Advances in Geographic Information Science

More information about this series at http://www.springer.com/series/7712

Jean-Claude Thill • Suzana Dragicevic
Editors

GeoComputational Analysis and Modeling of Regional Systems

Springer

Editors
Jean-Claude Thill
Department of Geography and Earth Sciences
University of North Carolina at Charlotte
Charlotte, NC, USA

Suzana Dragicevic
Department of Geography
Simon Fraser University
Burnaby, BC, Canada

ISSN 1867-2434 ISSN 1867-2442 (electronic)
Advances in Geographic Information Science
ISBN 978-3-319-59509-2 ISBN 978-3-319-59511-5 (eBook)
DOI 10.1007/978-3-319-59511-5

Library of Congress Control Number: 2017944436

Printed on acid-free paper

This Springer imprint is published by Springer Nature
The registered company is Springer International Publishing AG
The registered company address is: Gewerbestrasse 11, 6330 Cham, Switzerland

Preface

This volume addresses numerous research topics of GeoComputation as one of the important scientific areas of research evolving from Geographic Information Science. The book is the culmination of a few years of work that began with a series of special sessions on GeoComputation organized at the 2012 North American Meetings of the Regional Science Association International (RSAI) in Ottawa, Canada. In addition, it complements the celebration of 21 years of the GeoComputation conference series as a major scientific forum dedicated to exchanging scientific advances in this field.

We would like to express our sincere thanks to all our colleagues and authors who have participated in this project. They responded to our invitation and have selflessly given their time and effort to passionately contribute to this edited volume with chapters that cover various aspects of GeoComputation research. The twenty chapters are arranged into three thematic parts: an overview of GeoComputation as a cross-disciplinary field of research and its relevance to the science of regional systems; various cutting-edge aspects related to agent-based and microsimulations modeling; and finally the use of heuristics, data mining, and machine learning approaches.

It was indeed a pleasure to work with the 45 contributing authors, and we are thankful for their patience during the extended book editing process. We are also thankful to all the reviewers, authors, and external researchers, who have contributed with thoughtful comments to the blind peer review process and have strengthened the overall quality and scientific rigor of this book. Our special thanks go to Taylor Anderson and Olympia Koziatek from the Spatial Analysis and Modeling (SAM) Laboratory, Simon Fraser University, Canada, for assisting with the book editing process. The editors' support for this book project would not be possible without funding from the Knight Foundation Endowment Fund at the University of North Carolina at Charlotte and the National Science and Engineering Research Council (NSERC) of Canada.

This edited volume represents a coherent body of knowledge rooted in cutting-edge scholarship covering both theory and several application domains that will be of interest to GeoComputation researchers, graduate and undergraduate students as well as GIS practitioners in industry and government agencies.

Charlotte, NC, USA Jean-Claude Thill
Burnaby, BC, Canada Suzana Dragicevic
2017

Contents

Part I
General

GeoComputational Research on Regional Systems

Jean-Claude Thill and Suzana Dragicevic

The genesis of GeoComputation is not well-defined. On the onc hand, Stan Openshaw is widely regarded as the father of this field of inquiry and he is credited for having coined the term GeoComputation [1–3]. On the other hand, early GeoComputational research is difficult to separate from the profusion of research that was an integral part of the strands of research in quantitative geography, mathematical geography, and computational geography that grew out of the quantitative revolution in Geography. Indeed, the late 1990s saw a flurry of research contributions incorporating the basic elements of computation, simulation, and data-driven thinking in the scientific understanding of events, phenomena and structures with a spatial perspective. These activities were also enabled by improvements in computer hardware, processing performance, data storage and analysis software solutions [4].

Defining GeoComputation has been undertaken by many authors, including [1, 3, 5, 6], and a number of more recent thought leaders. For our discussion, it is appropriate to adopt Openshaw's [3] view that it is "concerned with the application of a computational science paradigm to study all manner of geo-phenomena including both physical and human systems." (p. 9). Crucial aspects of this definition are threefold: objects of interest are geographic or spatial or can in some way be described within this frame of reference; computationbased on

J.-C. Thill (✉)
Department of Geography and Earth Sciences, University of North Carolina at Charlotte, Charlotte, NC 28223, USA
e-mail: jean-claude.thill@uncc.edu

S. Dragicevic
Department of Geography, Simon Fraser University, Burnaby, BC, Canada
e-mail: suzanad@sfu.ca

J.-C. Thill, S. Dragicevic (eds.), *GeoComputational Analysis and Modeling of Regional Systems*, Advances in Geographic Information Science,
DOI 10.1007/978-3-319-59511-5_1

actual or simulated data is paramount; and the data-driven approach may serve as a temporary solution for the dearth of knowledge—including theories—of the subject matter under investigation.

While GeoComputation, Geography and Geographic Information Systems and Science share many similarities, there are distinct differences that uniquely separate them. Geography focuses on methodological traditions that are either more aligned to post-modern social-theoretical views or, alternatively, stricter neo-positivist positions that leave little room for a data-driven mind-set in the inquiry process. Geographic Information Science focuses on the science of data or information, with substantive emphasis on matters of data formatting, data storage, data representation, visualization, generalization, and other related topics. However, over the years Geographic Information Science has successfully operationalized various quantitative approaches such as spatial analysis, modelling, and simulation into its investigation toolbox for multiple computing platforms [7–10]. GeoComputation on the other hand takes the spatial data as the starting point and attempts to "make sense" of these inputs and to leverage them to disentangle the complex and often multi-scalar, dynamic, and non-linear relationships and influences that permeate the data representations of the world. While Geographic Information Science continues to struggle to handle matters in quality, uncertainty, incompleteness, and trustworthiness, GeoComputation methods are designed to be robust to these considerations, which makes them well-suited to many empirical contexts where the data is "dirty" as well as to forecasting purposes where the data is by nature ill-conditioned. GeoComputation, Geography and Geographic Information Science are complementary and mutually reinforcing [11–13].

GeoComputation has also some clear lineage with Regional Science. Emerging in the 1950s, Regional Science had established itself as the field of study of the region, as the spatial expression on many social, economic, and political structures that frame the operational relationships that maintain the functional cohesiveness of regions [14]. Regional Science has always espoused a neo-positivism approach to research. Just as Regional Science was prompt to adopt the emerging paradigm embedded in Geographic Information Science [14–16] to study the complex structures of urban and regional systems, it embraced GeoComputation a few years later, owing to the tremendous leap in scientific knowledge afforded by this new mode of scientific inquiry.

Regional systems are systems that operate among a number of objects, entities and agents in ways that may be thematically diverse; but the net outcome of the interactions and functional relationships among them is some form of order. This order extends over space (such as the Christallerian system of cities, towns and villages), and may change over time (for instance the emergence of polycentric urban systems, or socio-economic convergence of regions comprising a country's economic space). It may also exhibit properties that persist over certain ranges of scales and granularities, and morph into others on other ranges. Thus, the study of regional systems and their dynamics is at the core of regional science.

GeoComputational techniques are also at the emerging forefront of research and applications dealing with Big Data analysis, Cloud Computing and High

Performance Scientific Computing [17]. This is particularly important since it is necessary to manage and process the data, whether extremely small or extremely large volumes, in a manner that would yield robust insights into the underlying structures and patterns to provide reliable decision-support.

This book aims at contributing to both Regional Science and GeoComputation through a collection of 20 unique research contributions by noted scholars of regional systems. All the chapters of this book are original pieces of research; some were featured at a series of special sessions organized by the editors at Regional Science conferences in the United States and Canada. Others were solicited by the editors to complement the themes selected as areas of emphasis of this volume. Each chapter was subjected to a rigorous peer-review process and was also reviewed by the editors on the volume.

The book is organized in three parts. The first part contains five chapters that discuss several fundamental themes of research that cut across all areas of regional systems application and across many families of GeoComputational techniques. These include discussions of open source codes fostering the spread of the GeoComputational paradigm, considerations of diversity (as opposed to conformity) in spatial decision support systems, and finally state-of-the art discussions of parallel and high-performance computing matters.

The second part contains five chapters and presents contributions that methodologically add to the strand of research on agent-based simulations and microsimulations in urban and regional contexts. This GeoComputational tradition has been one of the most effective at interfacing bottom-up computational principles with the fundamental theories of behavioral, social, and economic sciences to advance understanding of the complex organizations of regional systems.

The third and final part contains ten chapters that leverage various heuristic methods (such as evolutionary algorithms), and techniques of data mining and machine learning to complement conventional methods of spatial analysis or as substitutes for such methods in order to alleviate the intrinsic limitations of these methods. The contributions of Parts 2 and 3 not only serve to highlight the diversity of GeoComputational techniques that can be advantageously applied to regional questions, but also the diversity of application areas, ranging from environmental impact assessment, travel behaviors, urban service provision, community health, and others.

The scholarship communicated through these Chapters speaks volume to the scientific merit of the GeoComputational analysis of regional systems. In an era when data collection is pervasive (crowd sourcing, volunteered geographic information, and so on), and when a large part of these data is georeferenced or geotagged, GeoComputation has a future that is brighter than ever. The ease of access to high-performance computing and cloud computing, the emergence of edge computing and quick expansion of open access coding for numerical and text analytics and visualization, and the embrace of public and public entities for "big data" make these exciting times indeed for GeoComputational scientists. We envision the contributions compiled in this book will have an enduring impact on the long-term expansion of GeoComputational research on regional systems.

References

1. Ehlen J, Caldwell DR, Harding S (2002) GeoComputation: what is it? Comput Environ Urban Syst 26:257–265
2. Openshaw S (1998) Towards a more computationally minded scientific human geography. Environ Plan A 30(2):317–332
3. Openshaw S (2000) GeoComputation. In: Openshaw S, Abrahart RJ (eds) GeoComputation. Taylor & Francis, New York, pp 1–31
4. Rees P, Turton I (1998) Geocomputation: solving geographical problems with new computing power. Environ Plan A 30(10):1835–1838
5. Gahegan M (1999) What is geocomputation? Trans GIS 3:203–206
6. Longley PA (1998) Foundations. In: Longley PA, Brooks SM, McDonnell R, Macmillan B (eds) Geocomputation, a primer. John Wiley, Chichester
7. Brunsdon C, Fotheringham S, Charlton M (2007) Geographically weighted discriminant analysis. Geogr Anal 39(4):376–396
8. Couclelis H (1998) Geocomputation and space. Environ Plann B Plann Des 25(SPEC. ISS): 41–47
9. Guan QF, Zhang T, Clarke KC (2006) GeoComputation in the grid computing age. In: Carswell JD, Tezuka T (eds) Web and wireless geographical information systems, proceedings, vol 4295. Springer, Hong Kong, p 237
10. Murray AT, Matisziw TC, Wei H, Tong D (2008) A geocomputational heuristic for coverage maximization in service facility siting. Trans GIS 12(6):757–773
11. Abrahart RJ, See L (eds) (2014) GeoComputation, 2nd edn. CRC, Boca Raton, FL
12. Atkinson P, Martin D (eds) (2000) GIS and geocomputation. Taylor and Francis, New York
13. Griffith D, Chun Y, Dean DJ (eds) (2017) Advances in geocomputation. Springer, Cham
14. Thill JC (2017) Regional science. In: Castree N, Goodchild M, Liu W, Kobayashi A, Marston R, Richardson D (eds) International encyclopedia of geography. Wiley-Blackwell, Hoboken, NJ
15. Fischer MM, Nijkamp P (1992) Geographic information systems and spatial analysis. Ann Reg Sci 26:3–17
16. Yeh AGO, Batty M (1990) Applications of geographic information systems in urban and regional planning. Environ Plann B Plann Des 17(4):369–374
17. Li S, Dragicevic S, Anton F, Sester M, Winter S, Coltekin A, Pettit C, Jiang B, Haworth J, Stein A, Cheng T (2016) Geospatial big data handling theory and methods: a review and research challenges. ISPRS J Photogramm Remote Sens 115:119–133

Code as Text: Open Source Lessons for Geospatial Research and Education

Sergio J. Rey

Introduction

The open source revolution continues to have major impacts on science and education and the field of spatial analysis is no exception. A number of overviews of open source spatial analysis and geographic information science have recently appeared[1] and my intent here is not to provide a similar comprehensive coverage of this area but rather to expand upon a particular set of themes I have raised previously [11]. I do so by drawing on the lessons learned in the development and evolution of the PySAL project [13] as it has intersected with my teaching and research activities.

My central claim is that while open source has attracted much interest in geospatial education and research its potential to enhance our activities has been constrained by a lack of understanding of how open source communities function and the differences that exist between these communities and those found in the academic and scientific worlds. In broad terms, the excitement around open source in academia is dominated by the cost advantages Free/Libre Open Source Software (FLOSS) offers to our teaching and research missions. While these are important and very real, by focusing on these we miss the forest for the trees. The true value of open source is its potential to revolutionize and fundamentally enhance geospatial education and research. I argue that this will only be possible if instead of seeing

[1]For overviews see [4, 8, 10, 11, 15, 16].

S.J. Rey (✉)
Center for Geographical Information Science, School of Geographical Sciences and Urban Planning, Arizona State University, Tempe, AZ, USA
e-mail: Sergio.Rey@asu.edu; srey@asu.edu

J.-C. Thill, S. Dragicevic (eds.), *GeoComputational Analysis and Modeling of Regional Systems*, Advances in Geographic Information Science,
DOI 10.1007/978-3-319-59511-5_2

open source code as only a tool to do research, we see the *code as text* and an object of research as well as a pathway to better geospatial education and research.

I begin with an overview of PySAL, discussing its origins, development, and current state, and I share several lessons I've learned as an open source developer living inside academia. Next, I focus on the role of open source in geospatial education. I then discuss the impacts of open source for research in geospatial analysis. I conclude the paper by identifying key areas of future concern and opportunity.

PySAL

Origins

PySAL's lineage can be traced back to two earlier projects, Space-Time Analysis of Regional Systems (STARS) [14] a package I developed with my students at San Diego State University, and PySpace which was Luc Anselin's project developed at the University of Illinois, Urbana Champaign [1]. STARS was designed to handle exploratory space-time data analysis while PySpace focused on spatial econometrics. Although they had different foci, the two projects relied on similar data structures (primarily spatial weights matrices), certain algorithms and statistical tests.

Collaboration between the project directors led to the realization that by pooling the development activities of our two teams, we could move these common features into a single library, rather than continuing to duplicate efforts. Additionally, such collaboration could allow for a more focused and optimized implementation of the core shared components and free up time for each of the respective projects to specialize on features that were unique to the individual package.

A second motivation for creating the library was that, at the time, spatial analysis was largely absent in the Python scientific computing community. There were some early efforts of packages for data integration (ogr), map projections (pyproj) and geoprocessing (shapely), but at the higher end of the spatial analysis stack there were no Python packages to support exploratory spatial data analysis and spatial econometrics. As Python was having major impacts elsewhere in scientific computing, and was starting to make in roads in GIS as reflected by ESRI's adoption of Python as a scripting language, we wanted to both contribute to further adoption of Python within the GIScience community but also fill the void of missing spatial analytical tools in the wider Python scientific computing portfolio.

Initial discussions about the design of the library laid out a comprehensive coverage of many areas of spatial analysis that not only included the feature sets in STARS and PySpace but an expanded vision to cover a broad set of methods, data structures and algorithms in the spatial analysis toolkit. Parallel to this coverage of the components in the library we also felt that the library should support a number of different types of uses. As a Python library PySAL could be used through imports at the Python interpretor to facilitate interactive computing at the command line. A second intended delivery mechanism was to use PySAL to develop rich desktop clients in the mode of GeoDa and STARS. Here the analytical engine of the application would be based on methods from PySAL, while advanced graphics toolkits, such as WxPython, could be used to implement fully interactive and dynamic graphics. A third way we envisioned the library being used was to build plugins or toolkits for other packages, for example ArcGIS, QGIS or GRASS. The fourth delivery mechanism we identified was to provide access to PySAL through distributed web processing services [9].

Early on in the implementation of the library we began to realize the advantages that adopting Python for this project would offer. Python is a dynamically typed scripting language which lends itself nicely to rapid prototyping of ideas which radically shortens the distance between the germ of an idea and its articulation in working code. Python also has a clean syntax which facilitates collaboration by making the implementation of algorithms and spatial analytical methods transparent as the code becomes text, a point I return to later.

Components

Figure 1 displays a schematic from the early design of PySAL. The key departure point for development of the library was the spatial weights module. Spatial weights are central to many areas of spatial analysis as they formalize the notion of neighbor relations governing potential spatial interaction between locations. Having efficient data structures to store, create, operate, and transform spatial weights is critical to the entire library and thus the weights module became the dominant focus early in PySAL's implementation.

A second building block in the library is the computational geometry module which provides a number of algorithms and data structures for spatial tessellations (Voronoi), minimum spanning trees, convex hulls, binning and R-trees which are required by several of the other modules in PySAL. For example, the construction of contiguity based spatial weights from shapefiles uses R-trees, or distance based weights using K-nearest neighbor algorithms relies on KD-trees.

The clustering module provides methods used to carry out spatially constrained regionalization as in the case of defining neighborhoods in geodemographics and urban analysis, or aggregating spatial units to satisfy some minimum threshold value when estimating disease rates in spatial epidemiology. The exploratory spatial data analysis module implements methods for global and local spatial autocorrelation analysis which includes enhancements to deal with rates and spatial smoothing. Methods for spatial dynamics form the fifth PySAL module and extend the class of Markov based methods from STARS to include LISA Markov chains, conditional and joint spatial Markov chains, and directional space-time indicators. Finally,

Fig. 1 PySAL components

methods for the testing, estimation, and validation of spatial regression models are contained in the spatial econometrics module.

The first official release of PySAL was in July 2010. We placed this release under the Berkeley Software Distribution (BSD) license since one of our goals for PySAL was to contribute the scientific computing stack in Python and BSD was the dominant and preferred license in this community. At the time of writing PySAL is in its 9th stable release (1.8), with the next formal release scheduled for January 2015. The project is housed at GitHub.[2] Since the first official release PySAL has been downloaded over 50,000 times and we are quite pleased with the reception of the library for the community. An important reflection of this reception is the inclusion of PySAL as a featured package in the leading Python distributions for scientific computing Anaconda Python Distribution [5] and Enthought Canopy [6].

[2]https://github.com/pysal.

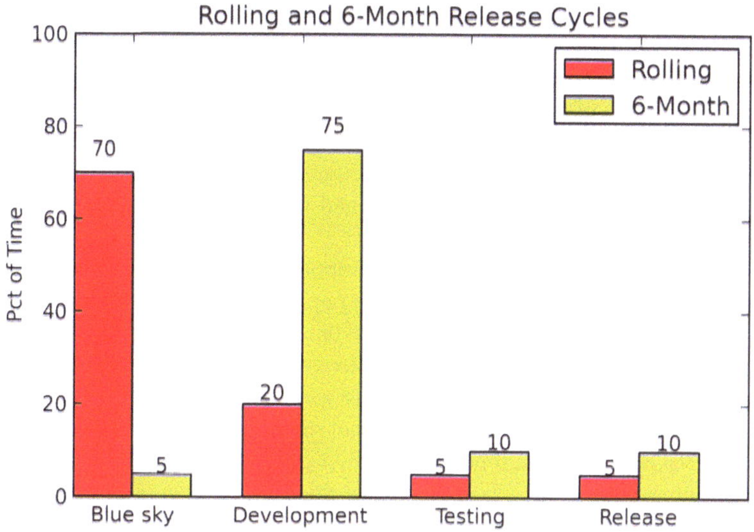

Fig. 2 Effort distribution under flexible and time based (release cycle) schemes

Lessons for Academic Open Source Developers

PySAL was born, and remains housed, in academia and the experience of managing an open source project inside of a university setting has been a rewarding and, at times, challenging experience. Because I expect more open source projects to find homes in academia in the future, I think it may be valuable to share a number of lessons that I've learned as an open source developer working inside academia.

Beginning with the first release (version 1.0) in July 2010 we adopted a 6 month release cycle. These cycles consist of several phases that begin with a 2 week period of considering PySAL Enhancement Proposals[3] (PEPs), followed by a developer vote on the PEPs that get priority for the current release cycle. The next phase of the cycle is the development component which lasts 14 weeks. The final month of the cycle is a feature freeze where attention shifts to bug fixes, testing, documentation and preparation of the release.

Prior to adopting the formal 6 month release cycle, work on PySAL and related projects (STARS, PySpace) followed no formal release schedule. We put out releases whenever we felt they were ready for public consumption, or in order to meet a deadline for a grant deliverable. I think this more flexible, less structured approach towards releases is more common in academic open source projects than the formal time-based scheme, and it is interesting to contrast the two (Fig. 2).

[3]We borrowed the PEP concept from the Python Enhancement Proposal model used in the development of Python: http://legacy.python.org/dev/peps/.

When operating under the informal release scheme, we (the developers) spent a large portion of the time engaged in design discussions about the new features, algorithms and related code architecture. These were very enjoyable discussions to be part of as they often generate a great deal of excitement and many fascinating ideas about what we could do. As an analogy to a research project, I would liken this to the phase where you are sitting with a potential collaborator discussing possible ideas over a beer. Here the sky is the limit and you can easily get caught up in the excitement of what is possible.

The excitement, however, can seem to evaporate when it comes time to actually write the proposal and, if successful, start to implement your grand ideas that not so long ago seemed so beautiful and seductive but now, in the face of the realities of pending deadlines and implementation challenges, take on a less enticing aura.

Given that the design and discussion phase was so engaging, while the implementation phase is much more like "work", it is not surprising that in the informal release scheme we spent a lot more time in the former and less in the latter. Unfortunately this meant that a lot of the ideas that drove the exciting design discussions never materialized in actual code.

When we shifted to the time-based release cycle we designed the lengths of the different phases to counteract this weakness. Basically we adopted the mantra of:

> You can do anything, but you can't do everything.
> *David Allen*

In other words, the blue sky discussions (i.e., you can do anything) were limited to the first 2 weeks of the cycle, and had to be formulated in a PEP which are essentially problem statements for the new feature. Voting on and prioritizing the PEPs for the release cycle reflected the recognition that we could not, in fact, do everything we might wish to do, and therefore we simply had to choose among all of our children which ones would be our favorites for the cycle.

The results of moving from the flexible to time-based release cycle have been very positive. The regular release schedule gives our users the ability to plan for any changes that they may want to make in light of new pending features in the library. For the developers it allows us to focus our energies on a subset of the features, resulting in more complete and better implementations of these in the release. Essentially we are sacrificing breadth for depth here.

In terms of the length of the release cycle, I think 6 months hits the sweet spot for an open source project inside academia, as we can align the ending of the cycle to coincide with the winter and summer breaks so that those release deadlines can be given full attention. Shorter cycles would mean release deadlines would compete against other deadlines during the academic year, while longer cycles (say 1 year) would slow the development of the project since new features could only be added once a year.

Our experiences with PySAL are likely similar to those of other projects that are housed inside academia. As is becoming increasingly recognized, much research code that is used in science is far from what could be called production quality software. Often it is written to get results for a particular project and then the researcher moves on to the next paper. What is typically missing are critical features such as documentation and testing that ensure the code could be used by others for purposes of replication and reproducibility.

Often the finger is pointed at the publish or perish dilemma as the main pressure leading to the rather poor state of research code. Less recognized, but arguably as important, is that most academics learn programming on the job rather than through a formalized sequence of courses in a degree program. As a result, even if the pressures to publish were absent and the researcher had time to document and test the code, they often lack the understanding of how to do so. To be fair, this lack of software engineering skills on the part of academic researchers could also be laid against many open source and proprietary projects outside of academia—many developers are self-taught. Equipping researchers with proper software development skills is a critical need that I return to below.

Lessons for Education

> It goes against the grain of modern education to teach students to program. What fun is there to making plans, acquiring discipline, organizing thoughts, devoting attention to detail, and learning to be self critical.
> *Alan Perlis*

Open source software and practices can have major empowering impacts on pedagogy. The free availability of the software offers a number of advantages in lab based courses. No longer are the students constrained to working in the school laboratory as they can now use the software installed on their own personal laptops, or home desktops, to complete exercises. This also allows for a greater degree of exploration and discovery by the student working by themselves and at their own pace.

These represent *potential* pedagogical wins for open source in geospatial education. My recent personal experience is that we still have far to go before these benefits are fully realized. During the fall of 2011 in my introductory course in GIScience, I decided to use QGIS as the software for the lab component in place of our traditional package of ArcGIS. This was something I had contemplated doing for quite sometime, but I always held back as the feature set and polish of QGIS were not yet at the stage where I felt comfortable doing so. By fall 2011, this had changed as the development of QGIS had reached an impressive state.

To my surprise this switch was less than well received by the students. Emblematic of the main complaint was the following comment I received on an anonymous teaching evaluation:

> I took this course as I heard we would be taught ArcGIS. I don't care about the science and the algorithms underneath the software, I want a job when I leave this class.
> *Anonymous student evaluation*

While there were a minority of students who told me they appreciated the introduction to an open source alternative, the vast majority of the students were not happy about the switch. In addition, I received push-back from some of my colleagues who were concerned that not covering ArcGIS threatened relationships with community internship partners that had been carefully cultivated over the years.

I was completely blindsided by these responses and felt a mixture of disappointment and puzzlement. In hindsight, I admit that these potential negative impacts never entered my decision making calculus. At the same time, while I now see that these are pressing concerns, they also raise some important questions regarding the role of geospatial education. On the one hand, the current demands in the labor market for students trained in ArcGIS reflects the reality that previous generations of students we have trained in this software are now in key positions in these agencies and companies. Additionally, many of these agencies have invested much time and resources in their GIS infrastructures and are understandably conservative regarding any changes. But, what about the future? Is our task to train students for today's labor market or to equip them with the skill sets and knowledge so that they are ready for, and can create, the future geospatial labor market?

A second general lesson for geospatial education concerns graduate education and the seemingly ironic situation of an embarrassment of riches in terms of freely available high quality programming tools for geospatial research on the one hand and, on the other, a general lack of desire to do any programming. I believe this stems from the challenges facing geography graduate students in that they not only need to acquire knowledge of substantive and methodological areas of the discipline but also somehow become proficient in programming. We have done a fairly poor job on the latter with solutions ranging from recommending introductory courses in computer science departments to learning on the job as part of a research project. The former is rather inefficient as my experience is geography students taking most introductory computer science classes come away without any idea of how to apply core concepts to geographical problems. The mentoring approach scores higher on this point, but does not really scale well.

There are several possible ways to address these issues. One approach I have adopted is to create a new course entitled "Geocomputation" that blends together both a primer on Python and open source tools, such as Git and text editors (Vim), together with formal lectures on a selection of spatial algorithms and their application to course projects. Open source tools, while very powerful, can have steep learning curves and a key motivation for this course is to flatten these curves. Thus far the course has been very well received as it equips the students with skill sets that are directly useful to their own thesis and dissertation research.

The course also offers a path to integrate PySAL into the curriculum as the library offers a rich set of possible topics to both use in lecture as well as to form core components of student projects. Stepping into an existing project gives the student hands-on experience with a large scale research code project, and rather than having to develop their own projects from scratch, they can choose from the ever expanding list of feature requests (and hopefully declining bug reports) for PySAL as their project topics.

It is here that I have seen the impact of the two key freedoms associated with FLOSS on geospatial education. The "free as in beer" freedom has already been alluded to since the students are free to download the software. This also is becoming increasingly important to educational institutions in an era of tightening budgets. The second freedom derives from the "free as in speech" aspect of FLOSS which means the code is now available for reading. Here seeing the code as text is enhanced in powerful ways by the free as in speech nature of FLOSS software and the use of Python. Students are empowered to think about the computational concepts and, due to the interpreted nature of Python and its clean syntax, they are unencumbered by the technical issues of compiling and linking that would be encountered in other languages hindering the learning process.

> You think you know when you can learn, are more sure when you can write, even more when you can teach, but certain when you can program.
> *Alan Perlis*

By coming to see the code as text, rather than as a black box, students' engagement with the fundamental concepts is deepened in a way that is simply not possible with closed source software. In my geocomputation course I endeavor to have the students come to see geospatial methods as not only tools they can use in their own research, but as possible subjects for research. For too long now the view in most geography departments has been that spatial analysis was something you use to do research, rather than something you do as research. We have only given scant attention towards nurturing the next generation of geospatial researchers who will produce the future advances in our fields. To facilitate the latter we have to affect a mind shift to see code as text.

There is, however, only enough demand at my own institution to offer a course like Geocomputation no more than every other year which leaves students entering our graduate program in the off year at a disadvantage as acquiring these skills early on in their studies is clearly desirable. A recent development in so called massively open and on-line courses (MOOC) offers some interesting possibilities in this regard. My experience in teaching PySAL workshops is that there is ample demand for these types of courses in the broader community. Offering such a course in the mode of a MOOC provides a mechanism to attract a staggering number of participants and could be a way to allow students at my own institution the possibility of taking the course each year. While there has been much debate about the impact of MOOCs on higher education, I am excited by the potential to reinvent the role of pedagogy at large research institutions as now it becomes possible to turn what could currently be seen as a boutique course into a staple offering.

Another attractive possibility for open source geospatial education can be seen in the Software Carpentry initiative [18]. Software Carpentry grew out of the recognition that most research scientists lacked basic software skills for scientific computing. Founded in 1988, the mission of Software Carpentry is to teach scientists basic lab skills for scientific computing through concentrated 2-day boot camps that cover topics such as shells, editors, code repositories and other technologies that help scientists become more efficient in their research computing.

Taken these lessons together I think that the reality of the situation of open source and geographic education is currently rather mixed. At the undergraduate level the impact has been much more limited than I would have originally believed, due mainly to the institutional factors raised above. The situation is more evolved at the graduate level. Here I've seen several instances where access to the source code in PySAL has enabled a motivated graduate student to gain a deeper understanding of a particular spatial analytical method. In hindsight, this mixed success may also suggest that a certain level of training and education may be required before the benefits of open source software can be experienced by students. I am optimistic that as the MOOC concept continues to evolve and Software Carpentry increases its outreach, our ability to engage more fully with the undergraduate population will be enhanced.

Lessons for Research

Analogies are often drawn between the logic of open source communities and the basic way science evolves. The notion of peer review is central to both arenas. Moreover, the ability to build on the contributions of others, as in the case of standing on the shoulders of giants, plays a central role in both open source and science. Finally, there are well accepted standards of behavior and norms in both communities. While these analogies have a ring of truth to them on the surface, closer inspection of each reveals subtle but important differences that may suggest the communities are more different than one might expect. Below I discuss a number of lessons the geospatial research community may draw from the open source world.

Open source has also had major impacts on research in GIScience. In the US this is clearly seen in research proposals to federal agencies as increasingly there are requirements that publicly funded projects include data and results management components so that subsequent research projects can replicate and extend funded projects. Having served on review panels for some of these agencies, a clear trend is that open source has been relied upon by many scientists to respond to these requirements. It should be emphasized that open source software offers clear advantages when it comes to replication as there are no longer any "black boxes" that conceal the implementation of a particular method or algorithm [19].

Peer review, while critically important to both science and open source development, works differently in these two communities. In the case of scientific journals, article reviews are typically done in a double-blind fashion to ensure candor from the referees. I've been an editor of a journal for 15 years, and have been impressed by the quality of reviews and the contributions these make towards often improving the original submission. Given that journal refereeing accounts for essentially zero in tenure and promotion cases, the fact that reviews are done at all, not to mention so well, is simply amazing to me. In the open source world by contrast, peer review is entirely open. The code commits a developer makes, the bug reports a user reports, feature requests, documentation contributions, and a host of other activities are all done in public view. This means that the individuals making those contributions are recognized and given credit. This is quite different from the scientists who spends several hours reviewing and, ultimately, improving a manuscript, since her comments only, and not name, are known to the author and wider scientific community. I think there are wonderful opportunities for academic publishing to learn from open source peer review processes.[4]

Another encouraging development can be seen in the evolving nature of the relationship between spatial scientists using open source code in their research and the development of that code. In the initial phase of adoption of open source in GIScience, the number of users of open source code dwarfed the number of developers of that code, and the intersection of users and developers was minimal. This situation is changing as there are important synergies between the two groups reflected in feature requests from users driving the development of the software. In other words, the distinction between user and developer is beginning to blur as users are coming to play more important roles in open source projects. Rather than being seen as end users or consumers, scientists adopting open source code in their research are increasing being viewed as collaborators in the open source development process [17].

This shift in collaboration will have major positive impacts on both the quality of future open source spatial analysis code as well as in the nature of the way geospatial research is conducted. One of the longstanding criticisms of open source code is that it can be "developer-centric" in the sense that only the developers understand and can make use of the code which is otherwise opaque to the end user. By integrating the research scientist into the development process, developers can be sensitized to the needs of the wider user community and improve the "user-centric" nature of the code. With regard to its impact on the practice of geospatial research and science, the open source model increases the likelihood that the scientific questions lead the way forward and the software itself is enhanced or modified to address these questions. This is in contrast to the proprietary world where the core programs themselves are not malleable. In the past this has led to the choice of research question being constrained by the functionality provided by the software.

The black box nature of proprietary spatial analysis software can mean that changes in APIs, data formats, and related design issues can break backwards compatibility yet, due to vendor lock-in, the costs of this breakage are largely borne by the community of users who are faced with the question of upgrading to the new version or finding an alternative. For the vendor, the clear gains are in a new revenue stream related to the upgraded version of the software. In some cases there

[4]A related development is the rise of open access journals and the open science movement. A full discussion of these is beyond the current scope, but can be found in [12].

may be legitimate debates as to whether the changes in the software reflect true enhancements to the software or not, but the impact is the same.[5]

This is not to say that similar changes in an open source project's code base do not happen. They can and they do. However, in our development of PySAL we have paid close attention to backward compatibility as we add new features and we are loath to break things. Moreover, users have access to the source code and can modify it to suite their own needs when there are changes in the code base. In the extreme case, the project could even be forked if development went in directions at odds with the wishes of our end users. Taken together I think that while open source projects have code bases that clearly evolve more rapidly than is the case of proprietary packages, the nature of community norms is such that the negative impacts of these changes are minimized.

To be sure these are all very positive developments. Yet, for the academic engaged in open source software development there are a number of challenges. A chief one regards the academic evaluation and promotion system which places heavy emphasis on scientific publications. Development, maintenance and documentation of an open source spatial analysis package requires a significant investment in one's time and this cuts into time that could go towards writing and submitting journal articles, books and proposals for funding. For a package that becomes widely adopted there is the possibility that scientists who use the package in their own research take care to cite the package, but my hunch is this is done less often than one would hope. There have been positive developments in this regard with journals such as *Journal of Statistical Software* that provide an outlet dedicated to developments in statistical software. It also reflects a shift in attitudes towards scientific software in that it is seen as scholarly work that should come under peer-review. In other words, the code is indeed viewed as text.

One often overlooked challenge that universities pose to open source developers is that these institutions are fairly conservative and slow to adapt to change. This was brought home to me in a very vivid way early on in my experience as an open source developer. As is customary when receiving a grant from a federal agency, I was called into a meeting with the chief technology officer (CTO) of my university when I received funding for the initial development of STARS from the U.S. National Science Foundation. The conversation went something like the following:

> CTO: "Has the software that is being developed in this grant been licensed?"
> Me: "Yes, it builds on code that I have placed under the GPL."
> CTO: "What is the GPL?"

I was amazed that in 2004 the CTO of a major research university had not heard of the GPL, but in hindsight I think it reflects the simple fact that institutional change occurs at a slower pace than technological change.

[5]For an example of this debate see the discussion and comments on the geodatabase thread at http://blogs.esri.com/esri/arcgis/2008/05/30/five-reasons-why-you-should-be-using-the-file-geodatabase/.

It can also be very difficult to secure funding in support of software development. In part this reflects the dominant view that production of analytical tools is not research, but rather something used in research. What gets funded is published research—text matters, code doesn't, and code isn't viewed as text. In this context one strategy we have adopted is to support parts of PySAL development through funding related to particular substantive projects. This requires having an infrastructure for the meta project that can keep the bigger picture in mind, while responding to the requirements of particular funded projects. In one sense the situation is not much different from that faced by most research active academicians attempting to juggle multiple on-going projects together with the next round of proposal writing. PySAL does, however, provide an overarching umbrella that can tie all these pieces together and, at least conceptually, allow one to see how future opportunities might be integrated into a research agenda.

An additional challenge for the use of open source code in geospatial research is that code itself is not enough. There is somewhat of a "build it and they will come" mentality at work in the all too common practice of new analytical methods developed as part of a research effort being made available as source code. In theory, this should allow other researchers to use the code and apply the new methods. In practice, however, there is often a great deal of heterogeneity in the quality of documentation accompanying the code as well as learning curves to install the code and any dependencies which may limit the dissemination and impact of the new methods. In other words, there can be a substantial gap between what is research code supporting a particular article, and production code that supports the use of the software by a wider audience. Here again, the incentive to the original creator of the new method was the creation of the method itself, the code is often a means to that particular ends. Indeed the competitive nature of academic research can result in a reluctance to release code since it may enable competitors to close the gap with the researcher-developer.[6]

There can also be a substantial gap between the amount and quality of testing that research code is subject to compared to the more extensive set of regression and integration tests that are viewed as a necessary part of an open source project. Those tests play a critical role in ensuring that new changes to the software do not introduce errors elsewhere in the code base, and this relies on the ability of the existing code to replicate a set of known results. Reproducibility is also a central pillar of the scientific process, yet it is highly ironic that much of the source code that generates new scientific results is rarely subject to even minimal software quality control measures. Adopting open source practices in the development of scientific research code could do much to improve the situation.

[6] See [3] for arguments as to why this reluctance may be misplaced.

Conclusion

Although the lessons outlined above treated development, education and research separately, this was for the purposes of exposition only. There are clearly strong potential synergies between these activities. At the same time, there are some challenges that can hamper our ability to exploit these synergies. One of our overriding goals in the development of PySAL has been to keep the level of code readability as high as possible, and here we have relied on the clear syntax of the Python language. We have always felt that the code can serve as a powerful source of information for students interested to learn the exact manner in which a spatial analytical method was implemented. While we have by and large kept to this goal, we have encountered tensions along the way. Keeping the code readable has required that we limit the number of third party libraries that PySAL requires. These libraries are often written in lower level languages such as C, C++, or Fortran and can offer substantial speed gains over pure Python implementations. At the same time the lower level code can be more difficult for the newly initiated spatial analyst to decipher. Faced with this trade-off, we have chosen pedagogy over speed.

As the number of open source spatial analysis projects within academia continues to grow, a difference in attitudes towards collaboration in the open source world versus academia is starting to emerge. The attitude of "not invented here" appears to be more prevalent in academia relative to what I have experienced in the broader open source community. In part, this reflects the pressures that open source academicians face in that citation of their work is critical for their own career advancement. This is, however, unfortunate as opportunities for combining these different tool sets through different forms of interoperability are lost.

The notion of a scientific work-flow has gained much attention in the cyberinfrastructure community, however progress in the implementations of architectures to support these work-flows faces a fundamental problem in that there are many areas of spatial analysis where we lack a consensus on the proper sequence of tools, or even choice of an individual tool. Paradoxically, the problem is not one of a scarcity of tools but rather abundance as users face a bewildering array of software packages. However, many of these are closed source which means their black box characteristic has hindered a deeper understanding of the methods enabled by the software. Open source provides a way to shed light on this area and will be critical in facilitating open discussions about methodological work-flows in spatial analysis [2].

Scientific geospatial analysis offers an important vetting framework—code can be evaluated for its scientific soundness through the formal peer review outlet of journals. As mentioned before, this can stand in contrast to the more open peer review process in open source where the comments of the community can reflect a heterogeneous mix of perspectives and levels of expertise. The rise of open source spatial analysis and tools has played a major role in the dissemination of these technologies beyond the halls of academia. As [7] has noted, this has shifted the educational mission from how to train professionals in the use of these technologies towards cultivating a more fundamental understanding of GIScience principles. In the end it is these principles that are paramount; the software and tools can be seen as a means to these ends. But how that software is built can have profound impacts on scientific and educational outcomes.

Acknowledgements A previous version of this paper was presented at the Open Source Geospatial Research and Education Symposium, Yverdon-les-Bains, Switzerland. I thank Ron Dorn, Olivier Ertz, Janet Franklin, and Stephane Joost for constructive feedback on this work. I have also benefited from the comments of the editors and anonymous reviewers. This research was partially supported by NSF Awards OCI-1047917 and SES-1421935.

References

1. Anselin L, Le Gallo J (2004) Panel data spatial econometrics with PySpace. Department of Agricultural and Consumer Economics, University of Illinois, Urbana-Champaign, IL
2. Anselin L, Rey SJ (2012) Spatial econometrics in an age of cyberGIScience. Int J Geogr Inf Sci 26:2211–2226
3. Barnes N (2010) Publish your computer code: it is good enough. Nature 467(7317):753–753
4. Brovelli M, Mitasova H, Neteler M, Raghavan V (2012) Free and open source desktop and Web GIS solutions. Appl Geomatics 4(2):65–66
5. Contiuum Analytics (2014) Anaconda scientific python distribution. https://store.continuum.io/cshop/anaconda/
6. Enthought (2014) Enthought canopy https://www.enthought.com/products/canopy/.
7. Goodchild M (2011) Twenty years of progress: GIScience in 2010. J Spat Inf Sci 1:3–20
8. Hu Q, Chen Y, Zhou Y, Xiong C (2009) An overview on open source GIS software with its typical applications. Geomatics World 1:2009–01. http://en.cnki.com.cn/Article_en/CJFDTotal-CHRK200901014.htm
9. Open Geospatial Consortium (2012) Web Processing Service. http://www.refractions.net/expertise/whitepapers/opensourcesurvey/survey-open-source-2007-12.p
10. Ramsey P (2007) The state of Open Source GIS. In: Free and Open Source Software for Geospatial (FOSS4G) conference
11. Rey SJ (2009) Show me the code: Spatial analysis and open source. J Geogr Syst 11:191–207
12. Rey SJ (2014) Open regional science. Ann Reg Sci 52:825–837
13. Rey SJ, Anselin L (2010) PySAL: A Python library of spatial analytical methods. In: Fischer MM, Getis A (eds) Handbook of applied spatial analysis. Springer, Berlin, pp 175–193
14. Rey SJ, Janikas MV (2006) STARS: space-time analysis of regional systems. Geogr Anal 38(1):67–86
15. Steiniger S, Hay G (2009) Free and open source geographic information tools for landscape ecology. Eco Inform 4(4):183–195
16. Steiniger S, Hunter AJ (2013) The 2012 free and open source GIS software map - a guide to facilitate research, development, and adoption. Comput Environ Urban Syst 39:136–150
17. Wang S, Wilkins-Diehr N, Nyerges T (2012) CyberGIS - towards synergistic advancement of cyberinfrastructure and GIScience: a workshop summary. J Spat Inf Sci 4:124–148
18. Wilson G (2006) Software carpentry: getting scientists to write better code by making them more productive. Comput Sci Eng 8(6):66–69
19. Yalta AT, Yalta AY (2010) Should economists use open source software for doing research? Comput Econ 35:371–394

Considering Diversity in Spatial Decision Support Systems

Ningchuan Xiao

Introduction

Many decision problems contain certain elements that are related to space [32, 35]. For example, to place a set of facilities, factors such as locations, distance, and connectivity among potential locations must be considered. Political redistricting is another example where space plays a significant role in determining the final plan that must satisfy restrictions such as spatial contiguity and compactness. We broadly refer to these as spatial decision problems.

Spatial decision problems are often difficult to solve due to many factors. Researchers have long recognized that spatial decision problems are often computationally intensive to solve [1]. This is because most spatial decision problems rely on a search algorithm to find feasible and optimal solutions from a huge set of potential solutions to the problem. The computational intensity of spatial decision problems often makes it impractical to find the optimal solution to the problem as the time used to search for the solution may become excessive. For many real world problems, even if the global optimal can be found, the solution is only optimal in the context of how the problem is simplified by removing factors that are otherwise difficult to be considered in the optimization model.

In additional to the computational burden, spatial decision problems often have multiple stakeholders who decide how the final decision should be made [38]. These stakeholders often have different goals to achieve regarding a specific problem and some of these goals are typically translated as the multiple criteria or objectives of the problems. For many problems that have multiple objectives, there may not exist a single solution that is deemed to be optimal by all stakeholders. To address

N. Xiao (✉)
Department of Geography, Ohio State University, Columbus, OH, USA
e-mail: xiao.37@osu.edu; ncxiao@gmail.com

J.-C. Thill, S. Dragicevic (eds.), *GeoComputational Analysis and Modeling of Regional Systems*, Advances in Geographic Information Science,
DOI 10.1007/978-3-319-59511-5_3

such decision problems, a variety of solution approaches have been developed [8]. The literature, however, seems to be less concerned with how to incorporate these approaches such that a better solution can be ultimately reached. A more interesting question is, if existing solution approaches can be collectively used to provide high-quality solutions, is it useful to develop new ones? Moreover, how can we successfully incorporate different perspectives of decision makers and stakeholders to generate more robust and reliable solutions that are satisfactory to a wider group of people?

The above questions are related to an interesting topic in social science: diversity, referring to a state of *difference* exhibited in a system and its components. Recent developments have demonstrated that effectively incorporating diversity may provide better solutions to highly complex problems in social and economic domains such as long-term prediction [19]. The purpose of this paper is to explore how the concept of diversity manifests in spatial decision making and how spatial decision making can benefit by incorporating diversity in the solution process. Although this paper is focused on decision problems from an optimization perspective, many concepts developed here can also be applied to other types of decision making problems. In the remainder of this paper, I first identify the kinds of diversity in spatial decision making, and then discuss a number of approaches to incorporating diversity into geographical problem solving.

Kinds of Diversity

Let $\mathbf{x}$ be a vector of decision variables. For a spatial decision problem, at least a subset of these decision variables have spatial references, often encoded as location indices. For example, we can have $\mathbf{x} = (x_1, x_2, \ldots, x_n)$ as indices to n locations and assign x_i to 1 if the ith location is selected for a design purpose (e.g., facility location) and 0 otherwise. We then assume $\mathbf{x}$ must be drawn from a domain denoted as $\mathbf{S}$ that defines all feasible solutions. The goal of solving a spatial decision problem is then to find an $\mathbf{x}$ such that a set of m objective functions, $\mathbf{f}(\mathbf{x}) = (f_1(\mathbf{x}), f_2(\mathbf{x}), \ldots, f_m(\mathbf{x}))^T$, can be optimized. Formally we write the problem in a generic form as

$$\begin{aligned} &\min \mathbf{f} \\ &\text{subject to } \mathbf{x} \in \mathbf{S}\,. \end{aligned} \tag{1}$$

Simon [24] suggested three steps that are commonly adopted in problem solving for a broad range of applications where decisions must be made. Starting at the intelligence step, the problem must be formulated so that alternative solutions can be found in the second step called design. In the third step called choice, a final decision must be made based on the alternatives identified. To solve a spatial decision problem, diversity is ubiquitous in all steps. For example, diversity occurs when the problem is interpreted and formulated by different stakeholders from different perspectives, solved using different methods, and presented to decision makers who have different preferences. Specifically in this paper, I examine diversity in spatial decision making from three perspective: (1) how solutions differ with respect to their decision variables and objective functions, (2) how the optimality of solutions differs and how their differences can be measured, and (3) how approaches to solving these problems differ.

Diverse Solutions

Solutions to a decision problem are typically described using two spaces: solution space and objective space. A solution space is formed by all the feasible solutions to the problem. Formally, a solution space is an n-dimensional attribute space where each dimension is one of the n decision variables, and we can denote it as a set of $\{\mathbf{x}|\mathbf{x} \in \mathbf{S}\}$. An objective space, however, is an m-dimensional space where each dimension is one of the m objective functions, denoted as $\{\mathbf{f}(\mathbf{x})|\mathbf{x} \in \mathbf{S}\}$. For spatial decision problems, a third space can also be identified: the geographic space of the solutions because each solution can be mapped and the spatial pattern shown on the map conveys meaningful messages that will be critical in the decision process [3]. Here we use a general notation of $g(\mathbf{x})$ to indicate the measure of solution $\mathbf{x}$ in the geographic space and therefore the geographic space can be denoted as a set $\{g(\mathbf{x})|\mathbf{x} \in \mathbf{S}\}$. Figure 1 illustrates the relationship between these three spaces.

The difference between solutions in the solution space can be captured using a distance measure such as the Euclidean distance

$$d_{ij} = \sqrt{\sum_{k}^{n} (x_k^i - x_k^j)^2}\,, \tag{2}$$

where x_k^i and x_k^j are the kth decision variable in solutions i and j, respectively. Using the measure in Eq. (2), the distances between the solutions in Fig. 1 are $d_{AB} = d_{BA} = 2$, $d_{BC} = d_{CB} = \sqrt{2}$, and $d_{AC} = d_{CA} = \sqrt{2}$.

The difference between two solutions can also be calculated in the objective space, again using a Euclidean distance:

$$d_{ij}^{\text{obj}} = \sqrt{\sum_{k}^{m} (f_k^i - f_k^j)^2}\,, \tag{3}$$

where f_k^i is the kth objective function value for solution i. In the hypothetical objective space in Fig. 1, it can be noted that $d_{AC}^{\text{obj}} < d_{AB}^{\text{obj}} < d_{BC}^{\text{obj}}$.

While the above two measures provide the numerical distances between solutions, one may argue that because the selected nodes in solutions A and B are adjacent in each case, they are more clustered than in solution C where the

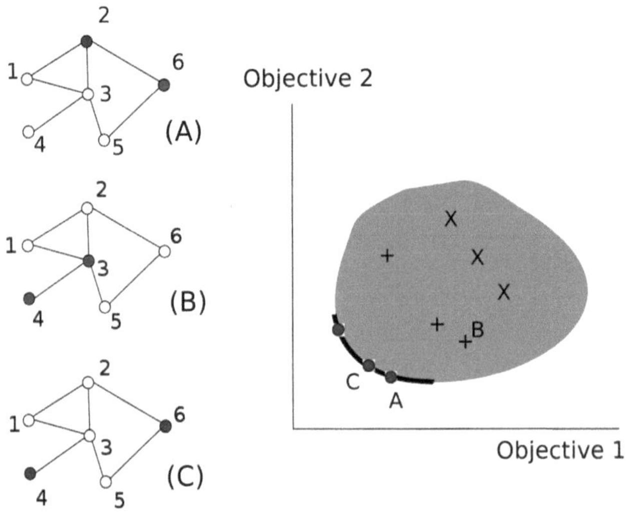

Fig. 1 A hypothetical spatial decision problem in which two nodes must be selected in a network of six nodes in order to minimize two objectives. Three possible solutions are illustrated as (A), (B), and (C). The number associated with each node is the index of the decision variable corresponding to that node. For each solution, its realization in the solution space is represented by whether a node is selected (*gray circle*) or not (*open circle*), or a set of values for the decision variables. For example, solution *A* is (0, 1, 0, 0, 0, 1). The geographic space realization is the network map shown in this figure, and each *dot* in the plot represents one of the hypothetical solutions in the objective space

selected nodes have no direct connections. Many measures can be used to reflect the geographic space of these solutions. Here we use a simple measure of the shortest distance or smallest number of edges on the path between selected nodes to illustrate the concept, and we have $g_A = 1$, $g_B = 1$, and $g_C = 3$. Accordingly, the distance in geographic space between these solutions can be simply calculated using the absolute difference between these measures:

$$d_{ij}^{\text{geog}} = |g_i - g_j|, \tag{4}$$

where g_i and g_j are the geographic measures of solutions i and j, respectively. In the three solutions in Fig. 1, we have $d_{AB}^{\text{geog}} = d_{BA}^{\text{geog}} = 0$, $d_{BC}^{\text{geog}} = d_{CB}^{\text{geog}} = 2$, and $d_{AC}^{\text{geog}} = d_{CA}^{\text{geog}} = 2$.

Diverse Optimality

The diversity in the objective space has two aspects. First, each solution can be identified using its objective function values as shown in Fig. 1 where the three dots marked as A, B, and C in the plot refer to the hypothetical values of the objective function values. The distance between these solutions in the objective space can therefore be simply calculated using the Euclidean distance between them.

Second, it is important to note that the multiple objectives for a problem reflect different and often conflicting goals. A consequence of such difference is the trade-off among alternative solutions, meaning there is no single solution that can be considered to be satisfactory with respect to all the goals. The trade-off among solutions can be formally understood using the concept of a domination. Here, we say a solution to a decision problem $\mathbf{x}_1$ dominates (or is better than) another solution $\mathbf{x}_2$ if and only if

$$\forall i f_i(\mathbf{x}_1) \leq f_i(\mathbf{x}_2) \wedge \exists i f_i(\mathbf{x}_1) < f_i(\mathbf{x}_2) .$$

In other words, solution $\mathbf{x}_1$ dominates solution $\mathbf{x}_2$ if $\mathbf{x}_1$ has at least one objective function value that is smaller (better) than that of $\mathbf{x}_2$, while all other objective values of $\mathbf{x}_1$ are not greater (worse) than that of $\mathbf{x}_2$. For a single objective optimization problem, there is typically only one solution that dominates all other feasible solutions. For a multiobjective problem, however, there often exists a set of solutions that are called non-dominated solutions, meaning they dominate all other solutions outside the set and each of these solutions does not dominate other members in the set. Solutions in this set are optimal and the set is often referred to as the Pareto front. In Fig. 1, the shaded area in the plot represents the objective space of the solutions, the thick curve represents the Pareto front and solutions on the curve are optimal (and therefore non-dominated) solutions.

A fundamental problem of (spatial) decision making is that the decision problem may be ill-structured because many social, economic, and environmental factors are difficult to be included in problem formulation [4, 25]. This feature suggests that the optimal solutions obtained based on the original problem formulation may become sub-optimal when new factors are considered as they often may be in real world applications. It is therefore important to understand the structure of the entire solution space instead of just the optimal ones, even if they can be found. We use the ranks of the solutions in the objective space to reveal this structure. Using the definition of dominance, we first give all the non-dominated solutions a rank of 1 (circles in the plot of Fig. 1). Then we increase the rank value to 2 and assign it to the non-dominated solutions in the remaining un-ranked solutions (pluses in Fig. 1). This process continues until all solutions are ranked.

After the ranking process is completed, we can measure the diversity of the objective space at different levels. First, we measure the between rank diversity of solutions using the inverse Simpson index [26]:

$$\frac{1}{\sum_{k=1}^{K} p_k^2}, \tag{5}$$

where p_k be the proportion of solutions that fall in rank k, and K is the total number of ranks in the solutions. The denominator is the probability that two random individual solutions have the same rank. If each solution has its own rank, we have $p_k = 1/K$ $(1 \leq k \leq K)$ and the between rank diversity is K. On the other hand, if all solutions are non-dominated (there is only 1 rank), we have a minimal between rank diversity of 1. For the 9 solutions in Fig. 1, the between rank diversity is $1/(\frac{1}{3^2} + \frac{1}{3^2} + \frac{1}{3^2}) = 3$.

Second, we can measure the diversity of solutions within each rank as the ratio between the number of solutions in the rank and a hypervolume of the solution space:

$$d_k = | \cup_{\mathbf{x} \in R_k} \mathbf{f}(\mathbf{x})| / \Pi_i^m (\mathbf{f}_i^u - \mathbf{f}_i^l)\,, \tag{6}$$

where the denominator is the hypervolume computed using the upper and lower bounds of each objective function values, $\mathbf{f}_i^u$ and $\mathbf{f}_i^l$, respectively, R_k is the set of solutions in rank k, and the numerator gives the number of unique individual solutions in rank k in the solution space.

Finally, while the above measures are aimed to provide a view for the solutions in the entire set or the ranked ones, diversity of solutions can also be measured at the level of each solution by examining the crowdedness of the neighborhood of that solution. Here we can borrow the concept of niche count from the evolutionary algorithm literature [8, 11] to measure the crowdedness around a solution:

$$n_i = \sum_{j=1}^{N} \mathrm{sh}(d_{ij}), \tag{7}$$

where n_i is the niche count of solution i, d_{ij} is the distance between individual i and j, which can be any of the distance measures discussed above (Eqs. (2), (3), and (4)) depending on what type of diversity is to be measured, and function $\mathrm{sh}(d)$ is defined as:

$$\mathrm{sh}(d) = \begin{cases} 1 - (d/\sigma_{\mathrm{share}})^{\alpha} & \text{if } d < \sigma_{\mathrm{share}} \\ 0 & \text{otherwise,} \end{cases} \tag{8}$$

where σ_{share} is a constant distance threshold that dictates the size of the neighborhood to be used for a solution, and α is a constant coefficient that reflects the weight given to a distance. In general, a high niche count suggests a high number of solutions exist around the given solution. We typically focus on solutions that have a low diversity of solutions in their neighborhood as suggested by low niche counts. Without knowing how solutions are exactly distributed in the solution space or objective space, it is often desired that each solution have a similar niche count (or local diversity).

At this point, one might reflect on why the above diversity discussion and diversity measures matter. The answer relies on how spatial decision problems are or can be solved. If there exists a magic tool that can return the exact solutions to a spatial decision problem, then none of the above discussion would matter much because the problem can be solved exactly, meaning we can find the optimal solution to the problem and therefore can make the decision consequently. For many real world decision problems, however, it is often impractical to solve the problem exactly, and it is important to find as many solutions as possible to enable an informed decision making process. More critically, the solutions found need to be diverse so that decision makers are not biased toward a certain subset of the solutions.

Diverse Toolboxes

Given the formulation of a spatial decision problem, the optimal solution can be obtained using an exact method. However, as discussed above, such an exact approach may become impractical when the size of the problem increases and additional factors must be considered in the decision process. It is critical, therefore, to explore a diverse set of solutions to the problem to enable a comprehensive examination of the solution space during the decision making process so that the decision makers can make their final choice. A second type of solution approach, called heuristics, can be used for this purpose. Heuristics are often more efficient compared with their exact counterparts, though they do not guarantee the global solutions to be found. The literature has generally suggested the effectiveness of heuristics in finding high quality solutions that are optimal or near optimal [5]. However, it has not been the focus of existing research to discuss how heuristic methods can be used to generate a diverse set of solutions to facilitate the decision process. In this section, I give a brief overview on the diversity of the solution methods. I will then discuss in the next section on how to utilize the diversity of these methods.

A large number of heuristic methods for spatial optimization problems have been developed in the past few decades. A traditional approach to developing such a method is problem-specific and lacks the flexibility of applying to other problems. The effectiveness of this type of heuristics is evident in the literature [5, 7]. For example, the vertex exchange method developed to solve the p-median problem in location-allocation models [31], though highly effective [22], cannot be directly used for other location allocation problems such as the center problems without significant modifications. Another example is the heuristic method that is specifically designed to solve political redistricting problems [33].

In general, traditional heuristic methods can be considered in different categories. A simple approach is to develop a greedy algorithm that construct a complete, feasible solution by assigning the values to the decision variables step by step. During each step, a decision variable is assigned so that the solution appears to be the best at that step, which can be simply achieved by minimizing the increase of the objective function value caused by assigning the new decision variable (assuming minimization is the goal). A greedy algorithm can often be strikingly simple to develop but the performance may not be satisfactory for many problems, especially when the problems contain many local optimal solutions.

Different from greedy algorithms, a local search algorithm starts from a complete solution to the problem that is called the current solution. The search algorithm can be used to create a neighboring solution by manipulating the current solution. The neighboring solution will be used to replace the current one if the former exhibit a better objective function value. Otherwise, the algorithm keeps searching for other neighbors. The algorithm stops when no better solution can be found. The vertex exchange algorithm for the p-median problem [31] is an example of local search where a neighbor solution is generated by swapping a selected vertex with other candidates.

In contrast to traditional approaches that are typically tailored to specific problems, a new set of heuristics is aimed to solve a wide range of optimization problems. These new methods, called metaheuristics collectively, include evolutionary algorithms, tabu search, simulated annealing, and ant colony optimization algorithms. A common feature of these algorithms is their root in natural processes. Evolutionary algorithms (EAs), for example, are derived from the natural selection theory [11, 12]. For an EA to find an optimal or near optimal solution to a problem, a set of solutions called a population is maintained at the same time. Each solution in an EA population is evaluated and consequently rated using a fitness function related to the objective functions of the problem. Solutions that exhibit high fitness function values often have a high chance to be used to create new solutions for the next iteration. In addition to their nature-inspired search mechanisms, metaheuristic methods also try to represent various optimization problems in a general and adaptive fashion. In EAs, for example, binary, integer, or real number strings have been used to represent solutions to numerical optimization problems in general [23], and geographic optimization problems in particular [35].

Embracing Diversity

Diversity can be incorporated into a spatial decision making process in a variety of ways. Before we start the discussion of specific incorporation strategies, let us stipulate the importance and therefore the benefits of recognizing and incorporating diversity in spatial decision analysis. First, the decision makers may wish to examine a diverse set of solutions such that important solutions, though may not be optimal according to the original mathematical formula, can be discussed and may be further modified. Second, a diverse set of solutions in the solution process can be used to maintain useful components of optimal solutions that otherwise may not exist in the "good" solutions chosen by the search algorithm. Here, I identify a number of technologies that can be used to promote or utilize diversity discussed above for the purpose of spatial decision making.

Encouraging and Maintaining Diverse Solutions

Several methods have been suggested in the evolutionary algorithm literature to maintain a diverse set of solutions. These methods try to balance two kinds of power in a search algorithm. First, a search algorithm is exploratory if it focuses on finding new solutions, especially solutions with new components that have not been found or included in those found so far during the search process. In EAs, a process called mutation is specifically designed to increase the exploratory power of search by randomly changing a portion of an existing solution with a hope of introducing new values which can then be combined with other solutions in order to construct better solutions. On the other hand, a search is exploitative if it tries to exhaustively use values in solutions found so far. In EAs, a crossover operation tries to combine two existing solutions to create new ones and therefore "exploits" current information that is already included in the two solutions. An exploitative operation tends to decrease the diversity of solutions while an exploratory one often increases the diversity.

Carefully balancing these two types of operations in a solution approach is critical for a successful search [6, 9, 37]. Some more recent work has also tested an adaptive fashion of using exploratory and exploitative search operations. For example, Tarokh [29, 30] suggest exploratory operations to be used more frequently if the lack of diversity is deemed in the current solutions. In EAs, the sharing method [11, 191] has been commonly used to reduce the chance of a solution to be selected if it is in a crowded neighborhood (measured in Eq. (7)). This concept is also used in EAs for multiobjective optimization problems where the fitness values of solutions in a crowded area in the objective space will be reduced so the solutions in less crowded areas have more chance to explore their neighborhood [8, 14].

Hybrid Solution Toolboxes

Solution approaches developed in the literature can be used in different ways. Though the common way of using these methods independently is useful, the overall performance can be improved if these methods are used collectively. One way of utilizing the diverse tools is to design a new process based on the components from existing methods. A method designed in this way can be called a hybrid method. For example, the concepts of vertex change and greedy algorithms are used to develop new and more effective hybrid methods to solve the p-median problem [21, 34]. While this type of hybridization is common in the literature [10, 16, 18, 20], a successful algorithm design may be ad hoc as many design aspects cannot be replicated in other problems.

Cooperative Methods

The recent literature has suggested another approach to incorporating different tools for problem solving. Hong and Page [13], for example, developed a general framework that includes a large number of problem-solving agents, each of which is a specific heuristic method that can be used to find a local-optimal solution to a problem. Each problem solving agent is evaluated using the average of the best solutions found. A subset of these agents is then selected to solve another set of random problems where each problem is solved sequentially, meaning one agent starts to solve it and then pass the final result to the next agent until all the agents are used. Their computational experiments on three different problem configurations suggested that a set of randomly selected agents outperformed the best agents on all cases.

In many real world problem situations, it has been observed that humans cooperate throughout the solution process and there have been different strategies in cooperating. In English, for example, it is often agreed that "two heads are better than one" [27, 28]. In this spirit, we can develop a new framework where problem solving agents work with each other through different cooperative (and sometimes non-cooperative) mechanisms, where some agents may prefer working alone while other may tend to solve a problem together with the others. There can be many cooperation strategies too. To illustrate various cooperation strategies, we discuss a recent development [36] in solving the p-median problem using two different approaches: a method called TB developed by Teitz and Bart [31], and a method called SA that is based on simulated annealing [15]. TB maintains a current solution and continuously replaces it with a better neighboring solution. TB stops when no better neighboring solution can be found. SA, however, uses a probability to accept a neighboring solution for replacement. while the acceptance probability for a better solution is always 1, SA also accepts solutions that are worse than the current one. The probability of accepting worse solutions decreases as the search progresses. SA terminates when no solution is accepted.

In this example [36], a total of seven modes of cooperation were implemented. First, TB and SA were two "work alone" modes where each ran separately and reported its own result. In addition to running these two methods independently, five cooperative strategies were also used. In a *relay* strategy, TB ran first and then the solution found by TB was used in SA; the process terminated after SA stops. A *sequential consensus* strategy was similar to relay, but the solution found by SA were passed on to TB again and the process repeated until no improvement can be made. To use a *compete* strategy, both TB and SA started independently and then, during each iteration of both methods, the current solutions were compared and the winner was used by both method for the next iteration. A *full cooperation* strategy depended on an exchange mechanism such that the two methods always exchanged their current solutions during each iteration. Finally, a *parallel consensus* strategy was developed so that both methods ran independently until they stopped and then they exchanged their best solutions found with each other; each agent then restarted

Table 1 Experiments on cooperation strategies for the p-median problem

	Number of optima found			Average deviation from optima		
Cooperation strategy	Best	Average	Worst	Best	Average	Worst
TB	25	5	5	0.07	0.32	0.78
SA	23	6	6	0.25	0.91	1.92
Relay	28	11	11	0.04	0.21	0.49
Sequential consensus	30	9	9	0.03	0.18	0.50
Compete	29	9	9	0.04	0.20	0.49
Full cooperation	30	10	10	0.05	0.17	0.39
Parallel consensus	34	15	15	0.01	0.11	0.24

using the solution from the other agent and continued the search process. Both kept exchanging solutions until not improvement can be made by any agents. A parallel computing environment was used to implement these methods.

Forty benchmark p-median problems [2] were used to test the above strategies. Each strategy was run 100 times for each problem. The best, average, and worst solutions generated in these 100 runs were used to report two summaries: the number of times these solutions were optimal for the 40 problems, and the average deviation from the known optima (Table 1). For example, the parallel consensus strategy found the optimal solutions to 34 of the 40 problems in the best case amid the 100 runs. The results clearly suggest that all the five cooperative strategies outperformed the two work-alone mode. Some strategies (e.g., parallel consensus) consistently outperformed the all other strategies, while some strategies (e.g., compete) may not necessarily outperform the other cooperative methods.

Extending the above experiment, we can consider each method as an agent that is equipped with a particular skill of solving some problems. An agent-based modeling framework, therefore, can be regarded as a platform to utilize the diversity of toolboxes in spatial decision making. In addition to such a toolbox perspective, agent-based models can also incorporate multiple players (decision makers) that have different belief systems and reflect different preferences to the decision problem. Simulation results of these models can be used by decision makers to learn interesting system behaviors.

Conclusions

The role of diversity has been recognized in many disciplines such as biology and sociology. In this paper, I attribute the importance of diversity in spatial decision making to the fundamentals of spatial decision making: multiple stakeholders with often conflicting goals, the ill-structured nature of the decision problem that leads to the need of exploring not only the optimal solutions but suboptimal solutions, and computational intensity of the solution approach. These characteristics entail the consideration of diversity for spatial decision making. This paper examines diversity in spatial decision making from three perspectives: solutions, optimality, and methods. The diversity of solutions can be identified and measured in the solution space, objective space, and their geographic space.

Considering diversity in spatial decision support systems is consistent with a postmodernist view [see, for example, 17] that adds to a computationally sophisticated environment of geocomputation. From a social or political point of view, promoting diversity in the decision process reflects a step toward a more appealing democratic process. It will be an informative debate to see if such an effort will provide us "better" decisions, the meaning of which may be beyond its methodological domain and of course is another aspect of diversity.

Acknowledgements An early version of this paper was presented at GeoComputation 2007 in Maynooth, Ireland.

References

1. Armstrong MP (2000) Geography and computational science. Ann Assoc Am Geogr 90(1):146–156
2. Beasley JE (1985) A note on solving large *p*-median problems. Eur J Oper Res 21:270–273
3. Bennett DA, Xiao N, Armstrong MP (2004) Exploring the geographic ramifications of environmental policy using evolutionary algorithms. Ann Assoc Am Geogr 94(4):827–847
4. Brightman H (1978) Differences in ill-structured problem solving along the organizational hierarchy. Decis Sci 9(8):1–18
5. Cooper L (1964) Heuristic methods for location-allocation problems. SIAM Rev 6:37–54
6. Crepinšek M, Liu SH, Mernik M (2013) Exploration and exploitation in evolutionary algorithms: a survey. ACM Comput Surv 45(3):1–33
7. Daskin MS (1995) Network and discrete location: models, algorithms, and applications. Wiley, New York
8. Deb K (2001) Multi-objective optimization using evolutionary algorithms. Wiley, Chichester
9. Eiben AE, Schippers CA (1998) On evolutionary exploration and exploitation. Fundam Inf 35(1–4):35–50
10. Estivill-Castro V, Murray A (2000) Hybrid optimization for clustering in data mining. In: CLAIO 2000, IMSIO, Mexico
11. Goldberg DE (1989) Genetic algorithms in search, optimization and machine learning. Addison-Wesley, Reading, MA
12. Holland JH (1975) Adaptations in natural and artificial systems. University of Michigan Press, Ann Arbor, MI
13. Hong L, Page SE (2004) Groups of diverse problem solvers can outperform groups of high-ability problem solvers. Proc Natl Acad Sci 46:16385–16389
14. Horn J, Nafpliotis N, Goldberg DE (1994) A niched Pareto genetic algorithm for multiobjective optimization. In: Proceedings of the first IEEE conference on evolutionary computation, IEEE World Congress on Computational Intelligence, vol 1. IEEE Service Center, Piscataway, NJ, pp 82–87
15. Kirkpatrick S, Gelatt CD, Vecchi MP Jr (1983) Optimization by simulated annealing. Science 220:671–680
16. Krzanowski RM, Raper J (1999) Hybrid genetic algorithm for transmitter location in wireless networks. Comput Environ Urban Syst 23:359–382

17. Macmillan B (1997) Computing and the science of geography: the postmodern turn and the geocomputational twist. In: Proceedings of the 2rd international conference on GeoComputation, University of Otago, Otago, New Zealand, CD-ROM
18. Nalle DJ, Arthur JL, Sessions J (2002) Designing compact and contiguous reserve networks with a hybrid heuristic algorithm. Forensic Sci 48(1):59–68
19. Page SE (2007) The difference: how the power of diversity creates better groups, firms, schools, and societies. Princeton University Press, Princeton, NJ
20. Preux P, Talbi EG (1999) Towards hybrid evolutionary algorithms. Int Trans Oper Res 6(6):557–570
21. Resende MGC, Werneck RE (2004) A hybrid heuristic for the p-median problem. J Heuristics 10:59–88
22. Rosing KE (1997) An empirical investigation of the effectiveness of a vertex substitution heuristic. Environ Plann B Plann Des 24(1):59–67
23. Rothlauf F (2006) Representations for genetic and evolutionary algorithms, 2nd edn. Springer, Berlin
24. Simon HA (1960) The new science of management decision. Harper and Row, New York
25. Simon HA (1977) The structure of ill-structured problems. In: Models of discovery. Boston studies in the philosophy of science, vol 54. Springer, Dordrecht, pp 304–325
26. Simpson EH (1949) Measurement of diversity. Nature 163:688
27. Surowiecki J (2004) The wisdom of crowds. Anchor, New York, NY
28. Tapscott D, Williams AD (2008) Wikinomics: how mass collaboration changes everything. Portfolio, New York, NY
29. Tarokh M (2007) Genetic path planning with fuzzy logic adaptation for rovers traversing rough terrain. In: Castillo O, Melin P, Kacprzyk J, Pedrycz W (eds) Hybrid intelligent systems. Studies in fuzziness and soft computing, vol 208. Springer, Berlin, pp 215–228
30. Tarokh M (2008) Hybrid intelligent path planning for articulated rovers in rough terrain. Fuzzy Sets Syst 159(21):2927–2937
31. Teitz MB, Bart P (1968) Heuristic methods for estimating the generalized vertex median of a weighted graph. Oper Res 16:955–961
32. Tong D, Murray AT (2012) Spatial optimization in geography. Ann Assoc Am Geogr 102(6):1290–1309
33. Weaver JB, Hess S (1963) A procedure for non-partisan districting. Yale Law J 73:288–309
34. Whitaker R (1983) A fast algorithm for the greedy interchange of large-scale clustering and median location problems. INFOR 21:95–108
35. Xiao N (2008) A unified conceptual framework for geographical optimization using evolutionary algorithms. Ann Assoc Am Geogr 98(4):795–817
36. Xiao N (2012) A parallel cooperative hybridization approach to the p-median problem. Environ Plann B Plann Des 39:755–774
37. Xiao N, Armstrong MP (2003) A specialized island model and its application in multiobjective optimization. In: Cantú-Paz E, et al (eds) Genetic and evolutionary computation — GECCO 2003. Lecture notes in computer science, vol 2724. Springer, Berlin, pp 1530–1540
38. Xiao N, Bennett DA, Armstrong MP (2007) Interactive evolutionary approaches to multiobjective spatial decision making: a synthetic review. Comput Environ Urban Syst 30:232–252

Parallel Computing for Geocomputational Modeling

Wenwu Tang, Wenpeng Feng, Jing Deng, Meijuan Jia, and Huifang Zuo

Introduction

In this study, we present the utilization of parallel computing capabilities for geocomputational modeling. Since its emergence in 1990s, geocomputation has been playing a critical role in bridging computer science and geography [1–3]. Geocomputation, as Gahegan [4] identified, is based on four themes in computer science to support geographic problem-solving: (1) *computer architecture and design*, (2) *search, classification, prediction and modeling*, (3) *knowledge discovery*, and (4) *visualization*. Computer algorithms and technologies from these themes are often intertwined to enable the resolution of complex geographic problems through geocomputational modeling. The advancement of these algorithms and technologies in computer science has pushed the development of geocomputation domains. However, gaps often exist between the development of algorithms and technologies in computer science and their applications in geography [1, 5]. Thus, it is necessary to retrospect the development of geocomputational modeling enabled by parallel

W. Tang (✉) • W. Feng • J. Deng • M. Jia
Department of Geography and Earth Sciences, University of North Carolina at Charlotte, Charlotte, NC 28223, USA

Center for Applied GIScience, University of North Carolina at Charlotte, Charlotte, NC 28223, USA
e-mail: WenwuTang@uncc.edu

H. Zuo
Department of Geography and Earth Sciences, University of North Carolina at Charlotte, Charlotte, NC 28223, USA

Center for Applied GIScience, University of North Carolina at Charlotte, Charlotte, NC 28223, USA

Department of Educational Leadership, University of North Carolina, Charlotte, NC 28223, USA

J.-C. Thill, S. Dragicevic (eds.), *GeoComputational Analysis and Modeling of Regional Systems*, Advances in Geographic Information Science,
DOI 10.1007/978-3-319-59511-5_4

computing. The focus of this study is on parallel computing, representative of computer architecture and design in geocomputation themes.

Recent increasing parallel computing applications are attributed to the blossoming of high-performance computing resources in cyberinfrastructure. Cyberinfrastructure, also referred to as e-Science, is the integration of "computing systems, data, information resources, networking, digitally enabled-sensors, instruments, virtual organizations, and observatories, along with an interoperable suite of software services and tools." ([6]; page 1). Cyberinfrastructure, as highlighted by NSF [6], consists of three key capabilities: high-performance and parallel computing, massive data handling and visualization, and virtual organization. High-performance and parallel computing is the key component of cyberinfrastructure that provides massive and transformative computing power for scientific discovery across alternative domains. Domain-specific problems that are infeasible for desktop computing can be solved by using tremendous computing power from cyberinfrastructure-enabled high-performance computing resources [7]. The use of cyberinfrastructure for enhancing problem-solving requires knowledge and skills from computer hardware, software, and specific science domains to best exploit the capabilities of cyberinfrastructure [6–9]. Of course, as cyberinfrastructure continues to develop, requirements on computer knowledge and skills tend to be relaxed. This will greatly urge the domain applications of cyberinfrastructure-enabled high-performance computing. There are a suite of representative cyberinfrastructure, including U.S. XSEDE (Extreme Science and Engineering Discovery Environment; see http://www.xsede.org), Open Science Grid (see http://opensciencegrid.org/), and DEISA (Distributed European Infrastructure for Supercomputing Applications; see http://deisa.eu). High-performance computing resources from these cyberinfrastructures are open to domain scientists for computationally intensive analysis and modeling.

The objective of this chapter is to discuss geocomputational modeling driven by parallel computing at the era of cyberinfrastructure. We organize the remainder of this paper as follows. First, we give an introduction to parallel computing. Then, we provide a detailed discussion on the applications of parallel computing on geocomputational modeling. We focus geocomputational modeling on four aspects: spatial statistics, spatial optimization, spatial simulation as well as cartography and geovisualization. We then use a case study to demonstrate the power of parallel computing for enabling a spatial agent-based model that is computationally challenging. Last, we conclude this chapter and propose directions for future research.

Parallel Computing

Current mainstream computing paradigm is dominated by multi-core and many-core computing, both of which are inherently associated with parallel computing architectures and technologies [10–12]. Multi-core machines are shared-memory computers based on CPU technology, which can be interconnected to form computer clusters (i.e., distributed memory architectures; see [10, 12]). Many-core computing is fueled by the emergence of NVIDIA many-core GPUs (Graphics Processing Units) for general-purpose computation [13, 14]. Multi- and many-core computing resources are often coupled together—i.e., heterogeneous high-performance computing resources—for the need of parallel computing. These parallel computing architectures serve the basis for cutting-edge cyberinfrastructure-enabled computing, for example, cluster-, Grid-, and cloud-computing (see [10, 15, 16]). In particular, high-performance computing resources are increasingly available on cloud computing platforms [15, 17]. Thus, how to effectively utilize these high-performance computing resources is of greater interest than their accessibility. The solution lies in parallel computing.

Depending on the way that data or information is communicated among processors, two generic types of parallel computing methods exist: message passing and shared memory (see [12]). In message-passing parallel computing, a processor communicates with others for the data required for its subsequent computation through sending and receiving messages. The requested data are encoded into messages on the sender side and then decoded on the receiver side. In terms of shared-memory parallel computing, processors use common address space to exchange data among themselves. Message-passing and shared-memory parallelisms dominate the parallel computing paradigm with a focus on inter-processor communication, which may induce significant overhead. Further, because of inter-processor communication, synchronization is often needed to coordinate concurrent operations among processors. A set of synchronization approaches exists, including barrier, lock, or semaphore [12]. On the other hand, there exist problems for which divided sub-problems do not exchange data or the exchange is light-weighted. In other words, processors will not (frequently) communicate for data from others. For this case, an embarrassingly parallel computing approach (also often referred to as a master-worker approach; see [12]) is the idealistic parallel solution. Because no or little communication among processors exists, high performance on computation is likely to be obtained.

Besides synchronization, a set of parallel strategies, represented by decomposition and load balancing, is often needed to efficiently parallelize a problem. Decomposition strategies support the partitioning of a problem into sub-problems according to characteristics of data (domain decomposition) or functions (functional decomposition) involved [10, 12]. Depending on the size of the sub-problems being partitioned compared with the original problem, decomposition can be fine- or coarse-grained. For spatial problems, spatial domain decomposition that takes into account spatial characteristics of the problems is often used for partitioning and alternative decomposition strategies reported (see [18, 19]).

In particular, Ding and Densham [18] presented detailed discussion on spatial domain decomposition strategies based on the regularity and heterogeneity of spatial domains. As a result, four types (regular versus irregular; homogeneous versus heterogeneous) of spatial domains exist to guide the decomposition. Regarding consideration of interactions or influence among spatial features, Ding and Densham [18] discussed a suite of parallel spatial modeling: local, neighborhood, region, and global (also see [20]). Ding and Densham [18] suggested that the consideration of spatial characteristics, represented by heterogeneity and dependency, is instrumental in the development of parallel algorithms for spatial problems. For example, spatially heterogeneous characteristics may create an unbalanced distribution of computation across spatial domains of a problem, which will require (more) sophisticated domain decomposition for effectively parallelizing the spatial algorithm. Spatial dependency may affect the choice of the size of neighboring regions, exerting a significant impact on the synchronization mechanism for a spatial problem parallelized using either a message-passing or shared-memory approach.

Once a problem is decomposed into a set of sub-problems, each sub-problem will be wrapped into a task assigned to an individual computing processor. The relationship between tasks and computing processors can be one-to-one or many-to-one. The workload assigned to computing processors may be unbalanced—i.e., load balancing (see [12]) is needed to efficiently utilize the parallel computing resources. Static and dynamic strategies [12] can be applied to achieve load balancing. For static load balancing, tasks are assigned to processors before parallel computing. Once the tasks are executed, there will not be re-assignment of tasks. Optimization algorithms can be used for static load balancing as it is naturally an assignment problem. Dynamic load balancing allows for flexibly reassigning or scheduling tasks among processors to achieve possibly more balanced workload.

To evaluate the performance of parallel algorithms, quantitative metrics based on computing time can be used. Performance metrics mainly include speedup, efficiency, and communication-computation ratio (see [12]). Speedup and efficiency are based on the comparison of execution time between a single processor and multiple processors (Eqs. (1) and (2)). Both speedup and efficiency are positively related to the computing performance of parallel algorithms. Communication-computation ratio is calculated as the ratio of communication time over computation time (Eq. (3)). Heavy communication overhead of a parallel algorithm usually leads to a high communication-computation ratio.

$$s = T_1/T_m \tag{1}$$

$$e = s/n_p \tag{2}$$

$$c = T_{comm}/T_{comp} \tag{3}$$

where s, e, and c are speedup, efficiency, and communication-computation ratio of a parallel algorithm. T_1 denotes the execution time of the sequential algorithm (i.e., on a single processor). T_m is the execution time of the parallel algorithm. T_{comm} is the time spent on inter-processor communication. T_{comp} denotes the time on computation.

Parallel Computing for Geocomputational Modeling

Geocomputational modeling serves as an abstraction of real-world geographic problems. Spatial statistics, spatial optimization, and spatial simulation are three pillars of geocomputational modeling that provide inductive or deductive problem-solving support. Further, geocomputational modeling is inherently related to cartography and geovisualization because of the need of visual presentation of relevant data that are geographically referenced. Thus, in this study, we focus our discussion in terms of the use of parallel computing for geocomputational modeling on four categories: spatial statistics, spatial optimization, spatial simulation, and cartography and geovisualization (Fig. 1). We use articles summarized in Table 1 to guide our discussion.

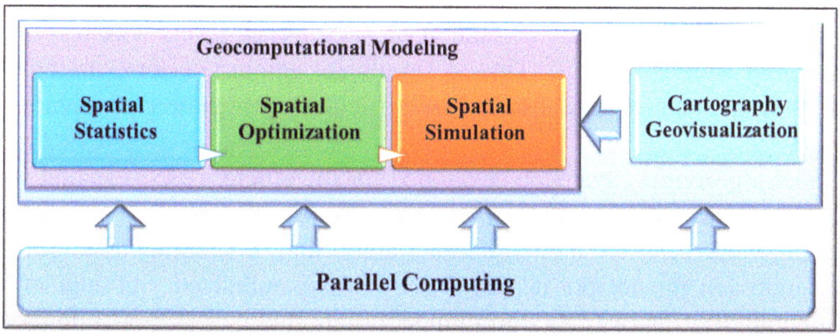

Fig. 1 Illustration of the use of parallel computing for geocomputational modeling

Table 1 List of literature of geocomputational modeling driven by parallel computing

Category	Citation
Spatial statistics	Armstrong et al. [21], Armstrong and Marciano [22], Cheng [23], Gajraj et al. [24], Guan et al. [25], Kerry and Hawick [26], Pesquer et al. [27], Rokos and Armstrong [28], Tang et al. [29] Wang and Armstrong [30], Widener et al. [31], Yan et al. [32]
Spatial optimization	D'Ambrosio et al. [33], Gong et al. [34], He et al. [35], Peredo and Ortiz [36], Porta et al. [37]
Spatial simulation	Abbott et al. [38], Deissenberg et al. [39], Guan and Clarke [40], Li et al. [41], Nagel and Rickert [42], Tang and Wang [43], Tang et al. [44], Tang [45], Uziel and Berry [46], Wang et al. [47]
Cartography geovisualization	Mower [48], Mower [49], Rey et al. [50], Sorokine [51], Tang [52], Vaughan et al. [53], Wang et al. [54], Wang [55]

Spatial Statistics

Spatial statistics provide a means of summarizing spatial characteristics of geographic data or inferring spatial patterns of interest based on first- or second-order properties of these data (see [56, 57]). Spatial statistics mainly comprise spatial autocorrelation analysis (e.g., Moran's I or Gerry's C), geostatistics (e.g., Kriging interpolation, semivariogram), and spatial pattern analysis (e.g., kernel density analysis, Ripley's K approach). Spatial statistics can be univariate, bivariate, or multivariate, thus facilitating the inference of spatial relationships within, between, or among spatial variables [56]. As geographically referenced data are increasingly available with respect to their size and type, spatial statistics provide necessary support for analyzing and understanding spatial characteristics (e.g., heterogeneity and dependence) in these data. Spatial statistics approaches involve comparisons of spatial entities in terms of distance, direction, geometry, or topological relationships [56]. These comparisons may operate at local or global levels with respect to the set of spatial entities [58, 59]. Thus, a significant amount of computation is often required for spatial statistics approaches, particularly when the geographic datasets are large.

Parallel algorithms have been developed for the efficient use of spatial statistics on high-performance computing resources. Armstrong et al. [21] presented their pioneering work in which a $G(d)$ statistic algorithm, functioning as a local spatial cluster approach for hotspot detection [60], was parallelized. Subsequent studies for the parallelization of $G(d)$ algorithm were reported (see [22, 30]). In particular, Wang and Armstrong [30] proposed a formal theory of spatial computational domain and applied it to parallelize the G algorithm. Spatial characteristics of geographic data were taken into account in the parallel algorithm to guide the efficient derivation of G values. Parallel computing efforts for other spatial statistics algorithms have been reported [28, 29, 31, 32, 61]. For instance, Yan et al. [32] developed a parallel Monte Carlo Markov Chain (MCMC) approach for efficient posterior sampling and applied it to parameterize a Bayesian spatiotemporal model based on Gaussian random field. Widener et al. [31] parallelized the AMOEBA (A Multidirectional Optimal Ecotope-Based Algorithm; see [62]) spatial cluster method using a message-passing approach. The computation of seeds required by the AMOEBA algorithm was partitioned and assigned to individual computing nodes. Tang et al. [29] presented a Ripley's K function approach accelerated through GPUs for spatial point pattern analysis. Acceleration factors, as reported by Tang et al. [29], can reach up to two orders of magnitude on a single Tesla Fermi GPU device and three (about 1501) when using 50 GPUs together.

With respect to geostatistics, Kriging interpolation is an approach that has been actively parallelized in the literature [23–27]. In Guan et al. [25] parallel work, fast Fourier transformation (to derive the covariance matrix) and the computation of weights for Kriging-based areal interpolation were parallelized within a message-passing environment. Guan et al. [25] examined their parallel areal interpolation algorithm on a high-performance computing cluster (about 5000

CPUs) and demonstrated that considerable speed-up was obtained. Pesquer et al. [27] proposed a row-wise decomposition approach to partition the computational load of ordinary Kriging, in which variogram fitting was automated, into a collection of worker nodes. Cheng [23] implemented a GPU-enabled parallel universal Kriging interpolation approach in which computationally intensive matrix-based operations (multiplication) were mapped to many-core architecture on GPUs. As Cheng [23] reported, the acceleration factor by using GPUs for universal Kriging is about 18.

Spatial Optimization

Spatial optimization is to search for optimal solutions from a set of alternatives that constitutes the solution space of a spatial problem of interest [63–65]. A spatial optimization algorithm is converged when its objective function (single- or multi-objective), constrained by a set of criteria, reaches maximum or minimum. Search approaches for optimization algorithms can be exact or heuristic [65]. Exact search enumerates and compares the entire set of solutions, guaranteeing for global optimum. Yet, exact search is only suitable for optimization problems that are relatively simple or small because of the brute-force search of solution space. Heuristic search, including deterministic (e.g., hill-climbing) and stochastic (e.g., simulated annealing, or evolutionary algorithms), introduces automated mechanisms that guide the convergence of the optimization algorithm. While heuristic search does not warrant global optima, it is well-suited to spatial optimization problems that are often sophisticated. Machine learning algorithms (e.g., decision trees, artificial neural networks, evolutionary algorithms, ant colony algorithm, and particle swarm algorithms; see [66–69]) have been extensively used to support heuristics search in optimization algorithms. These machine learning algorithms emulate the behavior of human or animals for intelligent problem-solving. The application of spatial optimization in geography, pioneered by Garrison [70], covers a suite of themes, including site search [71], location analysis [72, 73], spatial planning [74, 75], and ecosystem management [76].

The complexity of geographic problems often leads to a large solution space. As a result, computationally intensive search may be needed in order to obtain (near) optimal solutions for geographic problems, demonstrating the need of parallel computing for spatial optimization. Alternative parallel spatial optimization algorithms have been reported. Peredo and Ortiz [36] developed a simulated annealing algorithm parallelized using a message-passing mechanism to search for spatial patterns that match targeted ones. A tree-based strategy was used to accelerate the computation associated with the acceptance and rejection of perturbed spatial patterns. Machine learning algorithms, for example, artificial neural networks and evolutionary algorithms, have been parallelized. Gong et al. [34] proposed a hybrid parallel neural network algorithm as a nonlinear regression approach for empirical land use modeling. Parallel strategies were applied for the training and validation of ensemble neural networks. For evolutionary algorithms, the computation of each chromosome is independent with each other. Thus, the population of chromosomes is usually partitioned into a collection of subpopulations each assigned to a computing element for parallel computation. Because of independence among chromosomes' computation, computing performance for parallel evolutionary algorithms is usually high. For example, D'Ambrosio et al. [33] used a parallel evolutionary algorithm for optimal parameter estimation of a debris flow model based on cellular automata, and PGAPack, a parallel evolutionary algorithm software package (see https://code.google.com/p/pgapack/), supported their work. He et al. [35] developed a loose coupling strategy that applied parallel evolutionary algorithms to calibrate two hydrological models. Further, Porta et al. [37] implemented parallel evolutionary algorithms for optimal land use allocation within three types of computing environments: multi-core (shared memory), computing clusters (message passing), and hybrid.

Spatial Simulation

Spatial simulation is an approach that explicitly represents and generates the artificial history of a geographic system [77–79]. Components and their interrelationships in geographic systems are abstracted and represented in spatial simulation models. There are three types of generic spatial simulation [78–80]: system models, cellular automata, and agent-based models. System models, with a foundation in general systems theory [81], employ a set of differential equations to represent macro-level relationships among state variables in a system of interest [82]. Because of the complexity of geographic systems, analytic solutions may not be obtained for these differential equations. Differential equations in system models are often solved using a numerical approach. Cellular automata are based on neighborhood interactions and transition rules to represent spatial dynamics in geographic systems [78]. Agent-based models (or individual-based models) rely on the concept of agents that allow for the explicit representation of decision-making processes of spatially aware individuals or their aggregates [83, 84]. Both cellular automata and agent-based models are bottom-up simulation approaches tailored to the representation of decentralized interactions among components in a geographic system. Besides the three types of generic simulation, there are domain-specific simulation models, for example, hydrological models [85], that have been developed for the study of dynamic spatial phenomena. These simulation approaches (generic and domain-specific) have a vast body of literature in terms of exploring the spatiotemporal complexity of geographic systems.

The representational and generative power of spatial simulation models creates high computational demands, which trigger the motivation of utilizing parallel computing. Costanza and Maxwell [86] detailed the development of a parallel system model for the simulation of coastal landscape dynamics using spatially explicit differential equations. Guan and Clarke [40] presented the parallelization of SLEUTH, a cellular automat model of urban growth, and the application of the parallel model into the simulation of urban development of the conterminous U.S. Alternative spatial domain decomposition strategies were implemented to partition and allocate computational workload into parallel computing architectures. In Li et al. [41] work, a spatial cellular automata-driven urban simulation model was parallelized with support from strategies of ghost zones (for inter-processor communication) and load balancing (by area or workload). Besides cellular automata, parallel agent-based models have received attention from alternative domain scientists [39, 43, 47]. Uziel and Berry [46] presented a parallel individual-based model to simulate the winter migratory behavior of Yellowstone elk. Regular and irregular spatial domain decomposition strategies were used to cope with the irregular shape of the landscape that elk interacted with. Likewise, Abbott et al. [38] implemented a parallel individual-based model of white-tailed deer in which the foraging and movement of deer on their landscape were partitioned and distributed among multiple processors via a message-passing mechanism. Nagel and Rickert [42] proposed a parallel agent-based simulation of traffic in Portland and used a graph partitioning approach to divide the transportation network in the study area for load-balanced parallel computation. Tang et al. [44] applied a message-passing approach to parallelize a land use opinion model on a supercomputer. Further, as the increasing availability and maturity of GPUs technologies, a suite of parallel spatial simulation models accelerated by using the many-core GPU power have been reported (A detailed review is in [45]).

Cartography and Geovisualization

Cartography and geovisualization enable the presentation of 2- or 3-D spatial data through visual forms (e.g., maps or animations). Cartography has a focus on principals and techniques of mapping [87], while geovisualization is extended from cartography with an emphasis on interactive mapping and on-the-fly visualization of spatial information [88]. Map projection, data classification, generalization, and symbolization constitute fundamental components of cartography and geovisualization [87]. The combination of these cartographic components supports the design of alternative types of maps, including choropleth, dasymetric, isopleth, and proportional symbol or dot maps. Cartography and geovisualization pose a computational challenge [88]. For example, Armstrong et al. [89] illustrated the use of genetic algorithms for optimizing class intervals of choropleth maps and underlined that the process of developing optimal data classification requires computationally intensive search.

Each component in cartography and geovisualization could be highly computationally demanding, for which parallel computing provides a potential solution. Parallel algorithms have been developed to accelerate line simplification (see [49, 53]) and label placement of maps [48]. Tang [52] parallelized the construction of circular cartograms on GPUs. To leverage the massive thread mechanism of GPUs, the construction process was divided to a large number of fine-grained sub-tasks, while synchronization required by iterations of cartogram construction was conducted at a kernel level. Compared with advanced CPUs, the GPU-based parallel cartogram algorithm obtained a speed up of 15–20. In order to accelerate Fisher-Jenks choropleth map classification, Rey et al. [50] examined three different parallel python libraries, PyOpenCL, Multiprocessing, and Parallel Python, on both CPU-based parallel python and GPU-based PyOpenCL. Their results indicated that satisfactory speedup with the parallelization for moderate to large sample sizes can be achieved and performance gains varied according to different parallel libraries. Advance in high-performance computing greatly encourages the study and application of parallel scientific visualization [54]. Visualization software platforms enabled by high-performance and parallel computing, for example, ParaView (http://www.paraview.org/) and VisIt (https://wci.llnl.gov/codes/visit/home.html) are available and hold promise for accelerating the geovisualization of large geographic data. Sorokine [51] presented a parallel geovisualization module that allowed for leveraging high-performance computing resources for rendering graphics in GRASS GIS. A large geo-referenced image was divided into many smaller tiles concurrently rendered by back-end computing clusters. Similarly, in the work by Wang [55], a map tiling strategy was used for parallel visualization of vector- and raster-based GIS data.

Case Study

Agent-Based Spatial Simulation

In this study, we use a parallel agent-based model of spatial opinion to illustrate the importance and power of parallel computing for geocomputational modeling. The agent-based model was developed and parallelized within GPU environment (see [90]) for detail. In this model, geospatial agents situated within their spatially explicit environments develop and exchange opinions with their neighbors. Each iteration, an agent searches stochastically for its neighbors using a distance-decayed probability function (Eq. (4)).

$$p_{ij} = d_{ij}^{-1/\alpha} \tag{4}$$

The probability (p_{ij}) that two agents (i and j) are peered for opinion exchange is dependent on the distance between them (d_{ij}). After determining which neighbor for communication, the agent will exchange opinion with its neighbor, driven by a bounded confidence model that Weisbuch et al. [91] proposed. In this bounded confidence model, the opinion of an individual is a continuous variable with a range of 0–1. In our model, agents' initial opinions are uniformly randomly distributed. In other words, agents are randomly distributed on their opinion space. Each agent updates its opinion using two parameters: opinion threshold and exchange ratio.

Opinion threshold determines whether the agent will conduct opinion exchange with its neighbor. If the opinion distance between the two agents is shorter than the opinion threshold, the agents will use exchange ratio to update their opinions based on the opinion distance between them. Otherwise, no opinion exchange activities will occur if the opinion distance is longer than the threshold.

To enable the opinion modeling at a large spatial scale, the agent-based opinion model was parallelized and accelerated using general-purpose GPUs (see [90]). NVIDIA CUDA (Compute Unified Device Architecture; see [13, 92]) was the computing platform used for this parallel computing effort. GPU-enabled general purpose computing is based on a shared-memory data parallelism with thread technologies. A large number of CUDA threads are available for concurrently executing the assigned computing tasks on the streaming processors of a GPU device. In this study, the population of geospatial agents was divided into a collection of sub-populations based on a 1D block-wise domain decomposition strategy (see [11, 18]). Each sub-population may consist of one or multiple agents and the associated opinion development process is handled by a CUDA thread. Because the number of threads allowed in CUDA-enabled GPUs is large, massive agents are supported in this parallel spatial model.

Experiment

We designed an experiment to examine how parallel computing accelerates and thus facilitates the agent-based modeling of large-scale spatial opinion exchange. The experiment is to investigate the impact of communication range on the spatial opinion exchange. In this model, the distance coefficient (α in Eq. (4)) in the distance-decayed neighborhood search determines the communication range (see [90]). We varied this distance coefficient from 0.2 to 1.3 at an interval of 0.1 (corresponding to 3–398 cells). Consequently, there are 12 treatments in this experiment (noted as T1–T12). The distance threshold of agents was set at 0.22 and the exchange ratio is 0.40. A raster landscape was used in this study, and the landscape size of the model is 2000×2000. Each cell is situated by an agent. For each treatment, we repeated the model run 100 times, in total 1200 runs required. GPU devices that we used in this study are Nvidia Tesla Fermi M2050 (including 448 cores). CPUs are dual Intel Xeon hex-core processors (clock rate: 2.67 GHz; memory: 12 GBs).

In this model, as agents communicate with their neighbors, their opinions tend to move towards each other in their opinion space. When their opinions are clustered within a small range, these agents reach consensus. In this experiment, we are interested in the consensus development of agents. So we used an index of entropy [93] to quantify the spatial opinion patterns over time. The entropy index allows for representing the diversity of spatial opinions: a high entropy index illustrates that the spatial opinion pattern is diverse. Otherwise, a low entropy index is associated

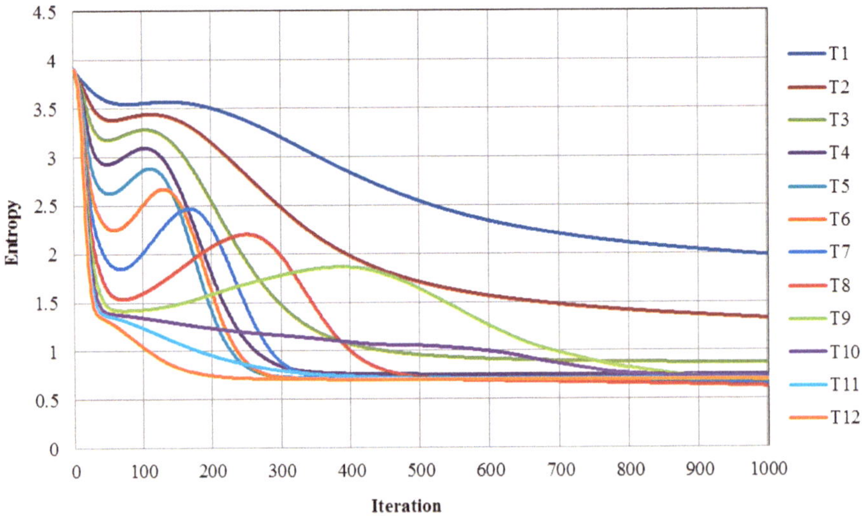

Fig. 2 Time series of opinion entropies over iterations (T1–T12: treatment 1–12)

with a homogeneous spatial opinion pattern—i.e., agent opinions are converged or consensus is reached.

$$en = -\sum_{i=1}^{n} p_i{}^* \log p_i \qquad (5)$$

where en is the entropy of an agent opinion pattern. p_i is the probability of opinion group i. n is the number of opinion groups. Figure 2 shows the time series of Shannon's entropy over 1000 iterations for the 12 treatments. For the first treatment, the communication range is short (threecells). The total number of possible neighbors that an agent exchanges opinion is small (79 neighbors). Thus, as agents communicate for exchanging opinions, entropy exhibits a gradually decreasing pattern. The averaged entropy at iteration 1000 is about 2.0. In other words, agents' opinions do not converge because of the limited communication range.

As increase in communicate range, entropy curves tend to reach minimum quickly. In most of the treatments entropy values converge within a range of 0.5–1.0. This illustrates that increment in communication range tends to increase the likelihood of communicating with more agents with diverse opinions. As a result, it is easier for agent opinion to converge for consensus. Of interest is the pattern of convergence iterations and entropies as communication ranges increase (see Fig. 3). For the first seven treatments, both convergence time and corresponding entropies tend to be lower when communication ranges increase. Yet, once communication range exceeds 40 cells (treatment eight and after), convergence time exhibits a wide range of variation (between iteration 1 and 1000). Most of the model runs

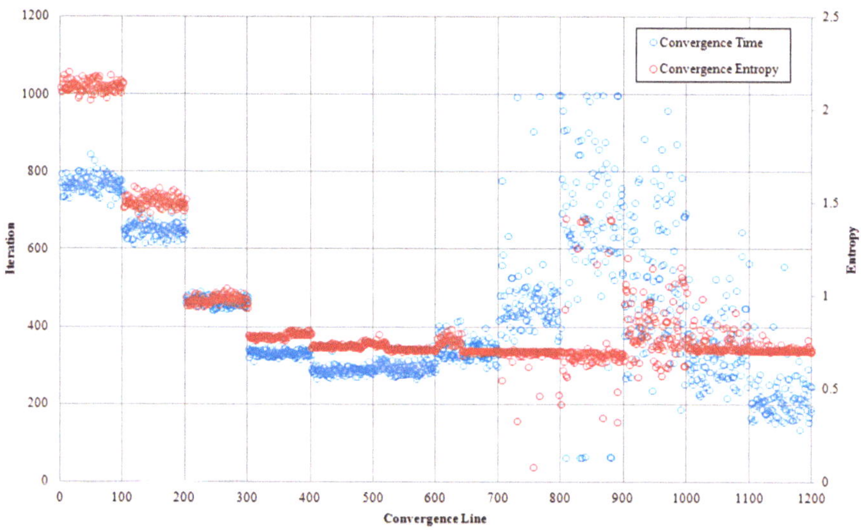

Fig. 3 Convergence iterations and entropies for model runs in the experiment

(except treatment T10) tend to converge at a small range of values for each treatment. Communication range of 40 cells is a critical threshold that triggers the state transition between stable and unstable convergence. This can be attributed to change in the interacting intensity (the amount of agent interactions per unit area) required for spatial opinion exchange. Before communication range reaches 40 cells, interacting intensity (given the same amount and types of agents) is high such that agents have sufficient opportunities to exchange their opinions for consensus. Yet, once the communication range exceeds 40 cells, interacting intensity required for opinion convergence tends to decrease (the number of interactions remains the same, but neighboring zones are enlarged). The decreased interacting intensity produces a form of diluting effect that introduces instability in the convergence of agent opinion.

We used acceleration factor, ratio of computing time on a single CPU over that on a GPU device (similar to speedup; see [90]; cf. [12]), to evaluate the computing performance of the parallel agent-based opinion model. Table 2 reports results of computing performance (including computing time and acceleration factor) of the 12 treatments. GPU computing time of each model run varies between 150 and 160 s (about 2–3 min), and corresponding CPU computing time falls within a range of 1600–1800 s (about half an hour per run). So the computing time required by this experiment is reduced from 600 h for a single CPU (0.5 h per run × 1200 runs; about 25 days) to 60 h for a single GPU. About 10–12 acceleration factors per GPU device per run were obtained. Because each run in this experiment is independent, we used 30 GPUs to concurrently execute these model runs to achieve further acceleration. The total CPU-based sequential computing time of this experiment requires 23.58 days. When using 30 GPUs together, it takes 6730.11 s to complete

Table 2 Results of computing performance of the agent-based modeling of spatial opinion exchange (time unit: seconds; Std: standard deviation)

	CPU time		GPU time		Acceleration factor	
Treatment	Mean	Std	Mean	Std	Mean	Std
T1	1773.32	142.28	158.93	10.69	11.23	0.89
T2	1685.88	116.88	158.37	10.49	10.84	1.04
T3	1655.66	100.26	157.95	11.99	10.58	1.09
T4	1670.81	117.33	156.04	12.51	10.66	1.12
T5	1652.46	105.76	159.71	9.69	10.24	0.87
T6	1644.99	86.87	158.53	11.02	10.52	1.18
T7	1675.75	120.79	153.85	11.54	11.04	1.27
T8	1659.36	105.97	164.69	3.15	10.11	0.71
T9	1679.23	112.68	156.38	12.56	10.89	1.15
T10	1660.30	105.06	155.97	10.54	10.46	0.94
T11	1677.63	119.56	159.75	10.10	10.39	0.90
T12	1756.70	141.86	160.18	11.12	11.14	1.41

the experiment. The corresponding acceleration factor (similar to speed up) for completing the entire experiment is 302.74 with respect to a single CPU. The influence of communication range on computing performance (both computing time and acceleration factors) is insignificant in this experiment.

Conclusion

In this study, we illustrated the power of parallel computing for geocomputational modeling and identified parallel strategies instrumental in tackling the associated computationally intensive issues. High-performance computing technologies are extensively available for domain-specific scientists in general and geographers in particular. Parallel computing strategies, represented by decomposition, synchronization, and communication, allow for best utilizing parallel computing architectures that high-performance computing is built on. In particular, for the parallelization of spatially explicit geocomputational modeling, spatial characteristics can be taken into account into parallel spatial algorithms to best leverage the high-performance computing capabilities of state-of-the-art cyberinfrastructure.

We focused our discussion on four categories related to geocomputational modeling: spatial statistics, spatial optimization, spatial simulation, and cartography and geovisualization. The first three approaches (statistics, optimization, and simulation) serve as the pillars of geocomputational modeling. These three approaches allow us to abstract and transform geographic problems into geocomputational modeling. The abstraction and representation of these problems in geocomputational modeling approaches make them computationally challenging. Parallel computing provides a potential solution to resolve the computational intensity of these geocomputational modeling approaches. Cartography and geovisualization support the visual presentation of geo-referenced data or information associated with geocomputational modeling. The use of parallel computing for the acceleration of cartography and geovisualization methods is needed when massive data are associated with, or produced from, geocomputational modeling.

Future research directions that we suggest for the applications of parallel computing in geocomputational modeling include (1) more elegant parallel spatial strategies for the best utilization of computing power in alternative high-performance computing resources, including heterogeneous multi- and many-core computing architecture; (2) more detailed investigation on the capability of parallel geocomputational modeling approaches (statistics, optimization, and simulation) for large-scale spatial problem-solving; and (3) parallel geovisualization technologies for the visual presentation of large GIS data and information (i.e., big data) associated with geocomputational modeling.

References

1. Armstrong MP (2000) Geography and computational science. Ann Assoc Am Geogr 90: 146–156
2. Longley PA (1998) Foundations. In: Longley PA, Brooks SM, McDonnell R, MacMillan B (eds) Geocomputation: a Primer. Wiley, New York
3. Openshaw S, Abrahart RJ (1996) Geocomputation. In: Abrahart RJ (ed) Proceedings of the first international conference on geocomputation. University of Leeds, Leeds, pp 665–666
4. Gahegan M (1999) What is geocomputation? Trans GIS 3:203–206
5. Openshaw S, Turton I (2000) High performance computing and art of parallel programming: an introduction for geographers, social scientists, and engineers. Taylor & Francis Group, London
6. NSF (2007) Cyberinfrastructure vision for 21st century discovery. Report of NSF Council. http://www.nsf.gov/od/oci/ci_v5.pdf
7. Atkins DE, Droegemeie KK, Feldman SI, Garcia-Molina H, Klein ML, Messerschmitt DG et al (2003) Revolutionizing science and engineering through cyberinfrastructure: report of the National Science Foundation Blue-Ribbon Advisory Panel on cyberinfrastructure. US National Science Foundation, Arlington, VA
8. Wang S (2010) A CyberGIS framework for the synthesis of cyberinfrastructure, GIS, and spatial analysis. Ann Assoc Am Geogr 100:535–557
9. Yang C, Raskin R, Goodchild M, Gahegan M (2010) Geospatial cyberinfrastructure: past, present and future. Comput Environ Urban Syst 34:264–277
10. Dongarra J, Foster I, Fox G, Gropp W, Kennedy K, Torczon L et al (eds) (2003) The sourcebook of parallel computing. Morgan Kaufmann, San Francisco, CA
11. Foster I (1995) Designing and building parallel programs: concepts and tools for parallel software engineering. Addison-Wesley, Reading, MA
12. Wilkinson B, Allen M (2004) Parallel programming: techniques and applications using networked workstations and parallel computers, Second edn. Pearson Prentice Hall, Upper Saddle River, NJ
13. Kirk DB, Hwu W-m (2010) Programming massively parallel processors: a hands-on approach. Morgan Kaufmann, Burlington, MA
14. Owens JD, Luebke D, Govindaraju N, Harris M, Krüger J, Lefohn AE et al (2007) A survey of general-purpose computation on graphics hardware. Comput Graph Forum 26:80–113

15. Armbrust M, Fox A, Griffith R, Joseph A, Katz R, Konwinski A et al (2010) A view of cloud computing. Commun ACM 53:50–58
16. Foster I, Kesselman C (eds) (2004) The grid: blueprint for a new computing infrastructure. Morgan Kaufmann, San Francisco, CA
17. Yang C, Goodchild M, Huang Q, Nebert D, Raskin R, Xu Y et al (2011) Spatial cloud computing: how can the geospatial sciences use and help shape cloud computing? Int J Digital Earth 4:305–329
18. Ding YM, Densham PJ (1996) Spatial strategies for parallel spatial modelling. Int J Geogr Inf Syst 10:669–698
19. Wang S, Armstrong MP (2003) A quadtree approach to domain decomposition for spatial interpolation in grid computing environments. Parallel Comput 29:1481–1504
20. Tomlin DC (1990) Geographic information systems and cartographic modeling. Prentice Hall, Englewood Cliffs, NJ
21. Armstrong M, Pavlik C, Marciano R (1994) Parallel processing of spatial statistics. Comput Geosci 20:91–104
22. Armstrong M, Marciano R (1995) Massively parallel processing of spatial statistics. Int J Geogr Inf Syst 9:169–189
23. Cheng T (2013) Accelerating universal Kriging interpolation algorithm using CUDA-enabled GPU. Comput Geosci 54:178–183
24. Gajraj A, Joubert W, Jones J (1997) A parallel implementation of kriging with a trend. Report LA-UR-97-2707. Los Alamos National Laboratory, Los Alamos
25. Guan Q, Kyriakidis P, Goodchild M (2011) A parallel computing approach to fast geostatistical areal interpolation. Int J Geogr Inf Sci 25:1241–1267
26. Kerry KE, Hawick KA (1998) Kriging interpolation on high-performance computers. Technical report DHPC-035. Department of Computer Science, University of Adelaide, Australia
27. Pesquer L, Cortés A, Pons X (2011) Parallel ordinary kriging interpolation incorporating automatic variogram fitting. Comput Geosci 37:464–473
28. Rokos, Armstrong MP (1996) Using Linda to compute spatial autocorrelation in parallel. Comput Geosci 22:425–432
29. Tang W, Feng W, Jia M (2015) Massively parallel spatial point pattern analysis: Ripley's K function accelerated using graphics processing units. Int J Geogr Inf Sci 29:412–439
30. Wang S, Armstrong M (2009) A theoretical approach to the use of cyberinfrastructure in geographical analysis. Int J Geogr Inf Sci 23:169–193
31. Widener M, Crago N, Aldstadt J (2012) Developing a parallel computational implementation of AMOEBA. Int J Geogr Inf Sci 26:1707–1723
32. Yan J, Cowles M, Wang S, Armstrong M (2007) Parallelizing MCMC for Bayesian spatiotemporal geostatistical models. Stat Comput 17:323–335
33. D'Ambrosio D, Spataro W, Iovine G (2006) Parallel genetic algorithms for optimising cellular automata models of natural complex phenomena: an application to debris flows. Comput Simul Nat Phenom Hazard Assess 32:861–875
34. Gong Z, Tang W, Thill J (2012) Parallelization of ensemble neural networks for spatial land-use modeling. In: Proceedings of the 5th international workshop on location-based social networks. ACM, Redondo Beach, CA, pp 48–54
35. He K, Zheng L, Dong S, Tang L, Wu J, Zheng C (2007) PGO: a parallel computing platform for global optimization based on genetic algorithm. Comput Geosci 33:357–366
36. Peredo O, Ortiz J (2011) Parallel implementation of simulated annealing to reproduce multiple-point statistics. Comput Geosci 37:1110–1121
37. Porta J, Parapar J, Doallo R, Rivera F, Santé I, Crecente R (2013) High performance genetic algorithm for land use planning. Comput Environ Urban Syst 37:45–58
38. Abbott CA, Berry MW, Comiskey EJ, Gross LJ, Luh H-K (1997) Parallel individual-based modeling of Everglades deer ecology. Comput Sci Eng IEEE 4:60–78
39. Deissenberg C, van der Hoog S, Dawid H (2008) EURACE: a massively parallel agent-based model of the European economy. Appl Math Comput 204:541–552

40. Guan Q, Clarke K (2010) A general-purpose parallel raster processing programming library test application using a geographic cellular automata model. Int J Geogr Inf Sci 24:695–722
41. Li X, Zhang X, Yeh A, Liu X (2010) Parallel cellular automata for large-scale urban simulation using load-balancing techniques. Int J Geogr Inf Sci 24:803–820
42. Nagel K, Rickert M (2001) Parallel implementation of the TRANSIMS micro-simulation. Parallel Comput 27:1611–1639
43. Tang W, Wang S (2009) HPABM: a hierarchical parallel simulation framework for spatially-explicit agent-based models. Trans GIS 13:315–333
44. Tang W, Bennett D, Wang S (2011) A parallel agent-based model of land use opinions. J Land Use Sci 6:121–135
45. Tang W (2013a) Accelerating agent-based modeling using Graphics Processing Units. In: Shi X, Volodymyr K, Yang C (eds) Modern accelerator technologies for GIScience. Springer, New York, pp 113–129
46. Uziel E, Berry MW (1995) Parallel models of animal migration in Northern Yellowstone National Park. Int J High Perform Comput Appl 9:237–255
47. Wang D, Berry M, Carr E, Gross L (2006) A parallel fish landscape model for ecosystem modeling. Simulation 82:451–465
48. Mower J (1993) Automated feature and name placement on parallel computers. Cartogr Geogr Inf Syst 20:69–82
49. Mower JE (1996) Developing parallel procedures for line simplification. Int J Geogr Inf Syst 10:699–712
50. Rey SJ, Anselin L, Pahle R, Kang X, Stephens P (2013) Parallel optimal choropleth map classification in PySAL. Int J Geogr Inf Sci 27:1023–1039
51. Sorokine A (2007) Implementation of a parallel high-performance visualization technique in GRASS GIS. Comput Geosci 33:685–695
52. Tang W (2013b) Parallel construction of large circular cartograms using graphics processing units. Int J Geogr Inf Sci 27(11):1–25
53. Vaughan J, Whyatt D, Brookes G (1991) A parallel implementation of the Douglas-Peucker line simplification algorithm. Softw Pract Exp 21:331–336
54. Wang L, Chen D, Deng Z, Huang F (2011) Large scale distributed visualization on computational grids: a review. Comput Electr Eng 37:403–416
55. Wang H (2012) A large-scale dynamic vector and raster data visualization geographic information system based on parallel map tiling [Thesis]. Florida International University, Miami, FL
56. Cressie NA (1993) Statistics for spatial data (revised edition). Wiley, New York
57. Ripley BD (2005) Spatial statistics. Wiley, Hoboken
58. Anselin L (1995) Local indicators of spatial association—LISA. Geogr Anal 27:93–115
59. Getis A, Ord JK (1996) Local spatial statistics: an overview. In: Longley PA, Batty M (eds) Spatial analysis: modelling in a GIS environment. Wiley, New York
60. Getis A, Ord JK (1992) The analysis of spatial association by use of distance statistics. Geogr Anal 24:189–206
61. Zhang J (2010) Towards personal high-performance geospatial computing (HPC-G): perspectives and a case study. In: Proceedings of the ACM SIGSPATIAL international workshop on high performance and distributed geographic information systems. ACM, San Jose, CA, pp 3–10
62. Aldstadt J, Getis A (2006) Using AMOEBA to create a spatial weights matrix and identify spatial clusters. Geogr Anal 38:327–343
63. Deb K (2001) Multi-objective optimization. In: Multi-objective optimization using evolutionary algorithms. Wiley, West Sussex, pp 13–46
64. Fletcher R (2013) Practical methods of optimization. Wiley, New York
65. Tong D, Murray AT (2012) Spatial optimization in geography. Ann Assoc Am Geogr 102:1290–1309
66. Bishop CM, Nasrabadi NM (2006) Pattern recognition and machine learning. Springer, New York

67. Bonabeau E, Dorigo M, Theraulaz G (1999) Swarm intelligence: from natural to artificial systems. Oxford University Press, New York
68. Goldberg DE, Holland JH (1988) Genetic algorithms and machine learning. Mach Learn 3: 95–99
69. Russell SJ, Norvig P, Canny JF, Malik JM, Edwards DD (1995) Artificial intelligence: a modern approach. Prentice Hall, Upper Saddle River, NJ
70. Garrison WL (1959) Spatial structure of the economy: II. Ann Assoc Am Geogr 49:471–482
71. Cova TJ, Church RL (2000) Exploratory spatial optimization in site search: a neighborhood operator approach. Comput Environ Urban Syst 24:401–419
72. Church RL (1990) The regionally constrained p-median problem. Geogr Anal 22:22–32
73. Murray AT, Gottsegen JM (1997) The influence of data aggregation on the stability of p-median location model solutions. Geogr Anal 29:200–213
74. Aerts JCJH, Eisinger E, Heuvelink GBM, Stewart TJ (2003) Using linear integer programming for multi-site land-use allocation. Geogr Anal 35:148–169
75. Scott AJ (1971) Combinatorial programming, spatial analysis and planning. Methuen, London
76. Hof JG, Bevers M (1998) Spatial optimization for managed ecosystems. Columbia University Press, New York
77. Banks J (1998) Handbook of simulation. Wiley, New York
78. Batty M (2005) Cities and complexity: understanding cities with cellular automata, agent-based models, and fractals. The MIT Press, Cambridge, MA
79. Benenson I, Torrens PM (2004) Geosimulation: automata-based modeling of urban phenomena. Wiley, London
80. Parker DC, Manson SM, Janssen MA, Hoffmann MJ, Deadman P (2003) Multi-agent systems for the simulation of land-use and land-cover change: a review. Ann Assoc Am Geogr 93: 314–337
81. Von Bertalanffy L (1972) The history and status of general systems theory. Acad Manag J 15:407–426
82. Costanza R, Voinov A (2004) Landscape simulation modeling: a spatially explicit, dynamic approach. Springer, New York
83. Epstein JM (1999) Agent-based computational models and generative social science. Complexity 4:41–60
84. Grimm V, Railsback SF (2005) Individual-based modeling and ecology. Princeton University Press, Princeton, NJ
85. Gassman PW, Reyes MR, Green CH, Arnold JG (2007) The soil and water assessment tool: historical development, applications, and future research directions. Trans Agric Biol Eng 50:1211–1250
86. Costanza R, Maxwell T (1991) Spatial ecosystem modelling using parallel processors. Ecol Model 58:159–183
87. Slocum TA, McMaster RB, Kessler FC, Howard HH (2009) Thematic cartography and geovisualization. Pearson Prentice Hall, Upper Saddle River, NJ
88. MacEachren AM, Gahegan M, Pike W, Brewer I, Cai G, Lengerich E et al (2004) Geovisualization for knowledge construction and decision support. Comput Graph Appl IEEE 24:13–17
89. Armstrong MP, Xiao N, Bennett DA (2003) Using genetic algorithms to create multicriteria class intervals for choropleth maps. Ann Assoc Am Geogr 93(3):595–623
90. Tang W, Bennett DA (2011) Parallel agent-based modeling of spatial opinion diffusion accelerated using graphics processing units. Ecol Model 222:3605–3615
91. Weisbuch G, Deffuant G, Amblard F, Nadal J-P (2002) Meet, discuss, and segregate. Complexity 7:55–63
92. CUDA (2016) CUDA. http://www.nvidia.com/object/cuda_home_new.html
93. Shannon CE (1948) A mathematical theory of communication. Bell Syst Tech J 27:379–423

High-Performance GeoComputation with the Parallel Raster Processing Library

Qingfeng Guan, Shujian Hu, Yang Liu, and Shuo Yun

Introduction

High-Performance GeoComputation

The recent advancements and wide adoption of spatial data collection technologies (e.g., high-resolution and hyperspectral remote sensing, Light Detection and Ranging [LiDAR], and global positioning system [GPS]) have led to the explosive growth of spatial data. Meanwhile, as geospatial science advances, a large variety of spatial analytical algorithms and spatial models have been developed in the last few decades. Big spatial data and complex spatial algorithms have been increasingly used in GeoComputational practices to solve complex spatial problems [1]. On the other hand, big spatial data and complex spatial algorithms often require massive computing power that vastly exceeds the capabilities of desktop computers. Therefore, high-performance GeoComputation is in need.

High-performance computing (HPC), usually referring to parallel computing, has been adopted in GIScience and GeoComputation since the 1980s. Openshaw explicitly points out that HPC is a significant component of GeoComputation [2]. The last three decades have seen a wide range of parallel geospatial computing studies, including transportation [3], land-use modeling [3, 4], spatial data handling and analysis [5, 6], least cost path [7], polygon overlay [8], terrain analysis [9–12], and geostatistics [13–17]. The recent developments of CyberGIS [18–20], spatial

Q. Guan (✉) • S. Hu • Y. Liu • S. Yun
National Engineering Research Center of Geographic Information System, China University of Geosciences, Wuhan, 430074, Hubei, China

Faculty of Information Engineering, China University of Geosciences, Wuhan, 430074 Hubei, China
e-mail: guanqf@gmail.com

J.-C. Thill, S. Dragicevic (eds.), *GeoComputational Analysis and Modeling of Regional Systems*, Advances in Geographic Information Science,
DOI 10.1007/978-3-319-59511-5_5

cloud computing [21, 22], and graphics processing units (GPUs) also stimulate the deployment of parallel computing in geospatial studies (e.g., [23–28]), as they provide easy-to-access HPC facilities and platforms.

Some characteristics of spatial data and algorithms must be taken into account when developing parallel geospatial algorithms [29]. For example, Guo et al. demonstrated the spatial distribution of features affect the spatial distribution of computational intensity for map visualization [30]. The amount of computing for visualizing an area where many features are concentrated is higher than that for visualizing an area where few features are sparsely scattered. Therefore, when decomposing the spatial domain of a dataset into sub-domains for multiple computing units (e.g., CPU cores) to process in parallel, the spatial distribution of computational intensity should be considered such that each computing unit is given approximately the same amount computing as others to achieve better performance. Spatially adaptive decomposition [16, 31], scattered mapping [32], and dynamic load-balancing techniques [33] can be used for such a purpose.

Because of the spatial autocorrelation (or spatial dependence), when processing a spatial unit, some algorithms (e.g., slope/aspect calculation and spatial interpolation) need to intake not only the spatial unit of interest, but also other spatial units within certain proximity. To parallelize such algorithms, the data must be carefully decomposed, such that a subset of data includes not only the part to be processed by a computing unit, but also the neighboring part (termed halo or ghost zone) required by the algorithm [32]. Furthermore, in some iterative algorithms (e.g., Cellular Automata), the neighboring part of a subset must be updated according to other subsets at each iteration, which usually results in communications between computing units [32]. Techniques for reducing the frequency and the amount of communications have been demonstrated in previous studies [4, 31].

As shown above, designing and implementing parallel GeoComputational algorithms requires not only the knowledge and skills of parallel computing, but also the understanding of the unique characteristics of spatial data and algorithms and their effects on parallel computing. The development complexity of parallel spatial computing can be extensively high for GIScientists and GeoComputation practitioners. Therefore, an easy-to-use programming library or middleware that is capable of facilitating the transformation from traditional sequential algorithms to parallel programs is expected to be greatly valuable to the GIS/GeoComputation community in the Big Data era.

Parallelizing Raster-Based Spatial Algorithms

Many GeoComputational algorithms for spatial analysis and modeling use raster data, for the following reasons: (1) a large proportion of spatial data are available in raster formats (e.g., remote sensing images, land-use and land-cover data, and digital elevation model); (2) the structure of raster data is simple, therefore easy to handle and process using computers; (3) the geographic coordinates of spatial units are indicated by row-column coordinates in raster data; and (4) raster data are suitable for representing continuous field and facilitating overlay analysis, which are commonly used in spatial problem solving.

Raster data are often organized as matrices of regularly shaped and sized cells, each one is associated with a value representing a certain attribute or condition at the cell's location. In many raster-based geospatial algorithms, the computation for a certain cell is independent from the computation for other cells, which means the computation is parallelizable. The parallelization strategy is usually straightforward because the matrix of cells can be decomposed into multiple sub-matrices to be processed in parallel (termed data parallelism). Note that computational independence does not mean data independence. Some algorithms require other cells when processing the cell of interest (e.g., focal [or neighborhood/moving-window based], zonal, and global operations), while still being parallelizable. Computational dependence exists in some algorithms. For example, in the flow accumulation algorithm, the flow accumulation of a cell depends on the flow accumulations of its upstream cells that must be computed before the current cell. Parallelizing a computationally dependent algorithm is harder, and usually requires special parallelization strategy specifically designed for the algorithm [34].

To facilitate the implementation of parallel spatial computing, some general-purpose programming libraries and environments have been developed. For example, Cheng et al. developed a set of general-purpose optimization methods for parallelizing terrain analysis using a Cellular Automata (CA) approach, which can also be used in the parallelization of other geospatial algorithms [35]. Qin et al. developed a set of parallel raster-based GeoComputation operators (PaRGO) for users to develop parallel geospatial applications on three types of parallel computing platforms [36].

The parallel Raster Processing Library (pRPL) is an open-source programming library designed for GIScientists and GeoComputation practitioners to easily implement parallel raster-based geospatial algorithms [31]. As a general-purpose parallel raster processing library, pRPL is primarily designed for data parallelism and can be used for a wide range of raster-based spatial algorithms. By encapsulating the underlying parallel computing details and providing easy-to-use interfaces, pRPL greatly reduces the development complexity of high-performance geospatial applications, hence enabling the utilization of advanced algorithms/models and big spatial data to solve complex geospatial problems.

This paper first gives a brief introduction to pRPL 2.0 (for more details about pRPL, please see [31, 37]), and then presents two showcases of using pRPL to implement high-performance spatial computing: slope/aspect calculation and Cellular Automata modeling. The experiments show that high-performance GeoComputation could be implemented with minimal parallel programming skills by using pRPL.

Key Features of pRPL 2.0

pRPL was developed using the C++ language based on the Message Passing Interface (MPI), a general-purpose framework for discrete-memory parallel programming [38]. The portability is guaranteed such that pRPL can be adopted on a wide range of HPC architectures, including multi-core CPU computers, computer clusters, computational grid, and cloud computing services. The rest of this section gives a brief introduction to pRPL 2.0.

Basic Components of pRPL

pPRL includes a hierarchy of data containers to hold raster data. A ***Cellspace*** is used to contain a matrix of cells, while a ***SubCellspace*** is used to contain a subset of cells within a *Cellspace*. pRPL 2.0 allows a data container to hold any C++ type of data (e.g., *int*, *short*, *long*, *float*, and *double*), as well as to retrieve and update a cell's value to/from a variable of any C++ type as long as the conversion between the cell's data type and the variable's data type is allowed. A ***Layer*** serves as a combining data container to hold a *Cellspace* and/or multiple *SubCellspaces*, and provides methods to add and remove *Cellspace/SubCellspaces*.

A data container also has an ***Info*** attribute component to hold the metadata, including the dimensions (i.e., numbers of rows and columns), data type, spatial reference information (e.g., datum, projection, cell size, and geospatial coordinates of the northwest corner), NoData value, and the minimal bounding rectangle (MBR) of a *SubCellspace*. A cell within a data container can be retrieved by either its row-column coordinates or geospatial coordinates.

To facilitate focal (a.k.a. neighborhood-based or moving-window) operations, pRPL provides a ***Neighborhood*** class. pRPL supports arbitrary neighborhood configurations, including not only the classical Von Neumann, Moore, and extended Moore configurations, but also user-defined ones that can be asymmetric and/or discontinuous. Also, varying weights can be associated with the cells within a *Neighborhood*, in order to facilitate some distance-decay algorithms.

A ***DataManager*** serves as a table of contents and maintains an indexing system for the management and manipulation (e.g., adding, removal, and retrieval) of multiple *Layers* and *Neighborhoods*. A *DataManager* facilitates the decomposition and mapping of data, i.e., dividing the *Cellspaces* into *SubCellspaces* and mapping them onto parallel processes[1]. It also controls the execution of a raster-processing algorithm, and coordinates the parallel processing automatically.

pRPL provides an intuitive programming guideline for users to implement application-specific algorithms (termed ***Transitions***). A basic *Transition* class is

[1]A process in parallel computing often represents a computational unit, e.g., a CPU or a CPU core.

provided by pRPL as a template, and a specific algorithm can be implemented as a customized child class by overriding the basic class. Writing a pRPL-based parallel program is like writing an ordinary sequential program and requires minimal parallel programming skills. pRPL automatically takes care of the underlying parallel computing details, and the users can therefore concentrate on the algorithms themselves.

Flexible Execution of Transitions

pRPL 2.0 provides three modes of executing customized *Transitions* to evaluate *Cellspaces/SubCellspaces*: EVALUATE_ALL, EVALUATE_SELECTED, and EVALUATE_RANDOMLY. The EVALUATE_ALL mode applies the *Transition* to process all of the cells within a *Cellspace/SubCellspace*. The EVALUATE_SELECTED mode uses the *Transition* to evaluate a set of user-selected cells. And the EVALUATE_RANDOMLY mode executes an iterative procedure that randomly selects a cell to evaluate at each iteration until a user-defined condition is satisfied. These three modes provide a great deal of flexibility and are able to implement a wide range of raster-processing procedures, such as scanning, tracing and sampling.

Multi-Layer Processing

pPRL allows a *Transition* to process multiple *Layers* of data, which is often the case in spatial analysis and GeoComputation. Not only multiple input *Layers*, but also multiple output *Layers* are allowed in version 2.0. As shown in Fig. 1, pRPL is able to synchronize the decomposition across *Layers*, such as to guarantee that the *SubCellspaces* on different *Layers* match with each other in terms of locations and dimensions. pRPL 2.0 also allows for varying decomposition configurations across *Layers*. This feature can be useful when the *Layers* have different spatial reference systems so the *SubCellspaces* on different *Layers* may refer to the same sub-region but have varying Minimal Bounding Rectangles (MBRs) within their original *Cellspace*. The *SubCellspaces* with the same ID on different *Layers* will be processed together.

Centralized and Non-Centralized Focal Processing

As mentioned above, pRPL supports not only local-scope and focal-scope processing, but also some zonal-scope and global-scope processing as long as they are parallelizable. For focal processing, both centralized (i.e., only the central cell of

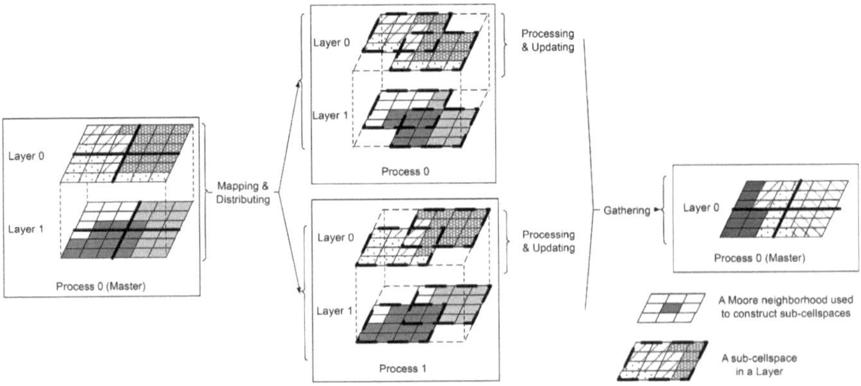

Fig. 1 Synchronized decomposition across layers

a neighborhood may be updated during the processing) and non-centralized (i.e., any cell within a neighborhood may be updated) algorithms are supported. Users only need to turn ON/OFF the "Only-Update-Centre" option of the *Transition*, and pRPL will automatically execute the special treatments according to the option for decomposition and parallel processing.

Flexible Domain Decomposition

pRPL provides multiple domain-decomposition methods for users, including regular row-wise, column-wise, and block-wise decomposition. Also, a spatially adaptive quad-tree-based (QTB) decomposition method is provided for cases when the computational intensity is extremely heterogeneous over space. The QTB decomposition iteratively divides the domain into four quadrats until all sub-domains have approximately the same workload (Fig. 2). Users must provide workload-calculation algorithms to the QTB decomposition according to their own raster-processing algorithms.

"Update-On-Change" and "edgesFirst" for Data Exchange

When focal *Transitions* are used, each *SubCellspace* contains not only the block of cells to be processed locally, but also a ring of "halo" cells serving as the neighbors of the edge cells. These halo cells are actually the replica of the edge cells of the neighboring *SubCellspaces*. Some iterative processing procedures (such as Cellular Automata) require executing one or more focal *Transition(s)* multiple times, which requires data exchange among parallel processes at each iteration

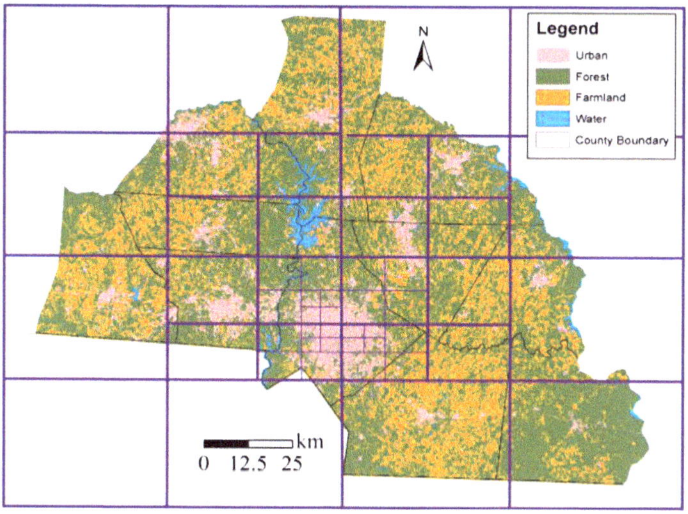

Fig. 2 QTB decomposition (the Greater Charlotte Metropolitan area, NC, 1992). Workload is based on the number of urbanized cells

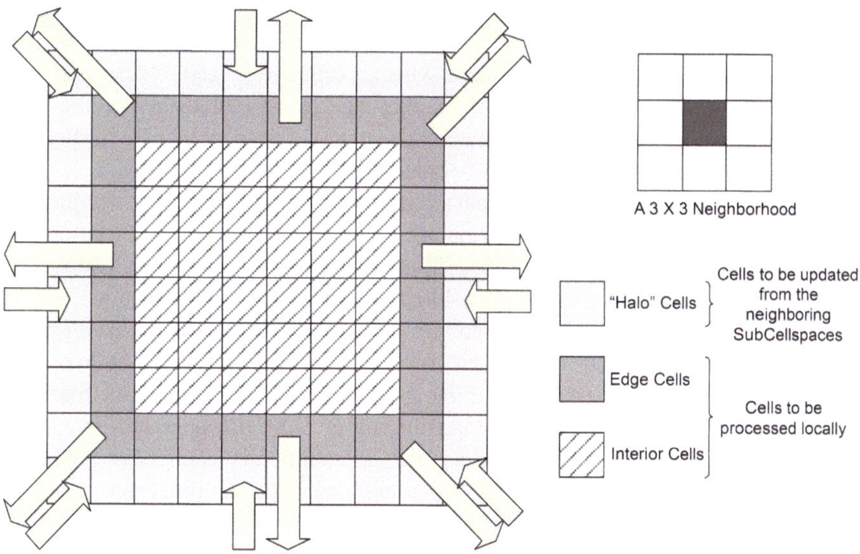

Fig. 3 Halo, edge and interior cells of a SubCellspace

so that the halo cells can be updated according to their origins (Fig. 3) [32, 39]. pRPL uses the "Update-on-Change" technique to help reduce the communication volume for data exchange among the participating processes, hence reduce the computing time. At each iteration, the edge cells of *SubCellspaces* are processed first, and only the changed cells are packed into data streams and transferred to other processes through non-blocking communications. During the data exchange, the interiors of *SubCellspaces* are processed. Such "edgesFirst" processing and non-blocking communication overlap the computation and data transfer, therefore reduce the waiting time for communication and improve the performance.

GDAL-based Centralized and Parallel I/O

pRPL 2.0 provides interfaces for algorithms to intake and output datasets in various raster data formats. Such flexible data I/O capability is implemented using the Geospatial Data Abstraction Library (GDAL, http://www.gdal.org), a general-purpose data I/O library that supports a variety of commonly used geospatial raster data formats, including GeoTIFF, Arc/Info ASCII grid, Erdas Imagine, SDTS, etc. Also, pRPL 2.0 provides two data I/O modes: centralized and pseudo parallel modes. In both modes, all the *Layers* needed by the *Transition* are first initialized. For input *Layers*, the master process reads the metadata (e.g., dimensions, data type, data size, NoData value, and spatial reference information) of the datasets and distributes to worker processes. For output *Layers*, the master process creates the datasets and distributes the metadata to worker processes. The metadata are then used to decompose the *Layers* (i.e., initializing the *SubCellspaces'* metadata). For regular decompositions (i.e., row-wise, column-wise and block-wise), the *Layers* can be decomposed without reading the actual cell data. For the QTB decomposition, however, the cell data must be read before decomposition such that the workloads of *SubCellspaces* can be calculated.

In the centralized mode, the master process is responsible for reading the *SubCellspaces* as needed and distributing them to worker processes before the computation, and gathering the *SubCellspaces* from worker processes and writing them to the output datasets. The non-blocking communication technique is used for *SubCellspace* transfer, similar to the data exchange procedure (see section "“Update-On-Change” and “edgesFirst” for Data Exchange"). In the pseudo parallel mode, all engaged processes read their assigned *SubCellspaces* from the input datasets directly in parallel, and write the output *SubCellspaces* into temporary datasets in parallel. The master process reads the temporary datasets and writes to the final output dataset. Both I/O modes assure that only one process writes a file instead of multiple processes writing to the same file simultaneously, because GDAL is not completely thread-safe (http://trac.osgeo.org/gdal/wiki/FAQMiscellaneous#IstheGDALlibrarythread-safe).

pRPL 2.0 also provides an option to use a writer process besides the master and workers. The writer process takes over the writing operations so the master can focus on coordinating the parallel processing. The writer dynamically receives output requests from other processes and writes subsets of data into the final output dataset.

Static and Dynamic Load-Balancing

pRPL 2.0 supports both static and dynamic load-balancing. In the static load-balancing mode, all *SubCellspaces* are mapped to processes before the computation. All processes read the input data of their assigned *SubCellspaces* through either centralized or parallel reading, and execute the *Transition* to evaluate the *SubCellspaces*. Once the computation of a certain *SubCellspace* is complete and output is needed, the output data is either transferred to the master/writer for output (i.e., centralized writing) or written into a temporary dataset (i.e., pseudo parallel writing).

The dynamic load-balancing mode uses the task-farming technique. A subset of *SubCellspaces* are first mapped to worker processes as their initial assignments. Whenever its assignment is near completion (i.e., only one *SubCellspace* is left to evaluate), a worker requests for more *SubCellspaces* from the master until a QUIT signal is received. The master dynamically maps *SubCellspaces* to workers in response to workers' requests until all *SubCellspaces* are mapped. The master or writer also dynamically receives output requests from workers, and either receives data from workers for centralized writing, or reads temporary datasets for pseudo parallel writing.

Showcases and Performance Assessments

To demonstrate the usability of pRPL and its performance, this section presents two showcases of high-performance geospatial computing implemented using pRPL 2.0: slope and aspect calculations, and Cellular Automata (CA) modeling. The slope/aspect calculations represent a variety of commonly used spatial analysis methods that are essentially local and focal operations. The CA modeling represents spatio-temporal dynamic simulations that require iterative execution of algorithms and frequent data exchange among parallel processes.

The experiments were conducted on a computer cluster composed of 106 computing nodes, each of which is equipped with four Opteron 2.1GHz 16-core CPUs and 256GB of RAM. The computing nodes are connected through a Quad Data Rate (QDR) Infiniband network at 10 Gigabit/s communication rate. The parallel programs were compiled using g++ compiler 4.7, OpenMPI 1.6, and GDAL 1.9, on the Scientific Linux 6.4 operation system.

Parallel Spatial Analysis—Slope and Aspect Calculations

Slope, measured in degrees or percentage, represents the maximum rate of change in value (e.g., elevation) from a cell to its neighbors. Aspect, measured clockwise in degrees from 0 (north) to 360 (north again), represents the downslope direction

Fig. 4 The 3 × 3 neighborhood for aspect calculation

1	2	3
4	0	5
6	7	8

of the maximum rate of change in value from a cell to its neighbors. Slope and aspect calculations are commonly used techniques in raster-based spatial analysis, such as terrain analysis, to identify surface change rates, surface facing directions, flow directions, and flat areas.

The algorithm of slope/aspect calculation often uses a 3 × 3 moving window (Fig. 4) to calculate the value of the central cell [40], and consists of the following steps:

1. Calculate the rate of change in the X direction:

$$dx = \frac{(Z[1] + 2 \times Z[4] + Z[6]) - (Z[3] + 2 \times Z[5] + Z[8])}{8 \times \mathit{CellWidth}} \tag{1}$$

where *Z[i]* indicates the value at *i*-th neighbor of the central cell in the input *Layer*.

2. Calculate the rate of change in the Y direction:

$$dy = \frac{(Z[6] + 2 \times Z[7] + Z[8]) - (Z[1] + 2 \times Z[2] + Z[3])}{8 \times \mathit{CellHeight}} \tag{2}$$

3. Calculate the slope:

$$slope = \text{ArcTangent}\left(\sqrt[2]{dx^2 + dy^2}\right) \times \frac{180}{\pi} \tag{3}$$

4. Determine if the cell is in a flat area

$$if\ slope = 0, then\ aspect = -1 \tag{4}$$

5. Calculate the aspect

$$aspect = \begin{cases} \frac{\pi}{2} - ArcTangent\left(\left|\frac{dy}{dx}\right|\right), & \text{if } dx < 0 \text{ } and \text{ } dy < 0 \\ \frac{\pi}{2} + ArcTangent\left(\left|\frac{dy}{dx}\right|\right), & \text{if } dx < 0 \text{ } and \text{ } dy > 0 \\ \frac{3\times\pi}{2} + ArcTangent\left(\left|\frac{dy}{dx}\right|\right), & \text{if } dx > 0 \text{ } and \text{ } dy < 0 \\ \frac{3\times\pi}{2} - ArcTangent\left(\left|\frac{dy}{dx}\right|\right), & \text{if } dx > 0 \text{ } and \text{ } dy > 0 \\ 0, & \text{if } dx = 0 \text{ } and \text{ } dy < 0 \\ \pi, & \text{if } dx = 0 \text{ } and \text{ } dy \geq 0 \end{cases} \tag{5}$$

6. Convert the aspect to compass direction value and correct to north:

$$aspect = 360 \times \frac{360}{2 \times \pi} + 180 \tag{6}$$

$$aspect = \begin{cases} aspect - 360, & \text{if } aspect \geq 360 \\ aspect, & \text{if } aspect < 360 \end{cases} \tag{7}$$

Even though the slope/aspect calculation is not a complex algorithm, when applied on a massive amount of data, it requires lengthy computing time. One can expect the computing time to be largely reduced using parallel computing.

Apparently, the slope/aspect calculation is a typical focal raster processing and can be easily parallelized using pRPL. In this study, a slope/aspect-calculation *Transition* class was developed based on the basic *Transition* class provided by pRPL, to implement the above algorithm, and a parallel program was developed to execute the customized *Transition* in parallel. The program intakes an input file (e.g., DEM), and is able to generate two output files (i.e., slope and aspect). A few parallel computing options are also provided, including the number of *SubCellspaces* to be generated by decomposition, load-balancing mode (static or dynamic), data I/O mode (no output, centralized I/O, or pseudo parallel I/O), and writer mode (with or without).

To test the performance of the program, a GeoTIFF file (1.93 GB) containing the DEM data of California was used as the input data (Fig. 5a). The input data includes 40,460 × 23,851 cells at 30-m resolution, and the elevation values are stored as *unsigned short integers*. The output data were written into GeoTIFF files (Fig. 5b and c) with the same dimensions and spatial reference as the input data, and the slope and aspect values are stored as *single-precision floating-point numbers*. Each output file's size was about 3.9 GB.

All experiments were conducted using the regular decomposition methods (i.e., row-wise, column-wise, and block-wise). Also, the scattered mapping technique (i.e., small-granularity decomposition and each process is assigned with multiple *SubCellspaces* that are scattered over the space) was used in the experiments, thus the chance for workload imbalance among processes was reduced [32].

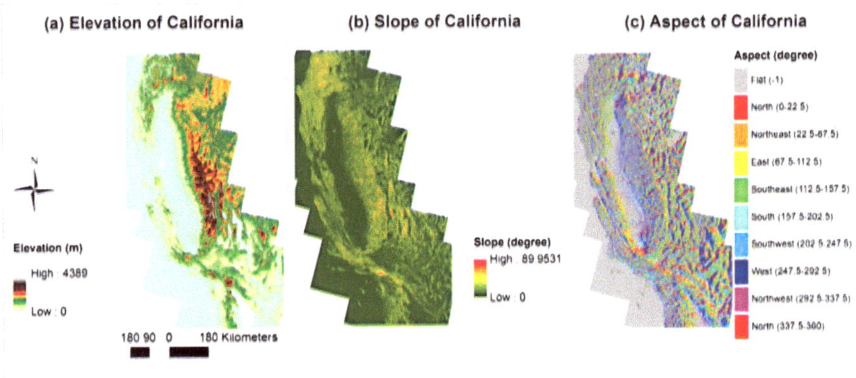

Fig. 5 Elevation, slope, and aspect of California

Our experiments showed that the column-wise and block-wise decomposition methods achieved a little lower speed-up than row-wise decomposition, because of the strip (i.e., row-wise) storage mode of the input GeoTIFF file. The following performance assessment focuses on the experiments using row-wise decomposition. Each *Layer* was decomposed into 1024 *SubCellspaces*, each of which consists of $40 \times 23{,}851$ or $39 \times 23{,}851$ cells to be processed locally, plus a ring of "halo" cells.

Without writing the output slope and aspect data, a sequential program took 4428.01 s (over an hour) to complete using one CPU core on the computer cluster. The parallel program greatly reduced the computing time by deploying multiple processes (i.e., CPU cores), as shown in Fig. 6. With 512 processes, using the dynamic load-balancing and parallel reading modes, the parallel program completed in 14.11 s, achieving a speed-up of 313.81[2]. Dynamic load-balancing outperformed static load-balancing in most cases, indicating the task-farming technique did improve the performance. The obvious exception occurred when four processes were used, in which case dynamic load-balancing was slower than static load-balancing. In the dynamic load-balancing mode, the master process is responsible for dynamically assigning *SubCellspaces* to the worker processes in response to their requests, and does not participate in the actual computation. When a small number of processes are used, isolating a process as the master means losing a significant proportion of the computing power, hence leading to poorer performance.

Figure 7 shows that parallel reading largely reduces the input time by allowing all processes to read their assigned *SubCellspaces* in parallel. When more than 16 processes were used, the reading time in the parallel mode was around 1 s, whereas in the centralized mode, the reading time was mostly above 10 s. When fewer processes were used, each process had to read a large number of *SubCellspaces*,

[2]Speed-up is a commonly used performance measurement of parallel computing, and is calculated as the ratio between the sequential computing time and the parallel computing time.

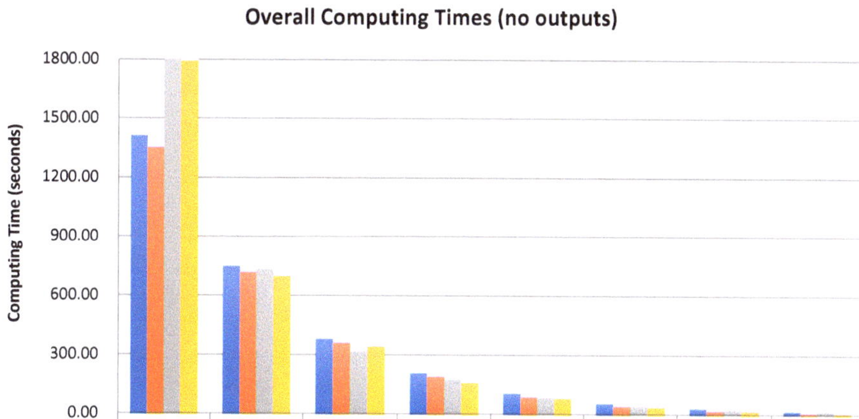

	4	8	16	32	64	128	256	512
SLB, CI	1418.75	754.39	386.37	213.50	110.04	59.41	36.03	21.16
SLB, PI	1360.85	725.81	367.92	196.52	94.80	48.25	25.15	14.15
DLB, CI	1811.05	742.38	326.61	184.56	91.85	49.18	30.98	21.67
DLB, PI	1801.71	707.42	352.70	168.44	90.10	46.09	25.16	14.11

Fig. 6 Overall computing times for no-output experiments (the X axis indicates the number of processes; *SLB* static load-balancing, *DLB* dynamic load-balancing, *CI* centralized input, *PI* parallel input)

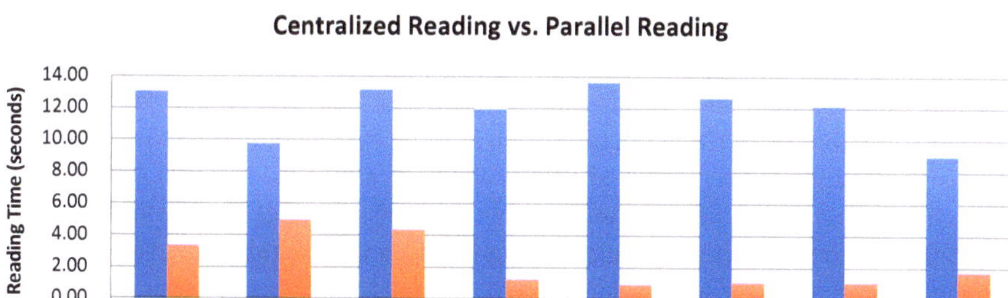

	4	8	16	32	64	128	256	512
CI	13.14	9.83	13.26	12.04	13.72	12.75	12.22	9.07
PI	3.48	5.06	4.47	1.33	1.04	1.18	1.18	1.82

Fig. 7 Times for centralized and parallel reading

leading to a longer reading time. Even so, parallel reading was still quicker than centralized reading.

When writing the output files, a sequential program took 4881.9 s to complete. With 512 processes including a writer, using the dynamic load-balancing and centralized I/O modes, the parallel program completed in 251.5 s, achieving a speed-up of 19.4. By comparing the results with that of the no-output experiments, we concluded that the writing operations took a large proportion of the overall computing time.

As shown in Fig. 8, in most cases, dynamic load-balancing outperformed static load-balancing when other conditions are the same. A writer process greatly reduced

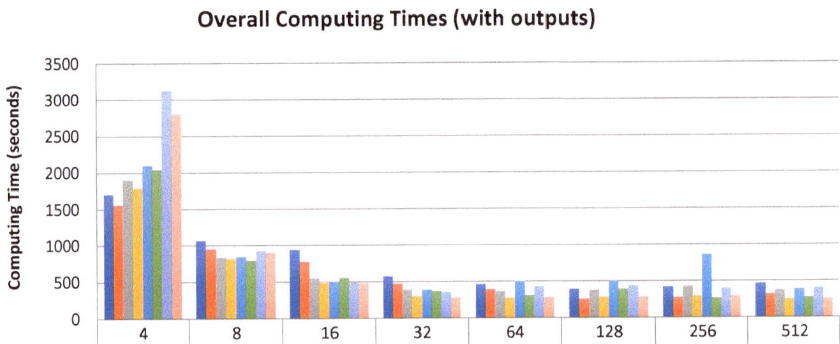

	4	8	16	32	64	128	256	512
SLB, NW, PIO	1709	1070.29	938.961	577.637	464.38	394.013	419.949	465.515
SLB, NW, CIO	1561.85	952.027	777.583	473.188	394.98	254.18	274.533	318.525
DLB, NW, PIO	1908.15	836.472	551.395	393.539	368.231	380.869	424.208	372.692
DLB, NW, CIO	1797.76	825.354	495.574	302.226	278.502	285.303	296.23	252.4
SLB, WW, PIO	2108.12	846.715	504.639	388.886	499.013	495.978	860.441	389.887
SLB, WW, CIO	2047.18	793.721	555.443	371.606	308.025	393.273	263.153	271.8
DLB, WW, PIO	3134.71	926.278	484.983	356.48	428.696	436.566	395.416	402.111
DLB, WW, CIO	2805.99	903.365	478.2	281.963	282.904	287.542	295.447	251.531

Fig. 8 Overall computing times for with-outputs experiments (*NW* no writer, *WW* with writer, *CIO* centralized I/O, *PIO* pseudo parallel I/O)

the overall computing time in the static load-balancing mode, while in the dynamic load-balancing mode, using a writer did not help improve the performance. This is because in the static load-balancing mode, the master also participates in the actual computation, and the writing operations have to be executed by the master after all its assignments have been finished if a writer does not exist. A writer, taking over the writing operations, can dynamically write *SubCellspaces* to the output file in response to other processes' requests, thus reduces the overall time. On the other hand, in the dynamic load-balancing mode, the master does not participate in the actual computation, and also dynamically writes data to the output files. Using a writer means losing a worker process, leading to poorer performance.

In almost all cases, parallel I/O yielded poorer performance than centralized I/O. As mentioned above, parallel reading could help reduce the time for data input. We therefore concluded that pseudo parallel writing was outperformed by centralized writing in these experiments. Pseudo parallel writing includes multiple reading and writing operations (i.e., the I/O of temporary files), which degrades the performance. Thus a better solution for parallel writing that allows processes to directly write subsets of data into the output datasets is needed, such as the parallel GeoTIFF I/O of the Terrain Analysis Using Digital Elevation Models (TauDEM, http://hydrology.usu.edu/taudem/taudem5/index.html) and the parallel netCDF parallel I/O of the pIOLibrary (http://sandbox.cigi.illinois.edu/pio/).

Parallel Spatio-Temporal Modeling—Cellular Automata

A classical Cellular Automata (CA) consists of a set of identically shaped and sized cells, each of which is located in a regular, discrete cellspace. Each cell is associated with a state within a finite set. The model evolves in discrete time steps, changing the states of cells according to transition rules, homogeneously and synchronously applied at every step. The new state of a certain cell depends on the previous states of a set of cells, which include the cell itself and its neighbors.

With the naturally embedded space and time properties, CA provides a straightforward approach for spatio-temporal dynamic simulations, and has been widely used in geospatial studies, such as land-use and land-cover change [41–44], wildfire propagation [45], and freeway traffic [46, 47].

In this showcase, we parallelized a classical CA model—the Game of Life (GOL), using pRPL. More complex CA models can be implemented using the same strategy (see [31]).

In the GOL, a cell can live or die depending on the condition of its 3 × 3 neighborhood, according to a simple transition rule. As a result, the living status of the cells can represent various spatial patterns throughout the course of iterations. The pseudo code of the GOL's transition rule is as follows:

```
FUNCTION Transition (cell, time_t)
 n = number of alive neighbors of cell at time_t
 IF cell is alive at time_t
   IF n ≥ 4
     THEN cell dies of overcrowding at time_t+1
   IF n ≤ 1
     THEN cell dies of loneliness at time_t+1
   IF n = 2 OR n = 3
     THEN cell survives at time_t+1
 ELSE (i.e., cell is dead at time_t)
   IF n = 3
     THEN cell becomes alive (i.e., born) at time_t+1
```

Similar to the slope and aspect calculation, the GOL's transition rule is a focal operation that can be easily parallelized. What makes the GOL more complex is that the iterative execution of the transition rule requires data exchange among the parallel processes at every iteration in order to update the "halo" cells of *SubCellspaces*. As mentioned in section ""Update-On-Change" and "edgesFirst" for Data Exchange", pRPL is able to automatically handle the data exchange while optimizing the performance. In this study, a customized *Transition* was developed to implement the GOL transition rule, and a GOL program to implement the simulation procedure. The GOL program first initializes a user-defined number of seeds (i.e., alive cells) that are randomly distributed over the space, iteratively executes the GOL *Transition* on all cells, and reports the number of cells alive at each iteration.

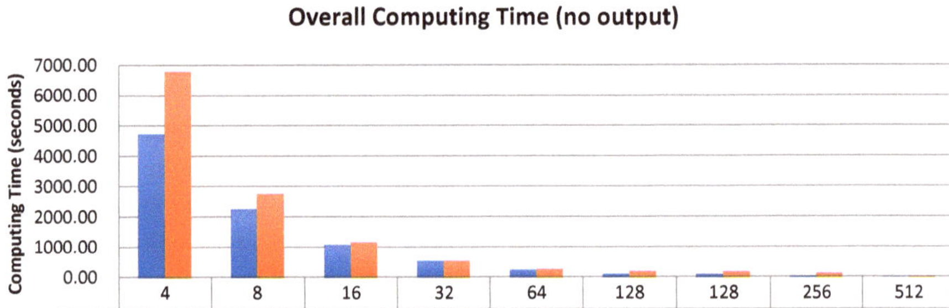

	4	8	16	32	64	128	128	256	512
SLB	4764.84	2300.02	1124.14	580.42	280.64	140.02	139.67	70.41	35.70
DLB	6846.83	2818.59	1212.62	589.10	314.82	240.91	229.71	178.05	54.01

Fig. 9 Overall computing times for GOL simulations

The parallel GOL program provides both dynamic and static load-balancing options. Dynamic load-balancing can only be used in the random seeding procedure, during which the master dynamically assigns *SubCellspaces* to workers in responses to their requests. Once complete, all *SubCellspaces* are assigned to workers. During the iterative execution of the GOL *Transition*, the assignments for workers stay static.

Without writing any output data, a sequential program took 12,242.7 s (about 3.5 h) to complete a 10-iteration simulation on a 20,000 × 20,000 cellspace with 80,000,000 initial seeds. All parallel experiments used row-wise decomposition, and the cellspace was divided into 1024 *SubCellspaces*. With 512 processes, the parallel program completed the same simulation in 35.7 s with a speed-up of 343 using static load-balancing, and 54 s with a speed-up of 227 using dynamic load-balancing (Fig. 9). The dynamic load-balancing (i.e., task-farming technique) yielded poorer performance than static load-balancing. This is mainly because the computational intensity of the random seeding procedure is quite different from that of the GOL *Transition*. The task mapping (i.e., assignments for workers) generated during the random seeding is not optimized for the GOL *Transition's* execution. Also, as pointed out by Guan and Clarke [31], a spatio-temporal model such as CA changes the spatial distribution of workload as it evolves because the cell values change at each iteration. Thus the task-farming technique is not suitable for such dynamic models.

When writing the final result, a sequential program completed the same simulation in 12,206 s. With 512 processes (no writer) and centralized writing, the parallel program completed in 90 s, achieving a speed-up of 135 (Fig. 10). Using a writer process yielded poorer performance in most cases. This is because the majority of the overall time was used for the iterative computation, and the writing only took place in the end. Pseudo parallel writing outperformed centralized writing when small numbers of processes were used (i.e., less than 128), which is different from the slope/aspect calculation experiments. This is because the output file of the GOL experiments was much smaller (800 MB), therefore the time for creation and writing was considerably shorter. When a small number processes were used,

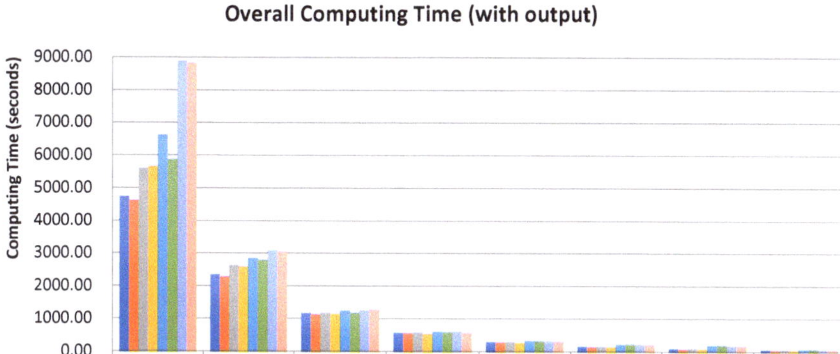

	4	8	16	32	64	128	256	512
SLB, NW, CW	4785.12	2388.58	1201.32	608.68	329.21	190.20	134.71	90.09
SLB, NW, PW	4667.26	2329.50	1174.56	604.09	328.93	191.87	127.72	91.48
SLB, WW, CW	5656.28	2684.94	1229.05	621.35	340.10	198.90	144.87	100.19
SLB, WW, PW	5720.81	2637.69	1210.24	591.25	326.54	198.07	142.45	102.69
DLB, NW, CW	6662.71	2887.50	1280.03	645.31	372.91	258.63	234.97	104.61
DLB, NW, PW	5923.02	2841.74	1221.97	638.74	373.29	271.93	242.76	108.32
DLB, WW, CW	8935.73	3129.91	1295.49	652.91	368.28	261.32	221.60	103.99
DLB, WW, PW	8874.42	3093.60	1324.20	623.56	355.52	264.44	230.41	105.35

Fig. 10 Overall computing times for with-output experiments (*CW* centralized writing, *PW* pseudo parallel writing)

writing temporary files to the shared disk space was more efficient than transferring the output *SubCellspaces* to the master/writer. However, when a large number of processes were used, the I/O channel of the shared disk space was saturated by the overwhelming number of concurrent writing operations.

Conclusion

Big spatial data and complex geospatial algorithms demand massive computing power that may largely exceed the capability of individual desktop computers. High-performance GeoComputation, by utilizing parallel computing technologies, provides promising solutions to overcoming the computational barriers and enables complex analytics and modeling using high-resolution data for large-area studies. Nevertheless, the high complexity of developing parallel geospatial algorithms has become a major bottleneck that discourages GIScientists and GeoComputation practitioners to exploit the advantages of high-performance computing in geospatial studies.

The parallel Raster Processing Library (pRPL), a general-purpose and open-source programming library, enables transparent parallelism by providing easy-to-use interfaces for users to parallelize application-specific raster-processing algorithms with minimal knowledge and skills of parallel computing. pRPL allows users to focus on their own algorithms, and automatically handles the underlying parallel computing, including domain decomposition, assignment mapping, algorithm execution, data exchange, load-balancing, and data I/O.

This paper presents two cases of high-performance geospatial computing implemented using pRPL: (1) the slope/aspect calculation, as an example of a wide range of spatial analytics that are based on local or focal operations; and (2) the Game of Life (GOL), a typical Cellular Automata model that represents spatio-temporal dynamic simulations.

The experiments showed that pRPL largely reduced the computing time. While parallel reading could effectively reduce the time for data input, the writing of large output datasets were found to be one of the main bottlenecks of performance. Therefore a true parallel writing mechanism is needed, which should be able to allow multiple processes to directly write data to the output datasets concurrently. Dynamic load-balancing outperformed static load-balancing in the slope/aspect calculation, indicating the task-farming technique did improve the efficiency for such non-iterative algorithms. However, it yielded lower performance in the GOL experiments, because the task-farming was only used for the random seeding procedure, not the iterative evolvement. Also, the task-farming technique is not suitable for spatio-temporal simulations because the spatial distribution of workload may change as the cell values change during the iterative procedure. Using a writer process was useful for improving the performance in the static load-balancing mode, because the writer is able to dynamically write subsets of data into the final output files.

In summary, pRPL provides a variety of options for parallel geospatial computing, and these options should be carefully chosen according to the characteristics and requirements of the algorithms and datasets, the parallel computing environments, and users' preferences.

References

1. Vatsavai R, Chandola V, Klasky S, Ganguly A, Stefanidis A, Shekhar S (2012) Spatiotemporal data mining in the era of big spatial data: algorithms and applications. Presented at the 1st ACM SIGSPATIAL international workshop on analytics for big geospatial data, Redondo Beach, CA, USA
2. Openshaw S (2000) GeoComputation. In: Openshaw S, Abrahart RJ (eds) GeoComputation. Taylor & Francis, New York, pp 1–31
3. Harris B (1985) Some notes on parallel computing: with special reference to transportation and land-use modeling. Environ Plan A 17(9):1275–1278
4. Li X, Zhang X, Yeh A, Liu X (2010) Parallel cellular automata for large-scale urban simulation using load-balancing techniques. Int J Geogr Inf Sci 24(6):803–820
5. Sandu JS, Marble DF (1988) An investigation into the utility of the Cray X-MP supercomputer for handling spatial data. In: Third international symposium on spatial data handling, Sydney, Australia, pp 253–266
6. Li B (1992) Opportunities and challenges of parallel spatial data analysis: initial experiments with data parallel map analysis. In: GIS LIS-international conference, San Jose, pp 445–458

7. Smith TR, Gao P, Gahinet P (1989) Asynchronous, iterative, and parallel procedures for solving the weighted-region least cost path problem. Geogr Anal 21(2):147–166
8. Wang F (1993) A parallel intersection algorithm for vector polygon overlay. IEEE Comput Graph Appl 13(2):74–81
9. Rokos D-K, Armstrong MP (1992) Parallel terrain feature extraction. In: Proceedings of GIS/LIS'92, Bethesda, MD, 1992, vol 2, pp 652–661
10. Puppo E, Davis L, DeDemthon D, Teng Y (1994) Parallel terrain triangulation. Int J Geogr Inf Sci 8(2):105–128
11. Kidner DB, Rallings PJ, Ware JA (1997) Parallel processing for terrain analysis in GIS: visibility as a case study. GeoInformatica 1(2):183–207
12. Zhao Y, Padmanabhan A, Wang S (2013) A parallel computing approach to viewshed analysis of large terrain data using graphics processing units. Int J Geogr Inf Sci 27(2):363–384
13. Armstrong MP, Marciano RJ (1997) Massively parallel strategies for local spatial interpolation. Comput Geosci 23(8):859–867
14. Cramer BE, Armstrong MP (1997) Interpolation of spatially inhomogeneous data sets: an evaluation of parallel computation approaches. In: Proceedings of GIS/LIS'97, Bethesda, MD
15. Kerry KE, Hawick KA (1998) Kriging interpolation on high-performance computers. In: Proceedings of the international conference and exhibition on high-performance computing and networking, pp 429–438
16. Wang S, Armstrong MP (2003) A quadtree approach to domain decomposition for spatial interpolation in grid computing environments. Parallel Comput 29(10):1481–1504
17. Guan Q, Kyriakidis PC, Goodchild MF (2011) A parallel computing approach to fast geostatistical areal interpolation. Int J Geogr Inf Sci 25(8):1241–1267
18. Wang S, Armstrong MP (2009) A theoretical approach to the use of cyberinfrastructure in geographical analysis. Int J Geogr Inf Sci 23(2):169–193
19. Wang S (2010) A CyberGIS framework for the synthesis of cyberinfrastructure, GIS, and spatial analysis. Ann Ass Am Geogr 100(3):535–557
20. Yang C, Raskin R, Goodchild M, Gahegan M (2010) Geospatial cyberinfrastructure: past, present and future. Comput Environ Urban Syst 34(4):264–277
21. Yang C et al (2011) Spatial cloud computing: how can the geospatial sciences use and help shape cloud computing? Int J Digital Earth 4(4):305–329
22. Yang C, Xu Y, Nebert D (2013) Redefining the possibility of digital earth and geosciences with spatial cloud computing. Int J Digital Earth 6(4):297–312
23. Tang W, Wang S, Bennett DA, Liu Y (2011) Agent-based modeling within a cyberinfrastructure environment: a service-oriented computing approach. Int J Geogr Inf Sci 25(9):1323–1346
24. Shook E, Wang S, Tang W (2013) A communication-aware framework for parallel spatially explicit agent-based models. Int J Geogr Inf Sci 27(11):2160–2181
25. Huang Q, Yang C, Benedict K, Chen S, Rezgui A, Xie J (2013) Utilize cloud computing to support dust storm forecasting. Int J Digital Earth 6(4):338–355
26. Liu Y, Sun AY, Nelson K, Hipke WE (2013) Cloud computing for integrated stochastic groundwater uncertainty analysis. Int J Digital Earth 6(4):313–337
27. Zhang J, You S (2013) High-performance quadtree constructions on large-scale geospatial rasters using GPGPU parallel primitives. Int J Geogr Inf Sci 27(11):2207–2226
28. Shi X, Ye F (2013) Kriging interpolation over heterogeneous computer architectures and systems. GISci Remote Sens 50(2):196–211
29. Ding Y, Densham PJ (1996) Spatial strategies for parallel spatial modelling. Int J Geogr Inf Syst 10(6):669–698
30. Guo M, Guan Q, Xie Z, Wu L, Luo X, Huang Y (2015) A spatially adaptive decomposition approach for parallel vector data visualization of polylines and polygons. Int J Geogr Inf Sci 29(8):1419–1440
31. Guan Q, Clarke K (2010) A general-purpose parallel raster processing programming library test application using a geographic cellular automata model. Int J Geogr Inf Sci 24(5): 695–722
32. Mineter MJ (1998) Partitioning raster data. In: Healey RD, Dowers S, Gittings B, Mineter MJ (eds) Parallel processing algorithms for GIS. Taylor & Francis, Bristol, PA, pp 215–230

33. Benedičič L, Cruz FA, Hamada T, Korošec P (2014) A GRASS GIS parallel module for radio-propagation predictions. Int J Geogr Inf Sci 28(4):799–823
34. Qin C-Z, Zhan L (2012) Parallelizing flow-accumulation calculations on graphics processing units—from iterative DEM preprocessing algorithm to recursive multiple-flow-direction algorithm. Comput Geosci 43:7–16
35. Cheng G, Liu L, Jing N, Chen L, Xiong W (2012) General-purpose optimization methods for parallelization of digital terrain analysis based on cellular automata. Comput Geosci 45:57–67
36. Qin C-Z, Zhan L-J, Zhu A-X, Zhou C-H (2014) A strategy for raster-based geocomputation under different parallel computing platforms. Int J Geogr Inf Sci 28(11):2127–2144
37. Guan Q, Zeng W, Gong J, Yun S (2014) pRPL 2.0: improving the parallel raster processing library. Trans GIS 18:25–52
38. Gropp W et al (1998) MPI: the complete reference, Vol. 2. The MIT Press, Cambridge, MA
39. Hutchinson D et al (1996) Parallel neighborhood modeling: research summary. CAM Press, New York
40. Burrough PA, McDonnell R, Burrough PA, McDonnell R (1998) Principles of geographical information systems, vol 333. Oxford university press, Oxford
41. Clarke KC, Hoppen S, Gaydos L (1997) A self-modifying cellular automaton model of historical urbanization in the San Francisco Bay area. Environ Plann B Plann Des 24(2): 247–261
42. Wu F, Webster CJ (1998) Simulation of land development through the integration of cellular automata and multi-criteria evaluation. Environ Plann B 25(1):103–126
43. Yeh AGO, Li X (2002) Urban simulation using neural networks and cellular automata for land use planning. In: Richardson D, van Oosterom P (eds) Advances in spatial data handling. The University of Michigan Press, Ann Arbor, pp 451–464
44. Silva EA, Clarke KC (2002) Calibration of the SLEUTH urban growth model for Lisbon and Porto. Comput Environ Urban Syst 26(6):525–552
45. Clarke KC, Riggan P, Brass JA (1995) A cellular automaton model of wildfire propagation and extinction. Photogramm Eng Remote Sens 60(11):1355–1367
46. Nagel K, Schreckenberg M (1992) A cellular automaton model for freeway traffic. J Phy I Fr 2:2221–2229
47. Benjamin SC, Johnson NF, Hui PM (1996) Cellular automata models of traffic flow along a highway containing a junction. J Phys A Math Gen 29(12):3119–3127

Part II
Agent-based Systems and Microsimulations

'Can You Fix It?' Using Variance-Based Sensitivity Analysis to Reduce the Input Space of an Agent-Based Model of Land Use Change

Arika Ligmann-Zielinska

Introduction

Land use system complexity originates from the interplay of key system drivers that form a web of reciprocal relationships resulting in nonlinearities, path dependence, and feedbacks across space, time, and scale [1–7]. One way to study a land system is by using an agent-based model (ABM), in which land operators (like developers, farmers, residents, businesses) are represented by distinct computational entities (called agents) that populate a common virtual environment, which reflects the target land system.

Uncertainty in ABMs is particularly important. These models require a large number of explanatory variables that describe the spatial, social, environmental, and socio-environmental components of the system, often combined using nonlinear functions and advanced algorithms. A common approach to addressing ABM uncertainty is to represent (a portion of) model inputs as probability density functions (PDFs) which are sampled many times with each sample used in a distinct model execution. This process, known as Monte Carlo simulation, produces distributions of outputs. Consequently, input uncertainty is propagated through the model and reflected in output distribution. While this uncertainty propagation (aka uncertainty analysis—UA) effectively quantifies the variability of results due to stochastic or ill-defined inputs, it does not reveal which of these inputs are instrumental in shaping output variability. Finding the drivers of output uncertainty can be useful in building a more parsimonious and transparent model with reduced parameter dimensionality.

A. Ligmann-Zielinska (✉)
Department of Geography, Environment, and Spatial Sciences, Michigan State University, Geography Building, 673 Auditorium Rd, Room 121, East Lansing, MI 48824, USA
e-mail: ligmannz@msu.edu

J.-C. Thill, S. Dragicevic (eds.), *GeoComputational Analysis and Modeling of Regional Systems*, Advances in Geographic Information Science,
DOI 10.1007/978-3-319-59511-5_6

It is therefore apparent that ABM output variability has to be subdivided to prioritize the influence of inputs, leading to input prioritization and model simplification [8, 9]. One approach to investigate model output variability is through sensitivity analysis (SA), which tests model response to changes in its inputs [10, 11].

Simplification has long been identified as one of the core steps in model development and application [12–23]. The principle of Occam's razor is often cited as a philosophical rationale for simplification. This principle states that, among competing explanations, we should adopt explanations with the lowest number of assumptions. Various methods of simplification have been proposed. In their seminal paper, Innis and Rexstad [19] provide a comprehensive review of 15 model simplification methods including optimization, filtering, sensitivity analysis, structure and logic tests, applying dimensionless variables (generalization), meta-modeling, analytical solutions, time constraints (e.g. a slowly changing system component may be represented as a model constant), and reduction of parameter space by eliminating collinearities. For their vegetation model, Moore and Noble [20] employ Bayesian analysis to generate a directed graph of a given size, leading to a simplified model at a specific resolution. The procedure is to discretize the input-output space in order to obtain a representative yet finite approximation of the otherwise continuous input-output data. Saysel and Barlas [22] use structure validity tests to generate models equivalent to their original system dynamics model of irrigation. Crout et al. [16] demonstrate a method of creating simpler models derived from a base model by systematically substituting its variables with constants—a procedure called model reduction by variable replacement. Finally, in their seminal paper on ABM simplification, Edmonds and Moss [18] propose a method of simplification that starts from an extensive model which accounts for the widest possible range of evidence and is simplified only when the evidence justifies it. They argue that simplification should be context-dependent. They also stress that, when simplifying a model, we should focus on retaining only the relevant behaviors in order to address a given problem. In this chapter, I postulate that these contexts can be embedded in model outputs. Consequently, I propose to utilize SA as a means of model simplification through factor fixing.

In the sections that follow, I propose a framework for ABM simplification by employing uncertainty and sensitivity analysis. UA is used to generate a distribution of outputs, which is further summarized using variance—a simple yet succinct measure of result variability. What follows is SA based on variance decomposition [24, 25], in which the variance is apportioned to model inputs, in order to quantify which of them and to what extent affect the variability of ABM results. Variance-based SA computes sensitivity indices that represent the fractional contribution of each input to output variance. The reported framework has very practical implications. By comparing the values of sensitivity indices, we can prioritize which inputs have a negligible effect on output variability (inputs with low values of sensitivity indices), and which inputs are the critical drivers of output uncertainty (inputs with high sensitivity). Each unimportant input can be fixed to some representative value (like a mean or a mode) leading to a reduction in ABM parameter space i.e., model simplification. Each influential input can be refined in order to improve ABM accuracy or reduce output variability in future model improvements. For example, finding critical model components allows for prioritization of input measurements, that is, efficient allocation of resources for future data acquisition [26]. While the reported framework is applied in the spatial ABM context, it can be easily used to evaluate the uncertainty of other complex systems models.

The UA and SA framework is demonstrated using an ABM of agricultural land conservation. The model emulates a process of farmer enrollment in U.S. Conservation Reserve Program (CRP) [27]. The model comprises two types of agents: farmers, who make decisions on their individual CRP participation, and the Farm Service Agency, which evaluates, selects, and accepts the enrollment offers made by farmers. A positive enrollment decision leads to the conversion of land use from row crop/pasture to fallow. The results of the ABM are maps of land use change, which are summarized using a number of metrics, from total fallow land acreage, through various measures of land use compactness and contiguity, to cost of land retirement. The distribution of each metric is used separately in variance-based SA, leading to alternative input prioritizations and potential model simplifications, depending on the type of output variable. Variance-based SA has been applied to spatial modeling in a number of previous studies [8, 24, 28–34]. What sets this framework apart is its emphasis on model simplification guided by a number of different outputs.

The chapter is organized as follows. Section "Comprehensive Uncertainty and Sensitivity Analysis of Agent-based Models of Land Use Change" sets the backdrop of this study by describing the UA and SA framework. Section "ABM of Agricultural Land Conservation and Model Setup" details the ABM of agricultural land conservation, the PDFs of data used in simulations, and the design of computational experiments. The results are presented in section "Results of the Original ABM", first by describing the UA and then by reporting the SA and the proposed model simplification. In section "Model Simplification and Discussion" I demonstrate the results of ABM simplification and provide some practical guidelines on choosing the right path to building a transparent model with the presented methodology. Section "Conclusions" concludes the chapter.

Comprehensive Uncertainty and Sensitivity Analysis of Agent-Based Models of Land Use Change

The roots of formal SA may be traced to engineering, scientific predictive modeling, and decision science [11, 30, 35, 36]. SA is frequently perceived as a tedious step in modeling that can be omitted without a significant loss of information about model performance. However, SA offers many benefits to improve the relevance of ABM to land use science and policy. Not only does it improve model validity by recognizing its critical inputs, but it also provides means of model

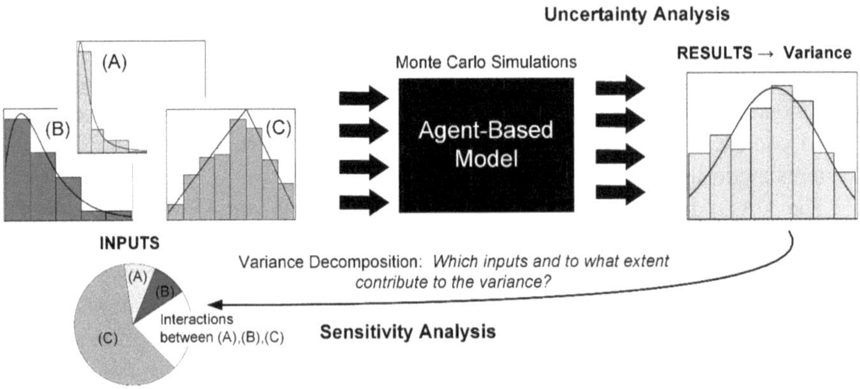

Fig. 1 A framework for model simulation coupled with uncertainty and sensitivity analysis

simplification (input reduction), which is especially valuable when dealing with highly dimensional ABMs. Finding important inputs allows for prioritizing data refinement to build better, operational versions of the model. SA can also assist in choosing the most accurate representation of the spatial system and, consequently, it can contribute to theory development [28]. For these reasons, I argue that SA plays an integral role in ABM development and application, requiring comprehensive methods that systematically examine model input and output uncertainty.

Framework

At the outset, it is important to distinguish between model uncertainty and sensitivity analysis (Fig. 1). By definition, models with varying inputs produce a number of output values, requiring a large number of model executions. The output values are then compiled into a probability distribution. This stage of modeling constitutes UA. The objective of UA is therefore to evaluate how the uncertainty of inputs propagates through the model and affects the values of its output variables. What follows is SA which evaluates how much each source of input uncertainty contributes to model output variability [11, 24].

The key issue in quantitative SA is to decide on the statistics summarizing the distribution of a given output. The selected statistic is then partitioned and distributed among inputs. Due to a number of reasons, variance is the most common statistic applied to evaluate the importance of inputs in shaping the variability of results [11, 37]. First, variance-based SA is model-independent, meaning that model functional complexity does not constrain the validity of SA. Second, variance has the capacity of capturing the influence on output variability of the full range of input variation, including the interaction effects among inputs. Finally, variance-based SA can deal with groups of inputs (e.g. one group of income inputs for all g agents in the model, rather than g number of separate income inputs—one for each agent) leading to computationally more efficient and analytically more useful SA. The procedure of variance decomposition is described in the following subsection.

Variance-Based Sensitivity Analysis

Variance-based SA decomposes the total unconditional variance (V) of the distribution of output Y caused by the changes in K model inputs and allocates each portion of V to individual input i (V_i) as well as i's interactions with other inputs (with increasing order effects):

$$V = V_i + V_j + \cdots + V_k + V_{ij} + \cdots + V_{ik} + V_{ijk} + V_{ij\ldots k} \qquad (1)$$

The decomposed variance is used to compute two sensitivity indices for every i. The first order sensitivity index (S_i or S for short) is a measure that quantifies the fractional contribution to output variance of i taken independently from other inputs [38, 39]:

$$S_i = \frac{V_i}{V} \qquad (2)$$

where $V_i = V[E(Y|X_i)]$ is a variance of the expected value of Y assuming a fixed i. A relatively high S_i denotes an input that substantially contributes to the variability of Y. Trivially, the sum of all $\mathrm{S_i}$ (S-sum) must be less than or equal to one. For additive models the S-sum equals one, meaning that all the variance of Y can be explained by the first-order effects alone. If this is the case, we can use a number of computationally less expensive techniques of SA like the correlation coefficients or the standardized regression coefficients [39]. In many complex models, however, the inputs interact with each other in a nonlinear manner. Spatial ABMs also exhibit such behavior [8, 33, 34]. The fractional contribution of nonlinear effects to the output variance is calculated by subtracting the S-sum from one. Nevertheless, this measure of interactions does not explain the partial interactions of individual inputs. The latter can be captured by using a total effects sensitivity index of i (ST_i or ST for short), which quantifies the entire fractional contribution to V of i including all of its interactions with other inputs [40, 41]:

$$ST_i = 1 - \frac{V\left[E\left(Y|X_{\sim i}\right)\right]}{V} \qquad (3)$$

where $V[E(Y|X_{\sim i})]$ is the overall contribution to Y's variance caused by all inputs but i. With ST, all first and higher order terms involving i are conveniently represented by a single index. Note that, for a large K, we would have to compute and interpret as many as $2^k - 1$ first and higher order measures of influence—an impractical and laborious task. Observe that the sum of all ST_i (ST-sum) must be greater than or equal to one. Whenever S-sum and ST-sum are equal (to one), no interactions affecting the variance of Y exist among model inputs. To obtain the sole interaction effects of i we need to calculate the difference between ST_i and S_i. More details on computing the S and ST can be found in Chap. 4 in Saltelli et al. [39].

Simple Example of Variance-Based SA

Consider a toy ABM of farmers growing corn on their land. Each farmer makes a decision on whether or not to grow corn based on its last season prices (P), the forecasted weather (W), and land productivity (L). The output variable of the model is the total area of grown corn with variance V_{Corn}, which can be broken down as follows:

$$V_{Corn} = V_{Corn}^{P} + V_{Corn}^{W} + V_{Corn}^{L} + V_{Corn}^{PW} + V_{Corn}^{PL} + V_{Corn}^{WL} + V_{Corn}^{PWL} \quad (4)$$

where $V_{Corn}^{P}$is the variance in corn area due to price, V_{Corn}^{PW} is the variance in corn area due to price and weather, and V_{Corn}^{PWL} is the variance in corn area due to the three inputs combined. Assume that the SA results in variances presented in Table 1.

The corresponding sensitivity indices are presented in Table 2.

Based on the computed indices we can conclude that P has the highest influence on corn area variability, followed by W, followed by L. Interestingly, the interaction effects of P and L are about the same (~5%), and yet the major first effect of P is six times that of L. Moreover, although W has a lower impact on output variance than P when treated singly (33% < 50%), its interaction effect is actually higher (8% > 5%). Only 8.7% of corn area variance can be attributed to input interactions (1–0.913).

The ABM simplification would not be advisable. Based on the value of S, input L is a potential candidate for fixing because only 8.3% of variance can be explained by L alone. However, considering the interaction effects, the role of L in explaining

Table 1 Total and fractional variances of corn area obtained from a simple farm ABM

V_{Corn}	V_{Corn}^{P}	V_{Corn}^{W}	V_{Corn}^{L}	V_{Corn}^{PW}	V_{Corn}^{PL}	V_{Corn}^{WL}	V_{Corn}^{PWL}
0.06	0.03	0.02	0.005	0.002	0.0005	0.002	0.0005

Table 2 Sensitivity indices for the example farm ABM

	S	ST	$ST - S$
P	$0.03/0.06 = 0.5$	$(0.03 + 0.002 + 2 \times 0.0005)/0.06 = 0.55$	0.05
W	$0.02/0.06 = 0.33$	$(0.02 + 2 \times 0.002 + 0.0005)/0.06 = 0.41$	0.08
L	$0.005/0.06 = 0.083$	$(0.005 + 0.002 + 2 \times 0.0005)/0.06 = 0.13$	0.047
Sum	0.913	1.09	

output variance increases to 11.9% (0.13/1.09). Consequently, we should always evaluate model simplification based on the values of ST rather than S.

Finally, P proves to be the most important input that drives the variance of corn area (~50% overall contribution). We conclude that, to improve the ABM accuracy, we should focus our data collection efforts on refining the PDF of price.

Sampling

Estimating the S and ST indices is computationally demanding. Based on the author's experience, eight inputs usually require more than 2000 model runs. For this reason, a number of sampling designs have been proposed and tested [25] with Sobol' quasirandom sampling outperforming other methods [42–45]. This method systematically probes the input space, leading to more uniformly distributed sample points in the multidimensional unit cube when compared to simple random sampling [46]. Consequently, Sobol' quasirandom sampling is used in the ABM experiments reported herein.

ABM of Agricultural Land Conservation and Model Setup

The ABM of agricultural land conservation simulates the enrollment in CRP—a government-sponsored agricultural land retirement program in the U.S. CRP was established in 1985 to conserve land production resources by reducing soil erosion [27, 47]. It was later amended to account for practices targeted toward improving water quality and maintaining biodiversity. The goal of CRP is to convert row crop and pasture lands back to natural land cover. Installation and practice of conservation activities is done on a volunteer basis. Farmers are paid for establishing and maintaining the conservation practices.

The ABM operates as follows (Fig. 2). Every time step (year), a farmer agent (FA) decides whether it is willing to enroll in the program. The decision is made based on individual values of selected sociodemographics including land tenure (owner or renter), farmer retirement (yes/no), value of production on the farm (in U.S. dollars), as well as the attitude to risk represented by FA's decision rule that also considers behavior of FA's neighboring agents. The first three decision criteria are simple statistical variables obtained from a survey [48]. The decision rule comprises a set of aggregation functions and a sub-model. The sub-model depicts a simple mechanism of social network influence, embodied in spatial interactions at an individual level, which allow for simulating the spatial diffusion of CRP participation through proximal agents [49]. The risk-based decision rule is defined using Ordered Weighted Averaging—OWA [50]. The OWA set is composed of rules spanning the continuum from completely risk-averse (all decision criteria must be above zero) to completely risk-taking (only one must be above zero). Criteria are combined using rank weights applied in various aggregation functions from the

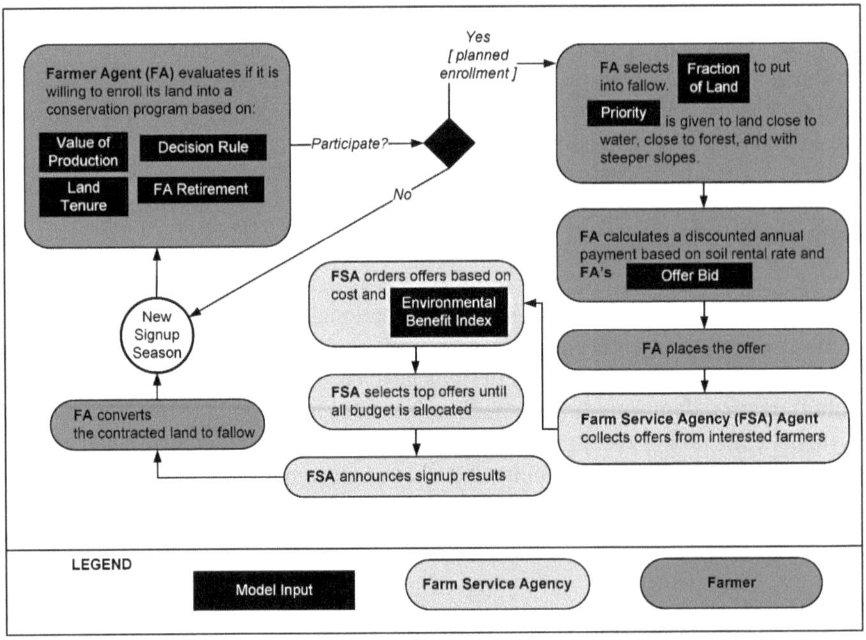

Fig. 2 The decision process in the ABM of CRP enrollment

most restrictive (in which all of the criteria need to be met) to the least restrictive (in which a high value of one criterion is enough to make the decision). OWA results in a standardized score, which is then compared to an empirically derived threshold value.

Once the decision is made, the FA evaluates what fraction of land and where should be selected for land retirement. Priority of selection is based on slope (steep slope is preferred), distance to water and forest (closer is considered better). The FA calculates the rental value expected from the land proposed for conservation, and applies a reduction (bid) to make its offer more competitive. The offer is then passed to the Farm Service Agency agent (FSA) which collects offers from all interested farmers and makes a final decision on admission to the conservation program. Due to a limited federal budget, CRP enrollment is established based both on cost of an individual offer and on a number of environmental factors which are jointly represented by an environmental benefits index (EBI) described in the following section [51]. Once the accepted offers are identified, FSA announces signup results leading to land use change from row crop or pasture to fallow land.

The ABM runs for 10 years. This allows for a simplified decision process since the minimum duration of a CRP contract is at least 10 years. Consequently, FAs who enroll in year one will not be able to withdraw before the end of the simulation and the ABM will not require algorithms for CRP withdrawal and return to agricultural production. The basic output of the model is a land use change map with a portion of farmland staying idle.

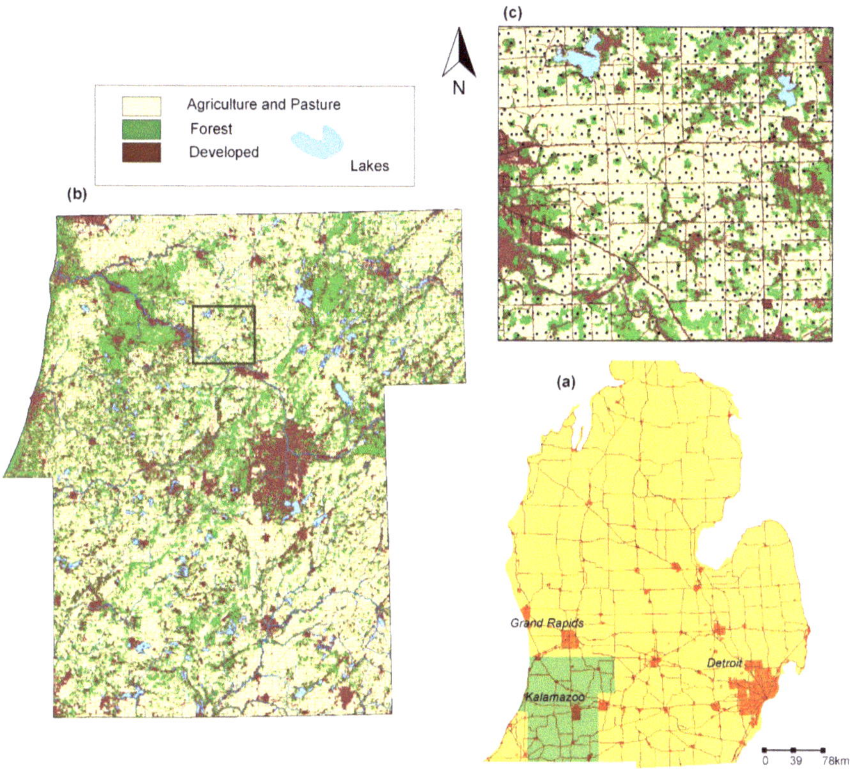

Fig. 3 Study area: (**a**) an overview map of Michigan with the study area highlighted, (**b**) a generalized land use map of the study area, (**c**) an inset map with a portion of study area at a larger scale where *dots* represent the locations of farmer agents

Study Area

The ABM is applied to study CRP enrollment in six counties (Allegan, Barry, Cass, Kalamazoo, St Joseph, Van Buren) in southwest Michigan, U.S. (Fig. 3). The area covers 9220 sq. km, with a substantial amount of agricultural land (51% of the total area). According to U.S. Census of Agriculture [52] there were 6689 farms in the area in 2007. Furthermore, about 3% of farmland was enrolled in CRP according to the census.

Data

The ABM inputs comprise seven constants and eight variables. Model constants include land use [53], slope [54], soils [55], soil rental rates for the selected counties [56], the total budget available to FSA [57], a threshold value that drives FA's

Table 3 Probability distributions of ABM inputs

Input name	Input description	Probability density function
RETIREMENT	Primary farm operator retired from farming (0: retired, 1: working)	D = [(0, .06), (1, .94)]
PRODUCTION	Total value of production on a farm	D = [(0, 0), (.2, .06), (.4, .06), (.6, .11), (.8, .15), (1, .62)]
TENURE	Ratio of owned to operated acres	D = [(0, .04), (.2, .14), (.4, .18), (.6, .14), (.8, .15), (1, .35)]
PRIORITY	Prioritization of land characteristics used in ranking the potential CRP locations	D = [six combinations with equal probability]
OWA	Farmer agent decision rule based on ordered weighted averaging	D = [17 combinations with equal probability]
LANDFRACTION	Fraction of farm parcel to set aside for conservation	U = (0, 1)
BID	Voluntary reduction by the farmer of the offer value below the maximum payment rate	D = [0%–16% of reduction with increments of 1, with equal probability of selection]
EBI	Environmental benefits index	D = [six spatial layers representing different scenarios of EBI distribution in the study area with equal probability of selection]

U uniform distribution, *D* discrete distribution (value, probability)

willingness to enroll [48], and a map with locations of land parcels eligible for CRP, delineated based on the common land unit specification [58]. The model assumes a separate decision maker per each virtual farming parcel, which amounts to a total of 26,095 farmer agents. Raster data was set to a common resolution of 30 m, resulting in 3540 columns and 3790 rows. All data was obtained for 2010—a year of the 41st CRP signup [51].

The eight variable inputs are depicted in Fig. 2 in black and their PDFs are summarized in Table 3. Notice that PRODUCTION, TENURE, and RETIREMENT are empirically-derived variables generated from data collected through the Agricultural Resource Management Survey [48]. Empirical data for the remaining variables was not available. It was therefore assumed that, for any ill-defined variable, all possible values can be selected with equal probabilities i.e., OWA, BID, EBI, and PRIORITY are represented as discrete uniform distributions and LANDFRACTION is described with a continuous uniform distribution (Table 3).

All model inputs are aspatial except for EBI. The EBI maps were obtained based on USDA guidelines for assessing environmental benefits from land conservation [51]. The list of potential benefits includes five factors (plus the rental cost which is evaluated separately) divided into operational sub-factors. Since the specification leaves some room for modification based on local geographic conditions, a number of alternative EBI realizations were generated. Figure 4 depicts the six EBI maps

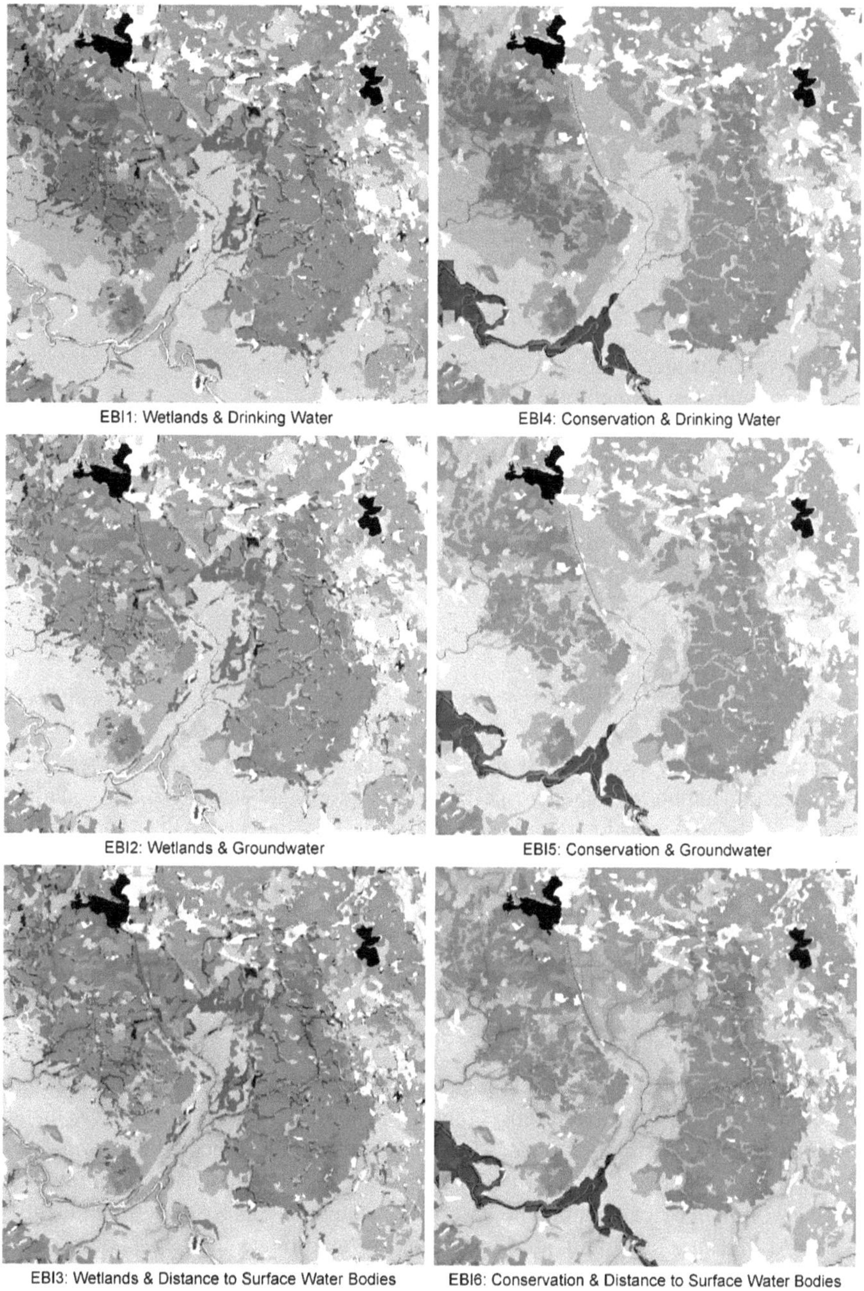

Fig. 4 Six alternative realizations of the environmental benefits index used in the ABM. Each caption describes the combination of factors used in its formulation. Each pixel represents a score on a range from 50 to 330 points established based on the USDA specification [51]. Higher values are depicted with darker shades of *grey*. For clarity, only inset (**c**) from Fig. 3 is shown

used in the reported simulations. The maps were produced by overlaying different combinations of the following six layers that embody three conservation objectives (names used in Fig. 4 are given in parentheses):

1. The wildlife objective maximizes expected wildlife benefits from conservation and is represented by two alternative maps:
   a. Wetland restoration priority zones (wetlands).
   b. Critical ecosystems for conservation (conservation).
2. The water quality objective minimizes inflow of soils, polluted runoff, and leaching and is operationalized as follows:
   a. Drinking water protection zones (drinking water).
   b. Groundwater vulnerability areas (groundwater).
   c. Safety buffers around streams, rivers, ponds, and lakes (distance to surface water bodies).
3. The erosion objective captures the vulnerability of soils to erosion and is embodied in one spatial layer—the soil erodibility index obtained from the SSURGO database [55].

Experiments and Outputs

The simulations were designed using the following protocol. After identifying the uncertain inputs and their respective distributions, N samples were generated using the Sobol' quasirandom sampling [45]. For the original (initial) model N = 2304 and the simplified model (reported in section "Model Simplification and Discussion") was evaluated with N = 1536. Monte Carlo simulations were run to generate numerous land use change maps, which were aggregated into scalars including the total area of fallow land (acres), the cost of land rent (cost), and nine fragmentation statistics of fallow land (spatial metrics) [59]. Out of the nine spatial metrics five were highly correlated ($|r| >= 0.9$), so the final set of output variables used in model evaluation was set to area, cost, and four spatial metrics i.e., the average nearest neighbor, the average radius of gyration, the largest patch index, and the average perimeter-to-area ratio (Table 4). The distributions of the six statistics were summarized using box plots. The outputs were then subjected to UA and SA. The UA was performed to quantify the variability of fallow land area resulting from CRP enrollment by calculating the descriptive statistics. The SA was run to identify the most and the least influential inputs and to investigate the dependence of the ABM outputs on input interactions by decomposing the scalars, apportioning them to inputs, and therefore determining the underlying causes driving the distribution of results. The resulting S and ST indices were visualized using pie charts (the ST indices were first normalized to sum up to one).

Table 4 Scalar outputs obtained from the ABM land use maps

Name	Definition	Unit
Acres	Total area of fallow land	Acre
Cost	Cost of land rent per year	U.S. dollars per acre
ANN	Average Nearest Neighbor: average distance between patches of fallow land measured along a straight line (a measure of patch isolation)	Meter
GYRATION	Average radius of gyration: a averaged measure of the extent of fallow land patches	Meter
LPI	Largest patch index: the percentage of the landscape comprised by the largest patch of fallow land (a measure of dominance)	Percent
PAR	Average perimeter-to-area ratio of fallow land patches (a simple measure of shape complexity)	Unitless

Definitions of spatial metrics based on McGarigal [60]

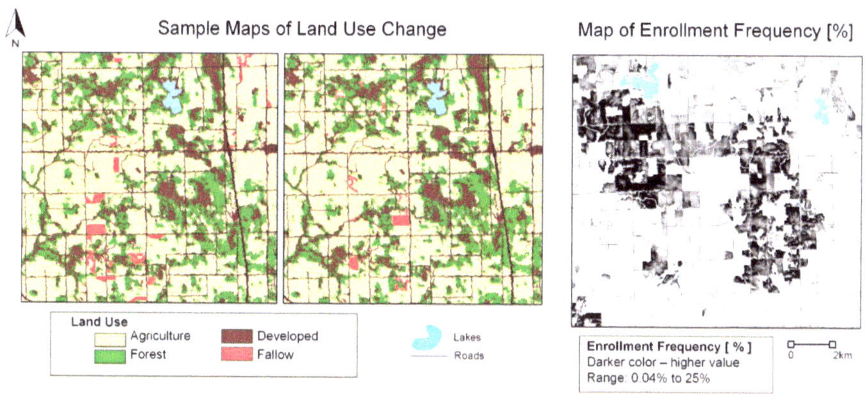

Fig. 5 Example output land use maps and the frequency map of agriculture-to-fallow conversion. For clarity only inset (**c**) from Fig. 3 is shown

The ABM was developed in Python (https://www.python.org/) and executed using the High Performance Computer Center at Michigan State University (http://icer.msu.edu/). The sensitivity indices were calculated with SimLab (https://ec.europa.eu/jrc/en/samo/simlab).

Results of the Original ABM

Figure 5 shows examples of output land use maps from two selected model executions. Observe that, compared to the input land use (Fig. 3), the maps contain one additional category of fallow land. Because FAs make CRP enrollment decisions on a pixel-by-pixel basis, most of the parcels at the end of simulation have

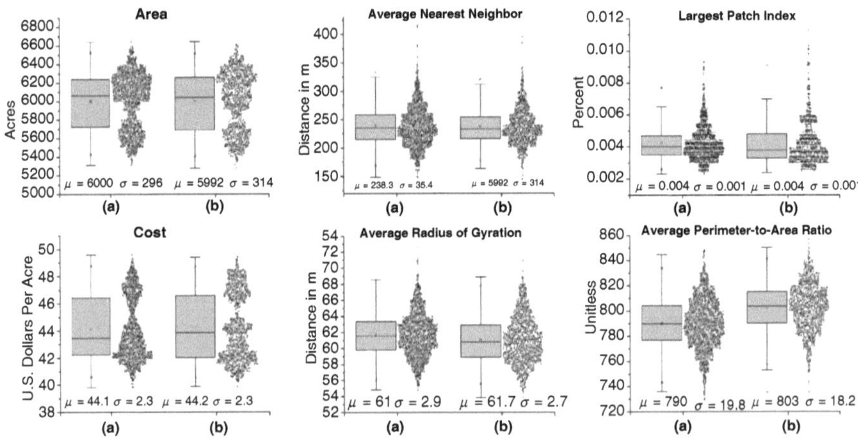

Fig. 6 Distributions of output scalars with their means and standard deviations: (**a**) for the original model with eight inputs, (**b**) for the simplified model with five inputs

only a portion of their land put to fallow. In addition to the land use output, I created a map of enrollment frequency by superimposing all output maps to calculate the percentage of times a particular pixel was converted to fallow. Clearly, considerable spatial heterogeneity in pixel enrollment can be observed, likely due to the complex interactions between inputs.

Uncertainty Analysis

The patches of fallow land were used to calculate the spatial metrics listed in Table 4. The distributions of the scalars and their respective means and standard deviations are rendered in Fig. 6 columns (a). Two observations can be made. First, the cost and area produce bimodal distributions whereas all the spatial metrics result in skewed normal-like distributions. The possible reason for the bimodal behavior is described in the following section. Second, cost is almost perfectly negatively correlated with area (Pearson's $r = -0.98$). This is not surprising given that the FSA operates with a fixed budget. Consequently, in locations with lower rental rates more land can be enrolled in the program (and vice versa).

Sensitivity Analysis

A reliable model simplification requires information about which inputs and to what extent contribute to output variability. By performing the decomposition of variances of the six scalars, we can identify inputs in the ABM that can be fixed

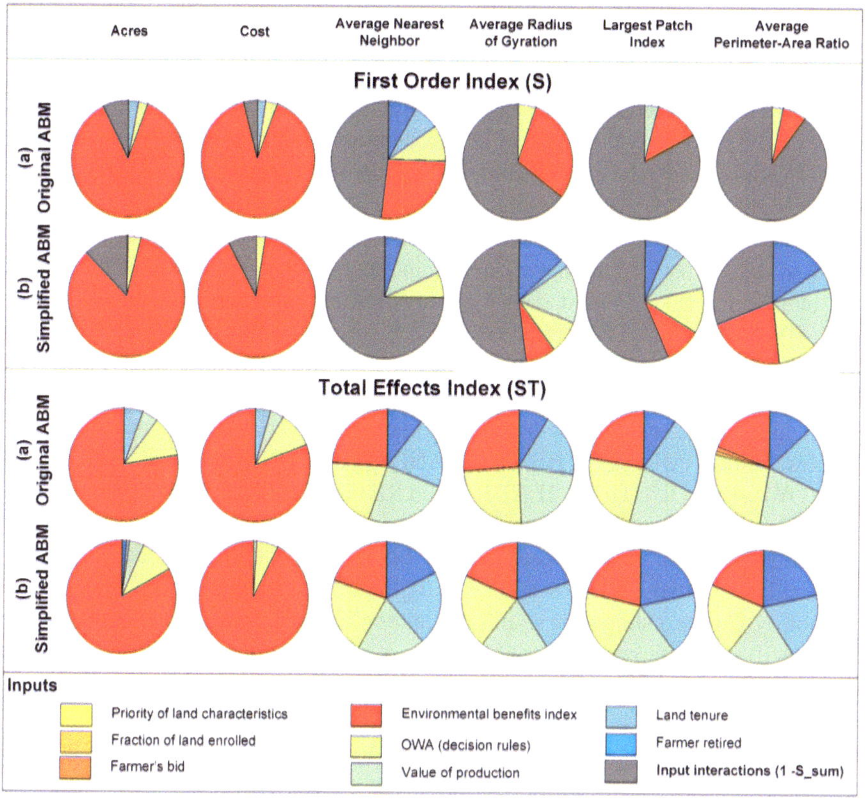

Fig. 7 Pie charts of the S and ST indices for the six output scalars: (**a**) for the original model with eight inputs, (**b**) for the simplified model with five inputs. Note that the ST values were rescaled

to constant values with minimal changes to the underlying output distributions. The results of SA—the S and ST indices—are depicted as pie charts in Fig. 7 rows (a).

Two distinct classes of sensitivities can be observed: the aspatial variables and the spatial metrics. The first group is characterized by a fairly linear input-output correspondence, as only 4% of cost variance and 7% of acres variance can be attributed to interactions (S-sum = 0.96 and 0.93 for cost and acres, respectively). Given that cost and acres are highly correlated, results of variance decomposition are quite similar. EBI is the input that predominantly affects the variability of both statistics (S_{EBI} = 0.9 for cost and 0.87 for acres). Three other inputs that play some role in describing the variability of these two statistics (especially in combinations with other inputs—see the ST pie charts), are OWA, TENURE, and PRODUCTION. Thus, based solely on the simple regional metrics of total area and cost of enrollment, we could easily fix RETIREMENT, PRIORITY, LANDFRACTION, BID. This post-processing analysis also indicates that, for these two variables, simpler linear methods of SA can be applied without a substantial loss of information.

The four spatial metrics comprise the second distinct group of sensitivities. This group is characterized by much larger interaction effects (from 48% for ANN to 90% for PAR). I hypothesize that such complex relationship between input and output variability can be attributed to the fact that the spatial metrics are sensitive to the configuration of various spatial layers present in the model like soils, forest, slope, and all the factors that build the EBIs. In this case, using only the S to identify the causes of output variance is ineffective and we have to resort to the ST indices to make a valid decision on ABM simplification. In all four cases EBI, OWA, PRODUCTION, and TENURE emerge as influential, which is similar to the inputs established for cost and acres. In addition to these four inputs, we can identify farmer's RETIREMENT as one more influential input. Assuming that ABM simplification should be based on all six output variables, the modified ABM involves setting PRIORITY, BID, and LANDFRACTION to constants, and leaving EBI, OWA, PRODUCTION, TENURE, and RETIREMENT as variable inputs. This results in a reduction of input dimensionality from eight to five dimensions.

Finally, to delve into the cause of the bimodal distributions of cost and area, I plotted these two variables against EBI, which is the major driver of output variability. Figure 8 shows the resulting scatter plot. As expected, the EBI layers are correlated with different combinations of cost and area. Moreover, they are a good candidate for explaining the bimodal variation. Under scrutiny, the cause may be both substantive and technical (data-related). Factors building the EBI of the 'high cost—low acre' cluster are a combination of a discrete raster (conservation) and two continuous distance rasters. The third 'conservation' combination (i.e., conservation and drinking water) is derived from two discrete layers of high priority zones (for drinking water and biodiversity, respectively). Consequently, the first (substantive) cause of the bimodal distributions is the presence of the conservation layer in EBI. The second (technical) cause is the data structure (i.e. a continuous raster) of the other layer used in combination with the conservation layer.

Model Simplification and Discussion

To evaluate the quality of ABM simplification, I performed UA and SA on the new ABM. The results are presented in Fig. 6 columns (b) and Fig. 7 rows (b). Visual comparison shows that all six output distributions maintained their shapes before and after the simplification (Fig. 6), with quite similar means and variances. I can conclude that the simplification is satisfactory for studying the general model behavior and the emergent land use patterns, but may be insufficient for more precise predictive purposes (in which case the use of ABM is generally not recommended anyway).

In addition to the qualitative visual examination of the equivalence of both models, I wanted to quantitatively compare the pre- and post- simplification

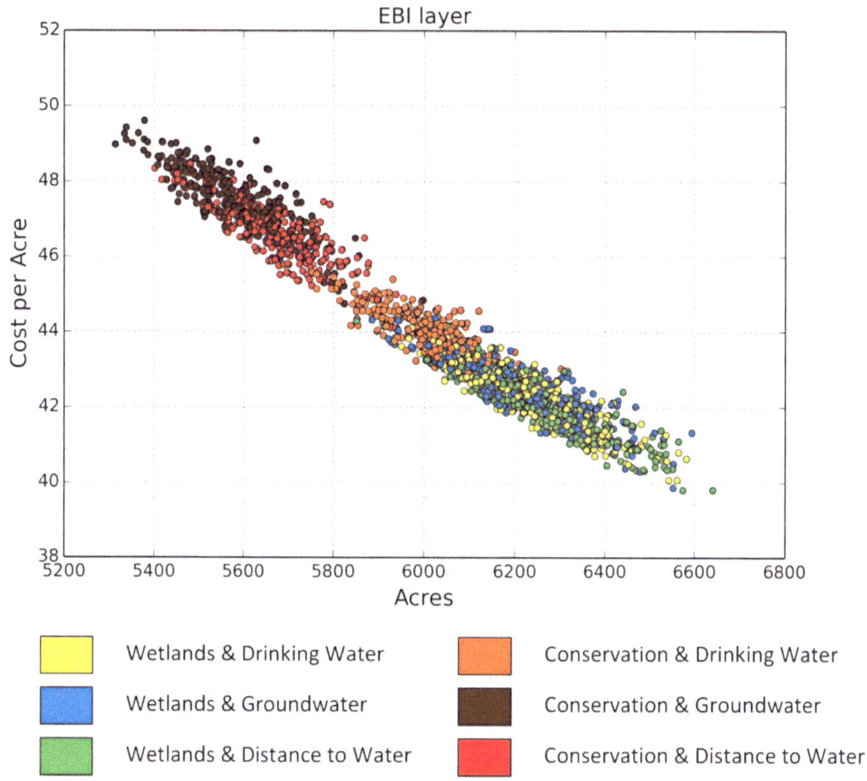

Fig. 8 Scatter plot of the bimodal distributions of Cost and Acres outputs obtained from the original model. The colors represent the corresponding EBI layers

Table 5 Results of the Kolmogorov–Smirnov test comparing the samples of the distributions of output variables before and after the ABM simplification ($\alpha = 0.05$, $N_1 = N_2 = 200$, $D_{critical} = 0.136$)

Output	D	p	Reject H_0?
Acres	0.08	0.5272	No
Cost	0.09	0.3767	No
ANN	0.09	0.3124	No
GYRATION	0.16	0.0144	Yes
LPI	0.11	0.1324	No
PAR	0.29	0.0000	Yes

distributions of the output variables. The Kolmogorov–Smirnov test was computed at $\alpha = 0.05$. To avoid too much power of the statistical test due to large N of both experiments, output values for each version of the model were sampled multiple times at N = 200 (a more likely number of Monte Carlo executions when dealing with complex, computationally demanding models). Table 5 shows the representative results of the statistical test ($D_{critical} = 0.136$).

The statistical results suggest that four distributions are not significantly different from their original counterparts. Moreover, while GYRATION does not pass the test, its distance value is fairly close to the critical one. The most observable difference is for PAR with a general shift of its distribution towards higher scores (Fig. 6). Since PAR has a number of limitations as a measure of landscape configuration [60], I conclude that, based solely on output distributions, the simpler model is a valid substitute for the original one.

The results of variance decomposition paint a slightly different picture of the simplification (Fig. 7). Since we discarded only the non-influential inputs, we would expect similar values of S and ST before and after simplification. While this notion is confirmed for cost and acres, the S pie charts for all four spatial metrics are visibly different. First, the interaction effects either substantially increase (by 27% for ANN) or decrease (by 59, 26, and 13% for PAR, LPI, and GYRATION, respectively). Second, the composition of influential inputs changed from the dominating EBI to various combinations of all five inputs. While somewhat puzzling, this behavior points to the dynamic role that the inputs play in this ABM. Simply put, the inputs are not passive explanatory variables of the dependent output. Rather, they are continually involved in many interactions during model execution changing their individual (independent) influence on the final variance. If we compare the ST indices before and after simulation (Fig. 7, bottom) we can conclude that the overall contribution of each input remains roughly unchanged. Thus, the ST indices offer a more comprehensive depiction of input influence on output variability and provide another argument for using total effect indices when evaluating complex spatiotemporal models.

Simplification—What for?

We may find it paradoxical that, in order to make a model computationally and structurally simpler, we need to resort to extensive computation. Since simplification is time consuming [21] a question could be raised whether this onerous exercise should even be contemplated. After all, the original model includes all known evidence and, by fixing some if its components, we risk reducing its predictive power and confine its applicability to a subset of problems. Modelers are more interested in avoiding the ‘sin of omission’ in model development—overlooking drivers that play a critical role in the system resulting in flawed simulations. Nevertheless, we should also recognize the dangers of ‘commission’—keeping a model over-parameterized may unnecessarily increase the uncertainty of the output, even when such uncertainty is not present in the target system [16]. Moreover, when a model is part of a decision-making process, its excessive complexity may be perceived as an attempt to obfuscate investigation of a problem [39]. Therefore, when applying models to controversial public policy problems, transparency should be pursued. Sensitivity analysis may therefore become instrumental in developing simplified, surrogate (yet credible) models, which are easier to understand and which are more efficient in subsequent scenario analysis and application.

Suggestions for Best Practices to SA-Based Model Simplification

I conclude with some practical guidelines for using SA to build simpler yet equally valid dynamic spatial models like ABMs. Those best practices are grouped by model accuracy, the role of interactions, and model scope and outputs.

Models are built for different purposes. We often start from 'quick and dirty' models for brainstorming about the target system and hypotheses building, and then gradually develop more detailed and empirically-rich models for case-based studies and policy evaluation. The latter may require more *accurate* results than the former. The presented method of SA is based solely on variance and its decomposition. Variance is a very elegant and succinct measure of result variability but it certainly does not reflect the whole distribution of the output. As the results in Fig. 6 and Table 5 show, it may lead to post-simplification distributions that quantitatively differ from their original counterparts. Using variance as the only statistics that drives model simplification may not be sufficient in studies requiring accurate predictions. An alternative approach to SA, called a moment-independent method, is more appropriate in such cases [61].

As shown in this chapter, the variability of model results can be affected by *interactions* among its inputs. Note that two types of interactions can be distinguished: the critical interactions and those that can be ignored during simplification. By definition, all complex models are imbued with nonlinearities and feedbacks that influence the behavior of their inputs. Not all of input interactions, however, manifest themselves in a given output variable. The example demonstrated herein clearly shows that some of the outputs are more affected by input interactions than others (compare cost versus ANN in Fig. 7). Thus, the type of output variable may have a significant effect on the metrics used in SA (e.g. standardized regression coefficients versus total effects indices). More complex outputs, like spatial metrics, may require evaluation of higher-order effects. Consequently, modelers should invest in methods that capture the full spectrum of input interactions, not just the major (first order) effects.

This leads to another important determinant of SA—the character of the *output variables* used to direct model simplification. Whenever a model is built to serve multiple purposes (like contrasting the economic objectives driving farmers' decision making with the ecological principles that guide FSA's decisions) one output variable may not be enough. Using multiple variables as inputs to variance-based SA leads to simplification that captures a wide spectrum of model behavior. For example, if I used only cost or acres as the output variable used to identify the non-influential inputs, I would end up with a model with fixed RETIREMENT—a clear driver of variability in land fragmentation. A possible method of simplification, utilized in this study, is to identify the influential inputs common to multiple output variables and fix only those inputs that prove unimportant in all outputs.

A related aspect of SA is the quality of data used to build input PDFs. Clearly, the distribution of outputs and the resultant variance decomposition are dependent on the type, shape, and other characteristics of the distributions of input variables.

In cases where the variable is ill-defined, it is prudent to assume a uniform PDF. By definition, a uniform PDF encompasses the whole spectrum of possibilities with equal probabilities. Consequently, if an input defined by a uniform distribution proves unimportant, it can be fixed to a representative value without the need for obtaining the empirical data in the first place. This outcome is very convenient when building complex models that require expensive data collection. In this case, it is best to start from building a rudimentary model that comprises the best available data, and then decide on input collection efforts based on the results of SA of the prototype model.

Conclusions

In this chapter, I presented an approach to model simplification that utilizes an extensive uncertainty and sensitivity analysis. I focused on identifying model inputs that can be set to constant values without significant changes in output distributions. Such model simplification is necessary since, as indicated by Crosetto et al. [29], irrelevant model inputs can degrade the overall model performance. Without model simplification the complicated and computationally demanding simulations may become infeasible.

I demonstrated that models of complex land systems are prone to interactions among inputs that need to be explicitly investigated to illuminate the uncertainty of the studied system.

I computed sensitivity indices for individual inputs and their combinations. The indices serve as quantitative representations of the drivers underlying model uncertainty. They prove useful in isolating the effects of the interconnected explanatory variables on the simulated emergent phenomena. Understanding functional dynamics embedded in ABM is critical if we want to realistically emulate social and ecological phenomena and build legitimate future scenarios for scientific and policy analysis.

Acknowledgements I would like to thank anonymous reviewers for providing a constructive feedback on the previous version of this manuscript. Financial support for this work was provided by the National Science Foundation Geography and Spatial Sciences Program Grant No. BCS 1263477. Any opinion, findings, conclusions, and recommendations expressed in this material are those of the author and do not necessarily reflect the views of the National Science Foundation. The funders had no role in study design, data collection and analysis, decision to publish, or preparation of the manuscript.

References

1. Agarwal C, Green GM, Grove JM, Evans TP, Schweik CM (2002) A review and assessment of land-use change models: dynamics of space, time, and human choice. U.S. Department of Agriculture, Forest Service, Northeastern Research Station, Newton Square, PA, p 61

2. An L, Linderman M, Qi J, Shortridge A, Liu J (2005) Exploring complexity in a human-environment system: an agent-based spatial model for multidisciplinary and multiscale integration. Ann Assoc Am Geogr 95(1):54–79
3. Brown DG, Verburg PH, Pontius RG Jr, Lange MD (2013) Opportunities to improve impact, integration, and evaluation of land change models. Curr Opin Environ Sustain 5(5):452–457
4. NRC (2013) Advancing land change modeling: opportunities and research requirements. The National Academies Press, Washington, DC
5. Parker DC, Manson SM, Janssen MA, Hoffmann MJ, Deadman P (2003) Multi-agent systems for the simulation of land-use and land-cover change: a review. Ann Assoc Am Geogr 93(2):314–337
6. Rindfuss RR, Entwisle B, Walsh SJ, An L, Badenoch N, Brown DG, Deadman P, Evans TP, Fox J, Geoghegan J, Gutmann M, Kelly M, Linderman M, Liu J, Malanson GP, Mena CF, Messina JP, Moran EF, Parker DC, Parton W, Prasartkul P, Robinson DT, Sawangdee Y, Vanwey LK, Verburg PH (2008) Land use change: complexity and comparisons. J Land Use Sci 3(1):1–10
7. Verburg PH, Kok K, Pontius JRG, Veldkamp A (2006) Modeling Land-Use and Land-Cover Change. In: Lambin EF, Geist HJ (eds) Land-use and land-cover change: local processes and global impacts. Springer, Berlin, pp 117–135
8. Ligmann-Ziclinska A (2013) Spatially-explicit sensitivity analysis of an agent based model of land use change. Int J Geogr Inf Sci 27(9):1764–1781
9. Tarantola S, Giglioli N, Jesinghaus J, Saltelli A (2002) Can global sensitivity analysis steer the implementation of models for environmental assessments and decision-making? Stoch Env Res Risk A 16(1):63–76
10. Longley PA, Goodchild M, Maguire DJ, Rhind DW (2010) Uncertainty. In: Geographic information systems and science. Wiley, Jefferson City, pp 147–177
11. Saltelli A, Chan K, Scott EM (2000) Sensitivity analysis. Chichester, Wiley-Interscience
12. Beck MB, Ravetz JR, Mulkey LA, Barnwell TO (1997) On the problem of model validation for predictive exposure assessments. Stoch Hydrol Hydraul 11(3):229–254
13. Brooks RJ, Tobias AM (1996) Choosing the best model: level of detail, complexity, and model performance. Math Comput Model 24(4):1–14
14. Collins A, Petty M, Vernon-Bido D, Sherfey S (2015) A call to arms: standards for agent-based modeling and simulation. J Artif Soc Soc Simul 18(3):12
15. Cox GM, Gibbons JM, Wood ATA, Craigon J, Ramsden SJ, Crout NMJ (2006) Towards the systematic simplification of mechanistic models. Ecol Model 198(1–2):240–246
16. Crout NMJ, Tarsitano D, Wood AT (2009) Is my model too complex? Evaluating model formulation using model reduction. Environ Model Softw 24(1):1–7
17. Eberlein RL (1989) Simplification and understanding of models. Syst Dyn Rev 5(1):51–68
18. Edmonds B, Moss S (2005) From KISS to KIDS—an 'anti-simplistic' modelling approach. In: Davidsson P, Logan B, Takadama K (eds) Multi-agent and multi-agent-based simulation, vol 3415. Springer, Berlin, pp 130–144
19. Innis G, Rexstad E (1983) Simulation model simplification techniques. Simulation 41(1):7–15
20. Moore AD, Noble IR (1993) Automatic model simplification: the generation of replacement sequences and their use in vegetation modelling. Ecol Model 70(1):137–157
21. Rexstad E, Innis GS (1985) Model simplification—three applications. Ecol Model 27(1):1–13
22. Saysel AK, Barlas Y (2006) Model simplification and validation with indirect structure validity tests. Syst Dyn Rev 22(3):241–262
23. Zeigler B (1976) Theory of modelling and simulation. John Wiley, New York
24. Lilburne L, Tarantola S (2009) Sensitivity analysis of spatial models. Int J Geogr Inf Sci 23(2):151–168
25. Saltelli A, Annoni P, Azzini I, Campolongo F, Ratto M, Tarantola S (2010) Variance based sensitivity analysis of model output. Design and estimator for the total sensitivity index. Comput Phys Commun 181(2):259–270
26. Saltelli A, Tarantola S, Campolongo F, Ratto M (2004) Sensitivity analysis in practice: a guide to assessing scientific models. Chichester, Wiley

27. USDA FSA (2012) Conservation Reserve Program Overview. http://www.fsa.usda.gov/FSA/webapp?area=home&subject=copr&topic=crp. Accessed 9 June 2012
28. Crosetto M, Tarantola S (2001) Uncertainty and sensitivity analysis: tools for GIS-based model implementation. Int J Geogr Inf Sci 15(5):415–437
29. Crosetto M, Tarantola S, Saltelli A (2000) Sensitivity and uncertainty analysis in spatial modelling based on GIS. Agric Ecosyst Environ 81(1):71–79
30. Gómez-Delgado M, Tarantola S (2006) GLOBAL sensitivity analysis, GIS and multi-criteria evaluation for a sustainable planning of a hazardous waste disposal site in Spain. Int J Geogr Inf Sci 20(4):449–466
31. Ligmann-Zielinska A, Jankowski P (2010) Exploring normative scenarios of land use development decisions with an agent-based simulation laboratory. Comput Environ Urban Syst 34:409–423
32. Ligmann-Zielinska A, Jankowski P (2014) Spatially-explicit integrated uncertainty and sensitivity analysis of criteria weights in multicriteria land suitability evaluation. Environ Model Softw 57:235–247
33. Ligmann-Zielinska A, Sun L (2010) Applying time dependent variance-based global sensitivity analysis to represent the dynamics of an agent-based model of land use change. Int J Geogr Inf Sci 24(12):1829–1850
34. Tang W, Jia M (2014) Global sensitivity analysis of a large agent-based model of spatial opinion exchange: a heterogeneous multi-GPU acceleration approach. Ann Assoc Am Geogr 104(3):485–509
35. French S (1992) Mathematical-programming approaches to sensitivity calculations in decision-analysis. J Oper Res Soc 43(8):813–819
36. Pannell DJ (1997) Sensitivity analysis of normative economic models: theoretical framework and practical strategies. Agric Econ 16(2):139–152
37. Saltelli A, Annoni P (2010) How to avoid a perfunctory sensitivity analysis. Environ Model Softw 25(12):1508–1517
38. Saisana M, Saltelli A, Tarantola S (2005) Uncertainty and sensitivity analysis techniques as tools for the quality assessment of composite indicators. J R Stat Soc A Stat Soc 168:307–323
39. Saltelli A, Ratto M, Andres T, Campolongo F, Cariboni J, Gatelli D, Saisana M, Tarantola S (2008) Global sensitivity analysis: the primer. Chichester, Wiley-Interscience
40. Homma T, Saltelli A (1996) Importance measures in global sensitivity analysis of nonlinear models. Reliab Eng Syst Saf 52(1):1–17
41. Saltelli A, Tarantola S, Chan KPS (1999) A quantitative model-independent method for global sensitivity analysis of model output. Technometrics 41(1):39–56
42. Beven K (2008) Environmental modelling: an uncertain future? Routledge, New York
43. Kucherenko S, Tarantola S, Annoni P (2012) Estimation of global sensitivity indices for models with dependent variables. Comput Phys Commun 183(4):937–946
44. Saltelli A (2002) Making best use of model evaluations to compute sensitivity indices. Comput Phys Commun 145(2):280–297
45. Sobol' IM (1993) Sensitivity estimates for nonlinear mathematical models. Math Model Comput Exp 1:407–414
46. MathWorks (2017) http://www.mathworks.com/help/stats/generating-quasi-random-numbers.html. Accessed 6 June 2017
47. Lambert DM, Sullivan P, Claassen R, Foreman L (2006) Conservation-compatible practices and programs: who participates? USDA Economic Research Report. United States Department of Agriculture, p 48
48. USDA (2011) Agricultural Resource Management Survey (ARMS). http://www.nass.usda.gov/Surveys/Guide_to_NASS_Surveys/Ag_Resource_Management/
49. Hägerstrand T (1968) Innovation diffusion as a spatial process. University of Chicago Press, Chicago
50. Yager RR (1988) On ordered weighted averaging aggregation operators in multi-criteria decision making. IEEE Trans Syst Man Cybern 18:183–190

51. USDA FSA (2011) Conservation Reserve Program Sign-up 41 Environmental Benefits Index (EBI) Fact Sheet. http://www.fsa.usda.gov/Internet/FSA_File/crp_41_ebi.pdf. Accessed 9 June 2012
52. USDA (2017) United States Census of Agriculture. http://www.agcensus.usda.gov/index.php. Accessed 6 June 2017
53. USDA NASS (2017) Cropland Data Layer (CDL). https://www.nass.usda.gov/Research_and_Science/Cropland/SARS1a.php. Accessed 6 June 2017
54. USGS (2013) National Elevation Dataset. http://ned.usgs.gov/. Accessed 15 Oct 2013
55. Soil Survey Staff (2013) The gridded soil survey geographic (gSSURGO) database for Michigan. http://datagateway.nrcs.usda.gov/
56. USDA FSA (2010) Notice CRP-663 sign-up 41 revised soil rental rates (SRR's) for 2010. United States Department of Agriculture Farm Service Agency, Washington, DC
57. USDA FSA (2013) Conservation Programs Reports and Statistics. http://www.fsa.usda.gov/FSA/webapp?area=home&subject=copr&topic=rns. Accessed 15 Oct 2013
58. USDA FSA (2004) Common land unit FSA handbook. United States Department of Agriculture Farm Service Agency, p 113
59. McGarigal K, Marks BJ (1995) FRAGSTATS: spatial pattern analysis program for quantifying landscape structure. General technical report. USDA Forest Service, Pacific Northwest Research Station, Portland, OR
60. McGarigal K (2014) Fragstats help. http://www.umass.edu/landeco/research/fragstats/downloads/fragstats_downloads.html. Accessed 7 May 2015
61. Borgonovo E, Castaings W, Tarantola S (2012) Model emulation and moment-independent sensitivity analysis: an application to environmental modelling. Environ Model Softw 34(0): 105–115

Agent-Based Modeling of Large-Scale Land Acquisition and Rural Household Dynamics

Atesmachew B. Hailegiorgis and Claudio Cioffi-Revilla

Introduction

Rural systems in most Sub-Saharan African countries are characterized by interdependent relationships between households dependent on subsistence agriculture and the biophysical system to which they are dynamically coupled. Rural households often rely on rain-fed agriculture, with climate variability directly affecting agricultural production. Their livelihood decisions are governed by the availability of and opportunity to use resources (human, social, environmental, or financial).

Recently, the rise in large-scale land acquisitions has become a major public issue altering the dynamics of rural systems and affecting the adaptive capacity of rural communities. Although most rural communities have developed adaptation mechanisms for prolonging their livelihood, the introduction of industrial-scale agribusiness enterprises (both national and international in origins) and the subsequent rapid change in land-use systems are challenging traditional ways of life. As a result, affected communities face a choice of migrating, protesting (including violence), or suffering hardship in situ.

The resilience of such regional systems under future socioeconomic uncertainty, and the complexity of the dynamics (i.e., heterogeneous actors/agents, nonlinear interactions, emergent micro-macro dynamics, and multiple spatio-temporal scales), pose a significant scientific challenge. Understanding the interaction between enterprises and rural communities, as well as the influence of commercialization of land on rural livelihoods and ecosystems, is key to improving policies for

A.B. Hailegiorgis (✉) • C. Cioffi-Revilla
Center for Social Complexity, George Mason University, 4440 University Drive, Fairfax, VA 22030, USA
e-mail: ahailegi@gmu.edu; atesbiz@gmail.com; ccioffi@gmu.edu

J.-C. Thill, S. Dragicevic (eds.), *GeoComputational Analysis and Modeling of Regional Systems*, Advances in Geographic Information Science,
DOI 10.1007/978-3-319-59511-5_7

enhancing the well-being of indigenous rural communities, and improving the prospects for economic development compatible with maintaining the sustainability of the ecosystems's functions and processes.

Rural Systems and Large-Scale Land Acquisition

The current surge in large-scale land acquisition in many developing countries has become a global (or at least a transnational) issue, attracting attention due to the scale and speed of acquisition [52]. The issue is especially felt in Sub-Saharan Africa—the area of the world with the highest hunger indices [53]—where a single enterprise could acquire nearly 500,000 hectares of land in a single purchase [20]. These land acquisitions involve diverse interest groups, both national and international, with diverse sources of investments, including privately owned, government-backed, and sovereign wealth fund investments.

Extent of Large-Scale Land Acquisition

Multiple lines of evidence yield the same overall trend in large-scale land acquisition in developing countries generally, and in particular Sub-Saharan Africa, including Ethiopia. The International Food Policy Research Institute (IFPRI) has estimated that nearly 20 million hectares of farmlands in developing countries have been acquired by enterprises since 2006 [53]. A World Bank study, based on more data sources including media reports, raised the figure to 57 million hectares [20]. Of these, more than two-thirds are in Sub-Saharan Africa [20]. Another empirical study by Cotula et al. [14] has shown that between 2004 and 2008, a total of 2.5 million hectares of land was acquired by national and international enterprises in five African countries: Ghana, Madagascar, Mali, Sudan, and Ethiopia. A recent study by The Oakland Institute, an independent policy think tank, reports that in Ethiopia alone the total amount of lands acquired by foreign and national enterprises increased from about 1.2 million hectares between 2004 and 2008 to nearly 3.6 million hectares as of January, 2011 [40].

Several processes drive large-scale land acquisition, the most common being demand for food and biofuel production. The empirical study by Cotula et al. [14] in five African countries indicated that, of the total 2.5 million hectares that were acquired by different domestic and foreign enterprises between 2004 and 2008, nearly 1.4 million hectares were for food and 1.1 million for biofuel production. In line with this, increased involvement by private enterprises and international organizations in the development of protected areas, nature reserves, ecotourism businesses, and construction of large-scale tourist complexes promote the conversion of productive lands into attractive tourist destinations and cause significant changes to land ownership and the traditional land-use system [54].

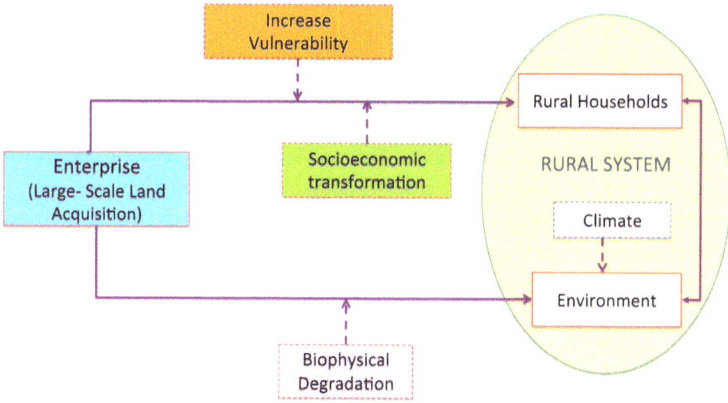

Fig. 1 Influence of large-scale land acquisition on rural systems. *Solid and dashed lines* indicate direct and indirect relationships between components, respectively. Source: adapted from [30]

Implications of Large-Scale Land Acquisition

The expansion of large-scale land acquisition can have various influences on a rural system, as shown in Fig. 1. Earlier studies have linked the activity of such enterprises to significant change in the socioeconomic dynamics of rural systems (e.g., [9, 20, 22, 23, 53], among others), showing that enterprises can contribute to a rural economy by creating jobs, transferring technologies, and developing infrastructure. In many cases, large-scale land acquisitions are oriented toward labor-intensive agriculture, thereby creating opportunities for a wage labor market in rural societies [20]. Wage labor enhances household incomes and diversifies livelihood options. It may also help households minimize dependency on subsistence farming, decrease vulnerability to climate change, and increase economic benefits [5]. Moreover, income growth can lead to a self-sufficient lifestyle and limit rural migration to urban centers [18].

The operation of more capital-endowed enterprises in a rural system can also increase rural infrastructure (e.g., roads, irrigation canals, bridges, storage facilities, transportation nodes) as enterprises seek to develop their business for maximum economic return. Infrastructure improvements can expose rural communities to broader opportunities, such as providing new or better links to markets and urban areas, minimizing travel costs, enhancing agricultural productivity, and improving the prospect for greater public services [10, 21]. Enterprises might also contribute to a rural system by increasing the production performance of land through the introduction of modern technology [13].

Although rural households' welfare might improve from the contributions of enterprises, large-scale land acquisitions also expose rural households to risks to their livelihood [7, 52]. Expansion of industrialized agricultural systems can deny the customary rights of rural households to utilize communal properties and lands, affecting indigenous adaptive capacity and increasing the vulnerability of households [41]. Considerable expansion of large-scale land acquisition can also cause dispossession and displacement of locals, as most of the lands are already occupied or used by rural households [2]. The policy of changing a rural system to a more modern capital-intensive agriculture system can neglect the social functions of land, which is more than a production-entity in traditional societies.

The environment is also stressed when land is converted to agriculture [29]. The conversion of "marginal" land—currently covered by forest or used for grazing—can accelerate land degradation and loss of biodiversity [37]. Even when large-scale farming is implemented in existing agricultural lands, the change from multi-cropping to mono-cropping causes an increase in vulnerability to drought and disease. Besides degradation and biodiversity loss, increased utilization of chemicals without proper treatment of effluent waters can cause environmental damage and increase health risks for humans, animals, and native vegetation.

Prior Agent-Based Modeling on Traditional Societies in Rural Systems

Several studies have been conducted to analyze the impacts of expansion of large-scale land acquisition in rural systems [45]. Earlier methodologies, such as statistical, equation-based, and systems models have been used to study the complexity of rural systems in different contexts [42]. However, these approaches have been criticized for their inefficiency in capturing complex interactions in human and biophysical systems. Earlier modeling approaches either over-simplify the representation of human actors or fail to capture temporal complexity, spatial complexity, and feedbacks [44].

More recently, understanding a rural system as a complex adaptive system has been a focus of attention, as this approach provides novel insights by capturing the complexity of interactions and dynamic feedbacks among system components [44]. The application of dynamic modeling could also help to explore the impact of climate change and large-scale land acquisition. The application of integrated models—such as agent-based models (ABMs) for capturing interactions among individuals and surrounding environment—is essential for understanding complex, dynamic, and nonlinear challenges faced by rural households when large-scale land acquisitions occur. An ABM provides a powerful computational laboratory for exploring and analyzing interesting scenarios focused on the local people's adaptive responses to different socioeconomic conditions and the resulting effects on their ecosystem [11, 15].

Several ABMs have examined interactions between rural households and their environments in developing countries, including assessments of consequences of household decisions on land use and land cover change (LUCC) [19, 33–35, 38,

47, 48]; households' migration behavior [24, 26, 32, 50]; vulnerability to climatic factors [1, 8]; adaptation to climate variability [12]; climate risk perception in land markets [25]; and diversification and adoption of new technologies [6, 31].

Although these and other prior models provide insights on complexity in coupled human and natural systems and the impact of human actions on the environment and vice versa, insufficient attention has been paid to the effect of large, industrial-scale enterprise actors, their interaction with local households, and how they affect rural landscape dynamics. Existing models focus mostly on individuals, households, and their interactions with biophysical environments and climate change.

A review of current research on coupled human and natural systems by Rindfuss et al. [47] identified shortcomings in terms of agent typology, scale of applications, and representation of feedbacks. Most of the models use only one type of agent, so agent heterogeneity is typically rendered through variation in household attributes, not in the characteristics of agents (e.g., households vs. enterprises) [43]. However, most systems of households consist of more than one type of agent. Competing actors with diverse objectives and goals interact with each other and with environments at different spatial and temporal scales [3]. The scale of intervention, type of interaction among actors, and factors affecting their decisions are usually different. Rindfuss et al. [47] have suggested that it is essential to consider the influence of diverse actors, their distinct characteristics, and the different factors affecting their decision-making to better understand how human and natural systems function.

Setting, Situation and Study Area

This study focuses on the South Omo Zone of Ethiopia, where large-scale land acquisition is an emerging public issue. The zone is rich in resources (fertile soils, rivers, irrigable lands) and has significant potential for increased agricultural and livestock production. It is comprised of 2.3 million hectares of arid and semi-arid lands in southern Ethiopia, with low and erratic rainfall, periodic droughts, and different types of vegetative cover and soils. The region borders Kenya in the south and South Sudan in the southwest, as shown in Fig. 2. Regional topography shows a distinct gradient along a northeast-southwest direction. Elevation in the northeast reaches 2500–3500 meters above sea level (MSL), while in the southwest it falls to 400–500 MSL. Vegetation cover varies along the elevation gradient. Lowlands are covered primarily with grasslands and woodlands, while highlands are covered with shrubs and trees. The Omo River dissects the zone running north to south, draining northern, higher-rainfall areas into Lake Turkana. The South Omo Zone is intersected by the Woito River on the southeast side, draining the northeast escarpments into the Chew-Bahir (also known as the "Salt Sea").

The local population consists of indigenous tribes living in a traditional system of subsistence agriculture. The total population of the zone is 569,448 inhabitants, according to the 2012 census, with 284,781 (50.01%) males and 284,667 (49.9%)

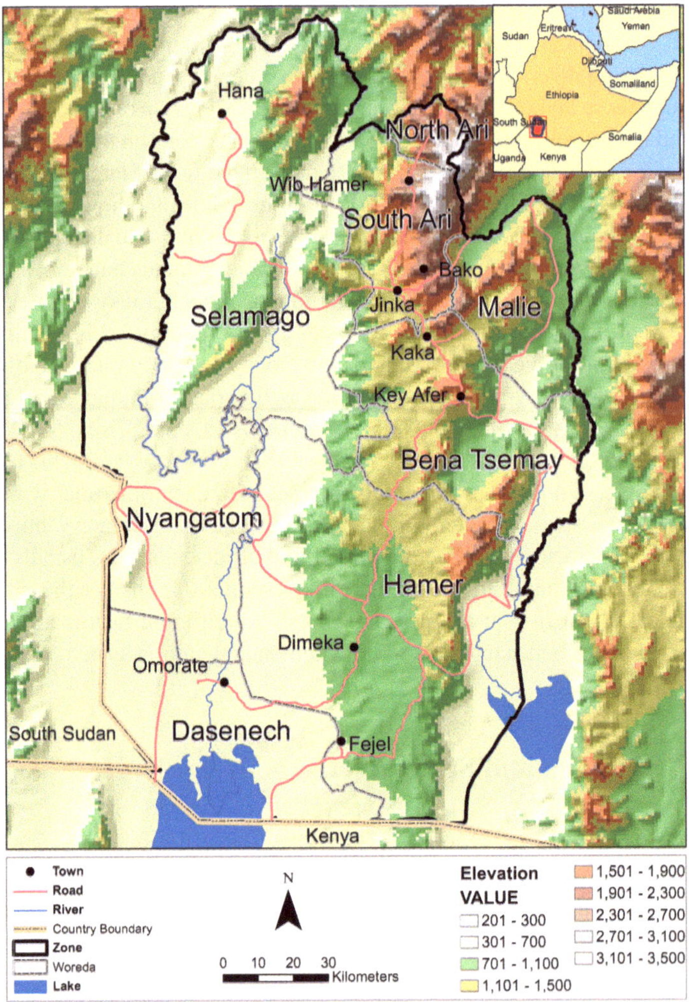

Fig. 2 Geographical location of the South Omo Zone of Ethiopia, also known as Debub Omo. Source: drawn by the first author, based on [30]

females [17]. The total number of households is 125,009 with an average household size of 4.6 persons, of which about 80% are male-headed households while the remaining 20% are female-headed households. The economic activity of the region is characterized by subsistence agriculture dominated by agro-pastoral and pastoral systems. Subsistence crop production is the traditional system, focused primarily on household consumption needs. Crop production is highly dependent on rain, although there are significant opportunities for irrigation and riverine crops. Livestock production occurs mainly in lowland areas, where moisture is a constraint. Lowland households realize 80% of their income from livestock [28].

South Omo is in the public spotlight due to a rising trend in large-scale land acquisitions. The government of Ethiopia is interested in increasing the socio-economic status of rural traditional communities and improving the agricultural sector by "commodification" of the land. Current policy entails transforming a rural society of many small farmers and subsistence agriculture into capital-intensive production enterprises for feeding a growing urban population [39]. Currently the federal land administration has assigned about 500,000 hectares of land in the South Omo Zone—which is about 20% of the total area of the region—to a variety of investment purposes [40]. The Oakland Institute [40] estimates that the amount of land already provided to large enterprises is 445,500 hectares, mainly along the Omo river, which is a significant ecosystem used by traditional pastoralists for coping and adaptation purposes. This trend in land acquisitions will likely bring major changes to the currently stable rural system—a hypothesis tested by our ABM. Growing pressures from national and foreign enterprises for large-scale agricultural production and ecotourism, and shifts in government policies, generate changes in socio-ecological dynamics, potentially affecting the adaptive capacity of rural households by limiting access to traditional resources.

The OMOLAND Model

Model Description

The OMOLAND model includes interrelated components of the South Omo Zone rural system. This ABM is designed to explore interactions and decision-making among different actors in the system: (1) rural households, whose livelihood is significantly coupled with climate and biophysical environments; (2) large-scale land enterprises, operating in the rural system; (3) climate, with variability that impacts actors and the biophysical environment; (4) actors, affecting their respective environments; and (5) the environment, producing feedback effects that influence actors' decision-making processes at different temporal and spatial scales. The representation of entities and their interactions uniquely distinguishes this model from previously implemented agent-based models of rural systems (e.g., [8, 19, 33, 48]). The model is implemented in MASON [36], an ABM simulation toolkit written in the Java programming language and primarily designed to facilitate the development of fast and efficient ABMs. MASON provides extensive libraries to integrate GIS data (vector and raster) [51].

Figure 3 illustrates the main components and relationships in OMOLAND using a UML (Unified Modeling Language) class diagram. The environment is the South Omo Zone, consisting of biophysical components (including natural and built systems) with spatial extent of 146.7 by 224.7 km and comprised primarily of heterogeneous parcels (built environment is minimal in this region, except for some roads). OMOLAND's spatial resolution is 1 hectare (100 by 100 m), based on the

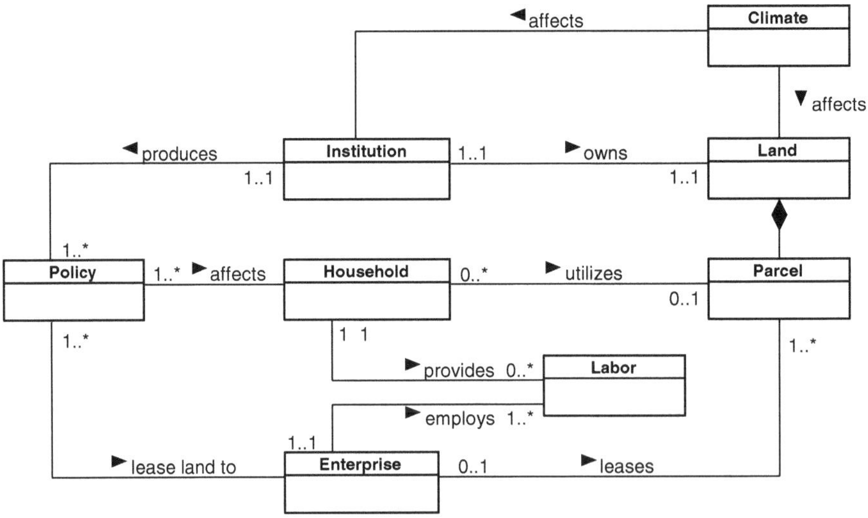

Fig. 3 High-level UML class diagram of the OMOLAND model in MASON. Source: adapted from [30]

average land-holding size of the rural households in the region. Each parcel has quality that is restored or depleted based on actions by households and enterprises. The biophysical system dynamically responds to climate and its variation. Such response will indirectly influence land-use choices by households and enterprise agents.

Climate in OMOLAND is represented by rainfall in terms of precipitation distribution patterns and variations, calibrated to the study area. Climate variation includes significant changes in weather patterns, shifting from normal to extreme events (e.g., drought, flooding). Climate determines the characteristics of biomass in the environment; i.e., type, growth rate, productivity, and annual number and length of growing period(s) or season(s).

Household agents represent individual households that live in a subsistence agricultural system (herding and/or farming) within the study area. Household agents are heterogeneous in their profile, livelihood choices, and decision-making processes. They have bounded rationality [49], lacking full knowledge of the environment, and making decisions based on information they have at hand and on their previous experience. However, households learn, imitate skills and techniques, and make adjustments to their livelihood. They are also social agents, cooperating among themselves and competing with others for resources. Each household has one or more family members. Each family member knows its employment situation.

Enterprise agents represent business actors operating large-scale agricultural production systems. They use much larger tracts of land in their possession and are heterogeneous, based on their land holdings and the number of employees they can hire at a given time. They also create jobs in the rural system and they interact with household agents in a labor market.

An enterprise agent is designed to capture two of the main influences of real-world enterprises in the rural systems. First, enterprises agents influence a biophysical environment by occupying a large tract of land and changing the land property from grazing land to farmland as lands are used for production of commercial crops through mechanized farming. The occupation and conversion of land has direct effects on the amount of vegetation production in the system. Conversion of grazing land to commercial farming reduces the total grazing area. This, in turn, has implications for livestock production, which solely depends on grazing. Moreover, occupation of land can also affect herding movements from place to place by fragmenting grazing areas. Second, enterprises affect the system by introducing a new livelihood option for rural households. These employment opportunities can diversify household income options. OMOLAND is designed to implement these two concepts.

Enterprises individually determine labor needs and announce openings to the public. Employment positions remain open until filled. When a position is filled, it is no longer searchable by the public and the enterprise does not employ additional labor. Positions are temporary. When a task reaches its time limit, the enterprise dismisses all employees and determines when to start a new task.

It is critical to point out that workers can also resign and leave an enterprise. A household may abandon off-farm activity at any given time and decide to return to agricultural activities, since in most rural systems off-farm activity is subsidiary to traditional farming and herding [27]. If a household member decides to leave an off-farm position, the enterprise will assess labor needs and immediately publicize employment positions as necessary [46].

In the OMOLAND model, each enterprise pays each worker an equal amount each day. Although skill or experience may affect the amount of money a worker can earn, this is not considered in the model. We believe that prior skill or experience are not yet valued in off-farm activity in South Omo, because the main opportunity is related to short-term labor activities.

The institution agent in Fig. 3 represents the government and is responsible for generating policies related to land use. This agent has overall knowledge of the entire area, assessing and designating land for different uses. For instance, the institution agent assigns lands to enterprises based on land quality. The allocation to enterprises can be either on lands that are occupied or unoccupied by rural households. The institution agent can also relocate household agents depending on demand for more lands from enterprise agents.

The temporal resolution of OMOLAND is a discrete time step, where 1 step = 1 day. Although such a temporal resolution is relatively fine, some processes occur only when necessary conditions are satisfied. For instance, crops can only instantiate and grow when a household agent sows crop seeds on his farmland. Similarly, a household member's age increases only once in a year. A full description of the OMOLAND model can be found in Hailegiorgis [30].

Model Sequence

The model sequence includes all components involved in the scheduling routine. Each procedure is activated by its generating actor or entity, and similar procedures are activated in the same order at each time step. The first routine concerns how climate affects land. This is updated by having rain fall on each parcel, which is followed by each parcel updating its soil moisture level. Equal amounts of rainfall generate equal amounts of soil moisture. In OMOLAND there is no overflow, inflow of water, or accumulation of soil moisture, as a simplifying assumption, so the updating mechanism is simple. If there is no rain on a given day, the amount of soil moisture added to a parcel is assigned as zero; otherwise, a parcel's soil moisture equals the amount of rainfall on the parcel. After updating rainfall, the vegetation subroutine is executed. Vegetation grows or decreases depending on a parcel's moisture, for each parcel where there is vegetation.

The second routine concerns household agents. In each time step, each household agent engages in livelihood activities, updates profiles, and assesses the success or failure of actions. The main sequential procedures of the household are predicting future climate conditions, analyzing adaptive response, selecting potential livelihood options, allocating resources for implementing livelihood-related activities, monitoring wealth status, updating profile, and updating memory. Routines are only executed at times when appropriate conditions are fulfilled. For instance, sequential procedures from predicting future climate conditions to determining livelihood options are executed once in a season. Each household predicts a date and amount of rainfall for the upcoming season. Based on the outcome of their action, each household makes an appropriate decision to either adapt or fail to adapt in response to the anticipated climatic condition of the season. Depending on the adaptation decision, each household determines the best livelihood or combination of livelihoods (herding, farming, or off-farming) that yields highest return. The household then allocates resources necessary for each livelihood in proportion to the share value of each livelihood. The household remembers its decision and allocation of resources for each livelihood option throughout the implementation of each activity. Implementation of an activity is carried out until each has either been discarded or completed. Memory update is executed at the end of each season.

The livelihood activity sequence of a household is scheduled in the following order: herding, farming, and off-farming. If a household engages in only one of the three livelihood options, it only implements the corresponding activity sequence. For instance, herding activities are invoked if the household has livestock. A household with livestock looks for high-quality grazing areas for its herds. A herder household also monitors its herd income in this sequence. Households with farmland implement farming activities, which include land preparation, planting, weeding, and harvesting. The implementation date of each activity is determined by when its necessary conditions are met. For instance, after the onset of rain, a household assesses if there is sufficient moisture to perform planting. Likewise, when a crop is ready for harvest, a household executes harvesting. At harvest time, each farmer household updates its income in proportion to yield harvested. Household agents can execute off-farm activities to earn extra income by seeking employment in one of the enterprises.

After the household routine, the herd sequence is invoked. Herds consume grass from their current location and move to an assigned location. They update their metabolic rate, food level, and size based on grass consumption.

Following the herd sequence, the crop sequence is invoked. A crop is activated only when it is planted. Similar to vegetation, crops respond to available moisture in their parcel, by either growing or decreasing. A crop updates its growth and production level at each time step.

The enterprise routine comes next. In each time step, enterprise agents manage the labor force by deciding whether to recruit or dismiss workers (daily laborers) and acting on their decisions. Each enterprise determines its labor requirements and allocates the resource to the task. If the current labor is more than required, the enterprise agent reduces the labor force to the required minimum level by dismissing excess workers. Conversely, if there is a need for labor, the agent searches for extra labor and hires to fill the labor gap.

The institutional sequence is invoked after the enterprise routine. The institution agent, which represents government, selects potential households needing capacity-building training or relief support, and provides such benefits as necessary.

Finally, an observer object, for managing data collection and statistics, is invoked and all the output is written to disk.

Policy Scenarios

The main aim of our scenario analysis is to explore and better understand impacts of large-scale land acquisitions on rural households by changing the scale and intensity of intervention of enterprise agents. Policy relevance is high, given the stakes and complexity of dynamic interactions among enterprises, households, climate, and environmental entities.

Scenario analyses using OMOLAND examine the impacts of enterprises on rural households. Analysis focuses on exploring whether enterprises increase the vulnerability of rural households or provide households an opportunity to diversify their livelihood options through off-farm activities. The following two main scenarios are analyzed: large-scale land acquisition without and with off-farm opportunities, corresponding to Scenario 1 and Scenario 2, respectively.

The main goal of scenario analysis is to assess issues related to commercial enterprises' contributions to providing additional off-farm opportunities to rural households. Although commercial enterprises increase employment opportunities, the probability that an individual person in a rural community will participate in such employment opportunities is not significant due to high competition (i.e., excess supply of labor). Moreover, such jobs are often short-term or seasonal and usually poorly paid [46]. Such factors discourage the engagement of rural households in off-farm activities, affecting their off-farm opportunities.

Scenario 1 depicts large-scale land acquisition with diverse spatial intensity and no opportunity for off-farm activity by rural households. In Scenario 2, enterprises offer labor employment opportunities to rural households [14], recruiting and dismissing employees depending on their labor requirements. Rural households search for such nearby off-farm opportunities and engage if they have extra time available from herding or farming, or if either of these two activities fails to provide sufficient household income.

Two trends in enterprise growth rates, "slow" and "fast," are explored in each scenario, corresponding to 2% and 5% annual growth rates, respectively—with the latter representing the current rate for the South Omo Zone.

Rural households and enterprises are considered as main agents in each scenario. The simulation runs for 18,250 steps. Since each step corresponds to a day, then 18,250 iterations are about 50 years. Monthly rainfall data from 1949 to 2009 is used as climate input. A set of 30 simulation runs is conducted for each scenario, using the same initial (default) parameter settings.

Results

It is important to discuss model verification before presenting results, following current standards in quality control. Verification is the process of ensuring that a simulation is implemented as intended by the conceptual model [4, 11, 16]. Verification of OMOLAND was performed by conducting code walkthroughs, debugging, profiling, and parameter sweeps. These tests insured that we made no logical errors in the translation of the model into code and there were no programming errors. No anomalies have been detected since the above verification procedures were carried out, so we feel confident that the model behaves as it is intended and it matches its design.

In this section we present results from simulation analysis of the two scenarios described in Section "Policy Scenarios".

Scenario 1: Without Off-Farm Opportunities

Figure 4 shows Scenario 1 ("no off-farm opportunity") results, assuming slow (2%) and fast (5%) rates in land acquisition expansion. As can be seen, the two figures are very similar, the only difference being a slightly lower final value for household totals in the figure on the right (b). When we explore the number of people that emigrate from the region, results in Fig. 5 also show that there is a significant difference between the 2% and 5% expansion rates, especially during the last 10 years of the simulation. However, emigration change is not linearly related to change in the expansion rate of large-scale land acquisition.

As shown in Fig. 5, at the end of the simulation, the number of emigrants reached 1 person per 9.7 hectares of land acquired by enterprises at a 2% ("slow") expansion

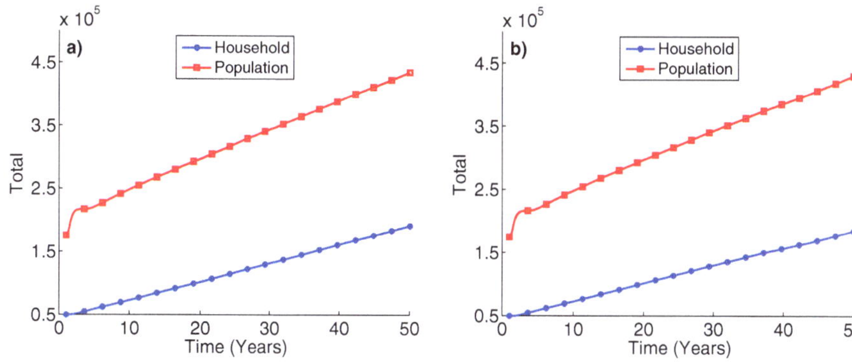

Fig. 4 Scenario 1 results: household and population growth without off-farm opportunities, assuming slow (**a**) and fast (**b**) rates of expansion in land acquisition, corresponding to 2% and 5% annual rates

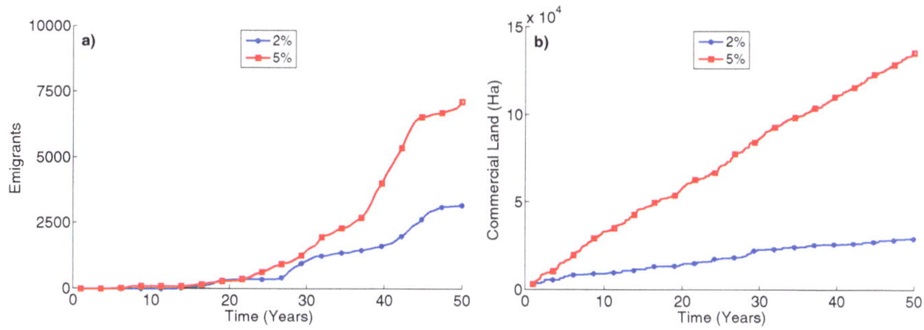

Fig. 5 Scenario 1 results: migration and land expansion without off-farm opportunity: (**a**) total number of emigrants and (**b**) area of land acquired by enterprise in hectares

rate, while at a 5% ("fast") rate of expansion the number of emigrants decreases to 1 person per 18.9 hectares. Note that, as a quantitative measure of social impact caused by enterprise expansion, this emigration effect is akin to a density measured in [persons]/[hectare]. This can also be interpreted as a displacement flow when time is added, or [persons]/[area][time].

The impact of expansion of large-scale commercial farming over time on crop and livestock production is shown in Fig. 6. Interestingly, increasing the rate of expansion of large-scale commercial enterprises does not significantly affect the per capita level of livestock and crop production in the region. Trends in livestock and production for the simulated period are similar under both rates (2% and 5%). However, in both cases, livestock production shows a slightly downward trend, whereas crop production ends with an opposite, slightly upward trend. This could be caused by the fact that most land assigned to enterprises is located in areas where livestock production is the dominant production system, such as in proximity to rivers.

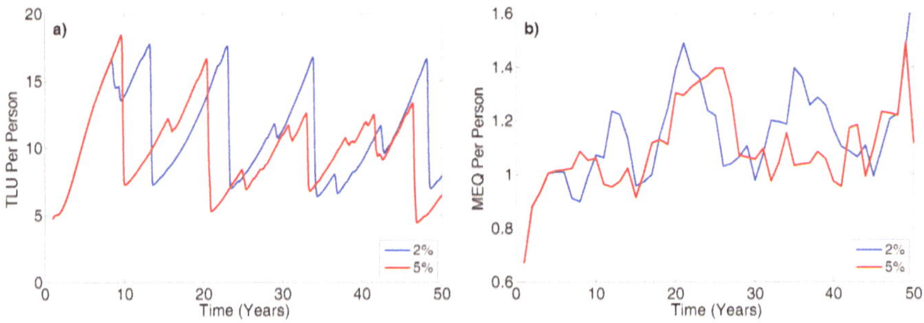

Fig. 6 Scenario 1: livestock and crop production with no off-farm opportunity: (**a**) number of livestock (TLU) per person and (**b**) crop (Maize Equivalent-MEQ in kilograms) per person

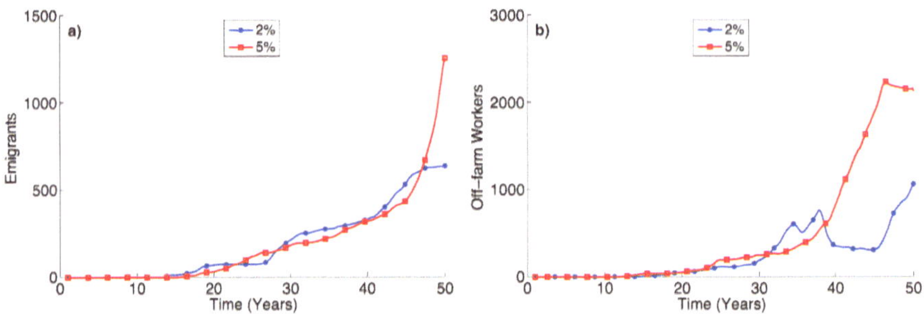

Fig. 7 Scenario 2: migration and off-farm activity with off-farm opportunity: (**a**) number of emigrants from the region and (**b**) number of persons engaged in off-farm jobs

Scenario 2: With Off-Farm Opportunities

In the second scenario, enterprises offer employment opportunities to rural households. In this case the OMOLAND simulation model provides different results in terms of the number of emigrants from the region and in patterns of livestock and crop production. Specifically, the total number of emigrants decreases significantly as compared to Scenario 1 as shown in Fig. 7a. This is mainly due to the number of people employed in off-farm jobs increasing as a function of time, following the expansion of large-scale commercial farming (Fig. 7b).

Although off-farm jobs offer additional income to households, such an opportunity does not entirely eliminate emigration, even under the "fast" rate of expansion (5% growth rate). This is because there are still persons who cannot sustain their livelihood under current climatic conditions and are forced to emigrate.

Scenario 2 also shows that off-farm opportunities affect livestock and crop production in opposite ways, as shown in Fig. 8. These results show that in Scenario 2 livestock production (TLU/person) and crop production (MEQ/person) show decreasing and increasing trends, respectively.

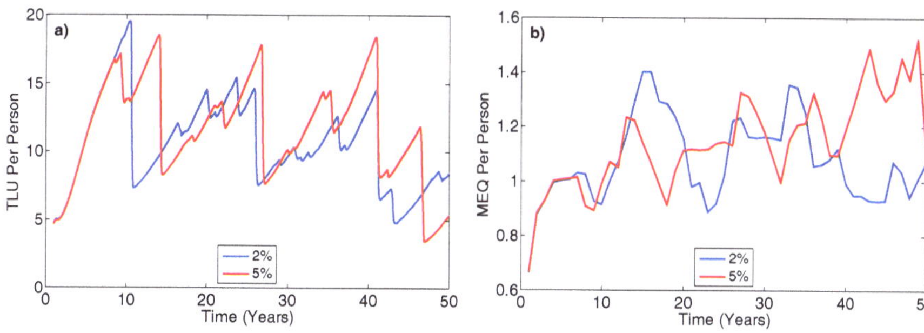

Fig. 8 Scenario 2: livestock and crop production with off-farm opportunity: (**a**) number of livestock (TLU) per person, (**b**) crop (Maize Equivalent-MEQ in kilograms) per person

Comparing the two scenarios, trends in both livestock and crop production are more pronounced in Scenario 2. Livestock production levels (20 TLU/person) reached 10 years from the start of the simulation in Scenario 1, decreased to less than 5 TLU/person by the end of the simulation in Scenario 2, while the same level was >7 TLU/person in Scenario 1. Similarly, crop production increased from 1 to 1.6 MEQ/person by the end of the simulation in Scenario 2, almost a 20% increase compared to Scenario 1.

Discussion and Conclusion

The OMOLAND model demonstrates two important points regarding the impact of large-scale land acquisition in the South Omo case, which can be comparable to other regions in developing countries. First, it suggests that land acquisition without providing new employment opportunities to local communities of herders or farmers can lead to catastrophic outcomes—potentially humanitarian disasters and crises—as more people are forced to migrate from the system [41]. Such events can create flows in internally displaced persons (IDPs) and, when national boundaries and border crossings are also involved, transnational refugee flows are another potential disaster. Although migration of rural households can occur simply as a result of extreme climate events (e.g., droughts or floods, both common in the region), results from the model clearly demonstrate that migration rates can be exacerbated as lands utilized by rural households are given to large-scale commercial enterprises.

OMOLAND also contributes to our understanding of the effect of off-farm employment opportunities on rural household livelihood. At first glance, it seems that simulation results align with the aspiration of government development policies or with those who highlight the economic contribution of enterprises in rural communities [52]. The emergence of additional sources of income influences the way in which rural people react to climate change and variability. As more persons work in off-farm jobs, particularly during times of drought, their vulnerability is reduced by the additional income they can generate through such jobs. However, close observation of these results indicates that emigration and displacement persist as more lands are controlled by enterprises.

Another result demonstrated by the model is the gradual transition of dominant rural livelihood from herding to farming as large-scale enterprises spread throughout the region. Galvin [27] argues that pastoral transitions in most East African countries are attributed to two major factors. First, grazing land is fragmented, due to numerous socioeconomic factors, such as changes in land tenure, agriculture, and institutions. Second, extreme weather events such as droughts are more common than elsewhere, due to climate change and variability. Although the OMOLAND model requires further development and analysis, the simulation results agree with real-world trends in terms of more households engaging in farming than in livestock production as grazing lands are converted into commercial farming.

Current trends of large-scale land acquisition are likely to continue over the next decade in most of Sub-Saharan Africa. Exploring the implications of commercially oriented farming enterprises, not only on rural households but also on the biophysical environment, is a promising potential extension of the model. For example, large tracts of rural lands recently have been given to a variety of enterprises with competing interests. These range from those interested in commercial farming (food crops and biofuel production) to others engaged in ecotourism. Greater demand for rural land will likely increase land value and, consequently, land marketing. It is important and feasible to further explore the dynamics of competition for land among different entities by incorporating into OMOLAND a more comprehensive land marketing mechanism. Such an integration of land markets or land transactions into the model could advance our understanding of rural land-use changes, socioeconomic transformations, and issues related to short- and long-term rural-urban dynamics. Although many future research directions are possible and arguably fruitful, this study contributes to investigating these basic and applied research questions by laying foundations for further rigorous work on complex dynamics in coupled human and natural systems.

Acknowledgements This work was supported by the US National Science Foundation through a Doctoral Dissertation Research Improvement (NSF-DDRI) grant (no. 112348), the Office of Naval Research (ONR) under a MURI grant to the GMU-Yale Joint Project on East Africa, and by the Center for Social Complexity at George Mason University. Thanks to the editor and reviewers of an earlier version of this chapter, and to Andrew Crooks, Alan Falconer, and Tim Gulden for comments and discussions. Only the authors are responsible for the content of this paper.

References

1. Acosta-Michlik L, Espaldon V (2008) Assessing vulnerability of selected farming communities in the Philippines based on a behavioural model of agent's adaptation to global environmental change. Glob Environ Chang 18(4):554–563

2. Aldrich S, Walker R, Arima E, Caldas M, Browder J, Perz S (2006) Land-cover and land-use change in the Brazilian Amazon: Smallholders, ranchers, and frontier stratification. Econ Geogr 82(3):265–288
3. An L (2012) Modeling human decisions in coupled human and natural systems: Review of agent-based models. Ecol Model 229:25–36
4. Balci O (1997) Verification, validation and accreditation of simulation models. In: Proceedings of the 29th conference on winter simulation. IEEE Computer Society, New York, pp. 135–141
5. Barrett C, Reardon T, Webb P (2001) Nonfarm income diversification and household livelihood strategies in rural Africa: concepts, dynamics, and policy implications. Food Policy 26(4):315–331
6. Berger T (2001) Agent-based spatial models applied to agriculture: a simulation tool for technology diffusion, resource use changes and policy analysis. Agric Econ 25(2–3):245–260
7. Berkes F, Folke C, Colding J (2000) Linking social and ecological systems: management practices and social mechanisms for building resilience. Cambridge University Press, Cambridge
8. Bharwani S, Bithell M, Downing T, New M, Washington R, Ziervogel G (2005) Multi-agent modelling of climate outlooks and food security on a community garden scheme in Limpopo, South Africa. Philos Trans R Soc, B 360(1463):2183
9. Booth D, Hanmer L, Lovell E (2000) Poverty and transport: a report prepared for the World Bank in collaboration with DFID, Overseas Development Institute (ODI)
10. Calderón C, Servén L (2004) The effects of infrastructure development on growth and income distribution, World Bank Policy Research Working Paper No. 3400
11. Cioffi-Revilla C (2014) Introduction to computational social science: principles and applications. Springer, London
12. Cioffi-Revilla C, Rogers JD Latek M (2010) The MASON HouseholdsWorld model of pastoral nomad societies. In: Takadama K, Cioffi-Revilla C, Deffaunt G (eds) The science of social simulation: the second world congress in social simulation. Springer, Berlin, pp 193–204
13. Cotula L, Vermeulen S (2011) Contexts and procedures for farmland acquisitions in Africa: what outcomes for local people. Development 54(1):40–48
14. Cotula L, Keeley L, Cotula S, Vermeulen R, Leonard J (2009) Land grab or development opportunity?: agricultural investment and international land deals in Africa. IIED, London
15. Crooks A, Heppenstall A (2012) , Introduction to agent-based modelling. In: Heppenstall A, Crooks A, See L, Batty M (eds) Agent-based models of geographical systems. Springer, Dordrecht, pp. 85–105
16. Crooks AT, Castle C, Batty M (2008) Key challenges in agent-based modelling for geo-spatial simulation. Comput Environ Urban Syst 32(6):417–430
17. CSA (2012) The 2007 population and housing census of Ethiopia: statistical report for southern nations, nationalities and peoples' region, vol 1. Central Statistics Agency of Ethiopia, Addis Ababa
18. De Haas H (2007) Turning the tide? why development will not stop migration. Dev Chang 38(5):819–841
19. Deadman P, Robinson D, Moran E, Brondizio E (2004) Colonist Household decision-making and land-use change in the Amazon Rainforest: an agent-based simulation. Environ Plann B 31(5):693–710
20. Deininger K, Byerlee D, Lindsay J, Norton A, Selod H (2010) Rising global interest in farmland: Can it yield sustainable and equitable benefits? World Bank-free PDF
21. Delgado C (1995) Agricultural diversification and export promotion in sub-Saharan Africa. Food Policy 20(3):225–243
22. Derbyshire H, Vickers P (1997) The sustainable provision of poverty focused rural infrastructure in Africa: a study of best practice, Department for International Development
23. Ellis F, Biggs S (2001) Evolving themes in rural development 1950s-2000s. Dev Policy Rev 19(4):437–448
24. Entwisle B, Malanson G, Rindfuss RR, Walsh SJ (2008) An agent-based model of household dynamics and land use change. J Land Use Sci 3(1):73–93

25. Filatova T, Voinov A, van der Veen A (2011) Land market mechanisms for preservation of space for coastal ecosystems: an agent-based analysis. Environ Model Softw 26(2): 179–190
26. Fontaine C, Rounsevell M (2009) An agent-based approach to model future residential pressure on a regional landscape. Landsc Ecol 24(9):1237–1254
27. Galvin KA (2009) Transitions: pastoralists living with change. Annu Rev Anthropol 38:185–198
28. Gebresenbet F, Kefale A (2012) Traditional coping mechanisms for climate change of pastoralists in South Omo, Ethiopia. Indian J Tradit Knowl 11(4):573–579
29. Geist HJ, Lambin EF (2002) Proximate causes and underlying driving forces of tropical deforestation. BioScience 52(2):143–150
30. Hailegiorgis AB (2013) Computational modeling of climate change, large-scale land acquisition, and household dynamics in Southern Ethiopia. Copyright - Copyright ProQuest, UMI Dissertations Publishing 2013; Last updated - 2014-02-28; First page - n/a. http://search.proquest.com/docview/1492669016?accountid=14541
31. Kaufmann P, Stagl S, Franks DW (2009) Simulating the diffusion of organic farming practices in two new EU member states. Ecol Econ 68(10):2580–2593
32. Kniveton D, Smith C, Wood S (2011) Agent-based model simulations of future changes in migration flows for Burkina Faso. Glob Environ Chang 21:34–40
33. Le QB, Park SJ, Vlek PL (2010) Land use dynamic simulator (LUDAS): a multi-agent system model for simulating spatio-temporal dynamics of coupled human–landscape system: 2. Scenario-based application for impact assessment of land-use policies. Eco Inform 5(3):203–221
34. Lim K, Deadman PJ, Moran E, Brondizio E, McCracken S (2002) Agent-based simulations of household decision making and land use change near Altamira, Brazil. Integrating geographic information systems and agent-based modeling: techniques for simulating social and ecological processes. Oxford University Press, New York, pp 277–310
35. Liu J, Dietz T, Carpenter SR, Alberti M, Folke C, Moran E, Pell AN, Deadman P, Kratz T, Lubchenco J et al (2007) Complexity of coupled human and natural systems. Science 317(5844):1513
36. Luke S, Cioffi-Revilla C, Panait L, Sullivan K, Balan G (2005) MASON: a multiagent simulation environment. Simulation 81(7):517
37. McLaughlin A, Mineau P (1995) The impact of agricultural practices on biodiversity. Agric Ecosyst Environ 55(3):201–212
38. Mena CF, Walsh SJ, Frizzelle BG, Xiaozheng Y, Malanson GP (2011) Land use change on household farms in the Ecuadorian Amazon: Design and implementation of an agent-based model . Appl Geogr 31(1):210–222
39. MoFED, Ethiopia (2010) Growth and transformation plan (GTP) 2010/11-2014/15
40. Mousseau F, Sosnoff G (2011) Understanding land investment deals in Africa: country report, Ethiopia, The Oakland Institute
41. Niamir-Fuller M (1999) Managing mobility in African rangelands, Food and Agricultural Organization and the Beijer International Institute of Ecological Economics
42. Parker DC, Manson SM, Janssen MA, Hoffmann MJ, Deadman P (2003) Multi-agent systems for the simulation of land-use and land-cover change: a review. Ann Assoc Am Geogr 93(2):314–337
43. Parker DC, Entwisle B, Rindfuss RR, Vanwey LK, Manson SM, Moran E, An L, Deadman P, Evans TP, Linderman M et al (2008) Case studies, cross-site comparisons, and the challenge of generalization: comparing agent-based models of land-use change in frontier regions. J Land Use Sci 3(1):41
44. Parker DC, Hessl A, Davis SC (2008) Complexity, land-use modeling, and the human dimension: fundamental challenges for mapping unknown outcome spaces. Geoforum 39(2):789–804
45. Perkins H (2006) Commodification: re-resourcing rural areas. SAGA Publications, Thousand Oaks

46. Richards M (2013) Social and environmental impacts of agricultural large-scale land acquisitions in Africa–with a focus on West and Central Africa, Technical report, Rights and Resources Initiative, Washington, D.C.
47. Rindfuss R, Entwisle B, Walsh S, An L, Badenoch N, Brown D, Deadman P, Evans T, Fox J, Geoghegan J et al (2008) Land use change: complexity and comparisons. J Land Use Sci 3(1):1
48. Saqalli M, Gérard B, Bielders CL, Defourny P (2011) Targeting rural development interventions: empirical agent-based modeling in Nigerian villages. Agric Syst 104(4):354–364
49. Simon HA (1996) The sciences of the artificial. MIT Press, Cambridge
50. Smith C, Kniveton D, Wood S, Black R (2011) Climate change and migration: a modelling approach. In: Williams CJR, Kniveton DR (eds) African climate and climate change. Advances in global change research, vol 43. Springer, Dordrecht, pp. 179–201
51. Sullivan K, Coletti M, Luke S (2010) GeoMason: GeoSpatial support for MASON, Department of Computer Science, George Mason University, Technical Report Series
52. Vermeulen S, Cotula L (2010) Over the heads of local people: consultation, consent, and recompense in large-scale land deals for biofuels projects in Africa. J Peasant Stud 37(4):899–916
53. Von Braun J, Meinzen-Dick RS, International Food Policy Research Institute (2009) "Land grabbing" by foreign investors in developing countries: risks and opportunities. International Food Policy Research Institute Washington, DC
54. Zoomers A (2010) Globalisation and the foreignisation of space: seven processes driving the current global land grab. J Peasant Stud 37(2):429–447

Spatial Agent-based Modeling to Explore Slum Formation Dynamics in Ahmedabad, India

Amit Patel, Andrew Crooks, and Naoru Koizumi

Introduction

In 2009, for the first time in history, more people lived in urban areas than in rural areas. By 2030, the global urban population is expected to be 59% of the total world population [1]. Most of this urban growth is expected to take place in developing countries and raise numerous developmental challenges. One of the most critical challenges is the lack of affordable housing for the urban poor, which results in increasing numbers of people living in slums. The issue of slums is compounded by the fact that it is both a large-scale global problem (e.g. a global crisis due to unprecedented magnitude in shelter deprivation) as well as posing localized problems for individual cities (e.g. the spread of infectious diseases). Currently, one-in-three urban residents (924 million people) live in slums globally, most of them in cities of the developing world. This number is projected to increase to two billion people by 2030 if adequate actions are not taken [2].

Many scholars and development practitioners recognize that the proliferation of slums is one of the most complex and pressing challenges that developing countries face today (e.g. [2, 3]). Inappropriate housing conditions for the urban poor is becoming an important concern for policymakers in developing countries since it is recognized that slums adversely affect the wellbeing of the entire city, raising wide concerns ranging from public health to that of safety [4]. The international development community has recognized the growth of slums as an important

A. Patel (✉)
University of Massachusetts Boston, Boston, MA, USA
e-mail: Amit.Patel@umb.edu

A. Crooks • N. Koizumi
George Mason University, Fairfax, VA, USA

J.-C. Thill, S. Dragicevic (eds.), *GeoComputational Analysis and Modeling of Regional Systems*, Advances in Geographic Information Science,
DOI 10.1007/978-3-319-59511-5_8

societal issue, as a response, Target 11 of the Millennium Development Goals (MDG) aimed to significantly improve the lives of 100 million slum dwellers by 2020 [5].

National and local governments in many developing countries have also called for slum up-gradation and slum improvement programs. The expansion of urban renewal programs and a greater focus on making cities slum-free has taken a front seat among policymakers in most developing countries. For example, in Kenya, their commitment to address the challenge of slums now appears in the national development agenda [2]. In India, the issue of slums has recently received significant political salience. For instance, the Jawaharlal Nehru National Urban Renewal Mission (JNNURM) highlighted tackling issue of slums as a key task [6]. Ms. Pratibha Patil, then president of India announced a policy targeted to make India slum-free within five years [7], which resulted in a massive housing program for slum dwellers called Rajiv Awas Yojana (RAY; [8]).

While policymakers have renewed their focus to address this challenge worldwide, slums are not a new phenomenon. Several policy actions in the past have attempted to resolve this issue. Slum policies evolved from "Site and Services" in the 1970s [9], to "Slum Redevelopment" in the 1980s [4, 10], to "Security of Tenure" in the 1990s [11] and "Slum-free cities" in 2000s [12]. Several indigenous policies have also been implemented within cities of developing countries. For example, in India, the city of Ahmedabad implemented the "Slum Networking Project" [13] and the city of Mumbai implemented the "Slum Redevelopment Scheme" [14]. Unfortunately, none of the past slum policies have proved to be a panacea to making cities slum-free. Many of these policies have been evaluated and have been found ineffective both in India and elsewhere (e.g. [15–18]).

It is evident that improved responses are required to address this challenge. Such a task is difficult, especially when there is a gap between slum policies and an understanding of slum formation and expansion processes. It is evident that slum policies have been either incremental or experimental. Both types of policies have been implemented without knowing their adverse implications for either an individual household at the micro-scale or slum formation patterns at the macro-scale. Ex-post analyses of these policies are in abundance (e.g. [19–21]) but attempts to understand the implications of these policies *ex-ante* are rare. In this sense, past slum policies were heuristic rather than evidence-based. We believe that part of the problem is the lack of research tools to evaluate proposed policy interventions with respect to slums [22, 23]. The model presented in this chapter is one such tool to conduct policy experiments in a simulated environment before slum policies are implemented in the real world.

Although our understanding of cities has increased throughout the twentieth century by incorporating ideas and theories from a diverse range of disciplines such as economics, geography, history, philosophy, mathematics and more recently computer science, it is now very clear that there are intrinsic difficulties in applying such understanding to policy analysis and decision-making [24]. This is especially challenging in the context of slums of the developing world where the availability of data and lack of prior research poses additional difficulties.

To gain a greater understanding of urban problems in absence of empirical data, researchers have recently focused on simulation approaches to model urban systems. Specifically, such approaches attempt to discover the basis of individual decision-making and its implications on urban systems [25]. One such approach is Agent-Based Modeling (ABM), which enables to simulate individual actions, study the resulting system behavior and aggregate spatio-temporal outcomes.

Traditionally, urban models focused on aggregate representation of cities and dealt with residential or employment distributions within cities. But emergent structures are not well suited to such traditional styles of urban modeling and raise questions on how to best handle them [25]. Slums display several properties of emergent structures and hence we, along with others [23, 26, 27], argue that bottom-up models provide a good alternative for developing new models of cities and especially for exploring slums. Both cities as well as slums are highly dynamic in space and time. Since formation of slums is a spatial phenomenon, integrating Geographical Information System (GIS) with ABM provides an appropriate framework to capture emergence of slums. Furthermore, such a framework provides an important medium for urban planning because the study and management of slums is affected by individual level locational and behavioral factors that are difficult to incorporate in traditional urban planning methods. Modeling also allows scientists to explore and test theories and practices about slums in a controlled computer environment to understand urban phenomena through analysis and experimentation, a traditional goal of science [25]. Urban modeling is also equally important to planners, politicians and communities to predict and invent urban futures [28]. Several policy measures regarding slums, such as slum upgrading and tenure formalization can be explored using such a model to generate various scenarios for urban futures, thus linking science to decision-making.

For making cities slum-free, it is important to understand how these slums emerge and evolve within cities. However, this is not a trivial task especially because cities are inherently dynamic, constantly evolving, undergoing changes, experiencing growth and decline, and restructuring simultaneously [29]. What makes it more complex is the fact that slums display the same dynamic properties as the cities within which they are located. Although advances in geosimulation methods have increased our understanding of urban systems in the developed world, these methods are rarely applied to study urban problems in the developing world such as slums.

In the remainder of this chapter, we first provide a review of geosimulation models of urban systems (see section "Modeling of Urban Systems") followed by the discussion on prior efforts to model slum formation (see section "Prior Efforts to Study Slum Formation using Geosimulation"). Then we present the conceptual framework for our model that integrates GIS and ABM (see section "A Geosimulation Approach to Model Slum Formation"). In section "Case Study: Ahmedabad", we describe the input data that we use as a basis to model slum formation in the city of Ahmedabad, India, and finally present the simulation results in section "Simulation Results". The chapter concludes with the challenges in linking GIS and ABM to study slums, and identifies future avenues of research.

Modeling of Urban Systems

Geosimulation is defined as a method of academic inquiry that simulates the systems by modeling adaptive collectives of interacting entities [30]. Unlike traditional top-down approaches, geosimulation studies the systems by dissecting them into logically justified components and it is characterized by a generative or bottom-up approach. The phenomena of interest (e.g. urban growth patterns, crowds, etc.) are therefore viewed as the product of multiple interactions between physically existing entities (e.g. households or pedestrians).

ABM provides an appropriate paradigm to think about dynamic urban systems, especially because it is well recognized that cities are complex systems [31] that emerge from the bottom-up rather than top-down [25]. Similarly, GIS is useful for representing systems of a geospatial nature and hence provides a useful medium to represent cities. However, it has been well established in the literature that GIS is not particularly well suited for dynamic modeling [32]. It is therefore important to merge the two methods since cities are both highly dynamic and geospatial in nature. In particular, urban models need to capture spatial changes such that if one or more location specific attributes or locations of activities themselves change, the outcomes of the model change [33]. This realization has led researchers to link GIS and ABM to model urban systems under the umbrella of geosimulation. The generative approach has been justified because planning and public policy do not always work in a top-down manner, instead aggregate conditions in cities emerge from the bottom-up from the interaction of a large number of entities at a local scale [25]. ABM is particularly well suited to model individual entities and GIS is appropriate to model locational aspects of cities. Before we discuss why it is appropriate to link ABM and GIS to study slums, we provide a brief review of models that link GIS and ABM to study urban systems.

Integrated simulations within the urban context are seen in a number of planning support systems in the developed world [34–36]. For example, the integration of Cellular Automata (CA) and GIS has been used to simulate urban dynamics, e.g. SLEUTH model by Clarke and Gaydos [37], which has been applied to several cities around the world [38]. Landis [39] developed a multi-scalar model named California Urban Futures (CUF) that predicts urban growth by integrating GIS and CA. The CUF evolved into two different models, the CUF II [40] and California Urban and Biodiversity Analysis Model (CURBA; [41]) that were both used to simulate policies and generate development scenarios. Engelen et al. [35] designed a model named Environmental Explorer (EE) to work as spatial support system for the assessment of socio-economic and environmental policies for the Netherlands.

However, most of the above-mentioned simulations were CA-based which has several limitations with respect to studying the inhabitants of cities. One of the principal limitations is the difficulty to adequately model mobile entities within CA models (e.g. households, pedestrians, vehicles). Another major limitation is the inability to apply heterogeneous behaviors to all cells within CA framework [42]. More recently, researchers have started to combine ABM and CA models to overcome these limitations and making them more 'agent-like' [25]. Examples of this combination of ABM and CA include Torrens [43] who combined CA with ABM to explore sprawl in Michigan. While Xie et al. [44] explored urban growth in China and used agents to explore pressure on land from developers. Such models place more emphasis on the individual decision maker than on the transition potential of cells within CA models or the aggregate flow of people within 'traditional' urban models.

The utility of exploring urban growth through the combination of CA and ABM is that land-use change has a temporal, spatial and behavioral component (e.g. why people want to live in a particular area). By combing CA and ABM, urban modelers are able to capture these three elements [45]. It has also been argued that the combination of CA and ABM provide a more decentralized view of exploring urban systems from the bottom-up [45], which some term as the cell space models [25]. The combination of CA and ABM allows urban modelers to explore human behavior and how such behavior impacts on urban growth patterns. For example, Wise and Crooks [46] explored how heterogeneous agents representing farmers, developers and buyers could influence the spatial pattern of residential development through interactions in the land market. Robinson and Brown [47] showed how lot-size zoning and municipal land acquisition strategies could reduce the impact of urban sprawl. While the majority of geosimulation models are applied in developed countries, there are a limited number of models that have been applied to developing countries, especially in the context of slums. In the next section, we discuss the past efforts to model slum formation using geosimulation methods.

Prior Efforts to Study Slum Formation using Geosimulation

While policy-oriented and theoretical literature on slums is abundant, the geosimulation approach for this phenomenon has been lagging. There are only a handful of attempts to model slums so far. Sietchiping [23] adopted the SLEUTH model in an unplanned urban context of Yaoundé to predict slum growth. However, as discussed above in section "Modeling of Urban Systems", CA models lack the human behavior aspects that ABM can provide. Barros [26] developed an ABM of slum formation and growth in Latin American cities to address this limitation of CA models. However, Barros' [26] model is rather abstract and lacks the explicit spatial representation of a city and hence cannot be used as a planning support tool. Xie et al. [44] developed a model that integrated remote sensing data and ABM to study emergence of Desakota (densely populated rural areas in the extended surroundings of large cities) in Suzhou region of China. The model incorporated interaction between global and local scale actions and included the supply-side of housing (i.e. behavior of developers). However, the model was limited to explain the emergence of Desakota, a phenomenon largely observed in the peripheral rural areas within China.

At the micro-scale, the Informal Settlement Growth Model by Young and Flacke [27], developed further by Augustijn-Beckers et al. [48], showed how housing patterns within a single slum can be simulated using simple rules of spatial change. Vincent [49] developed a spatially explicit ABM for a single ward of the city, thus limiting its capability for citywide planning. Overall, past slum models are not useful for policymaking either because they are either abstract or do not model the city as a whole. Our model attempts to bridge this gap by incorporating human behavior in a spatially explicit environment of an entire city and thus could be useful for citywide planning and policymaking.

A Geosimulation Approach to Model Slum Formation

In order to address the shortcomings of the previous simulation efforts discussed above, a holistic approach to model slum formation is required. Here we present a conceptual geosimulation framework, which captures the important processes pertaining to slum formation dynamics that builds upon Patel et al. [50]. Specifically, our framework aims to capture three important processes, which lead to the formation of slums within a city. First, population growth, both natural and thorough migration, drives overall housing demand within a city. Second, households' residential location choice behavior drives housing demand at specific locations. Third, the spatial configuration of the housing market shapes the availability of housing at specific locations. These processes are modeled because previous literature and studies have shown that they influence the formation of slums within a city. Our framework, presented in Fig. 1, is split into three different modules to capture these three important processes: (1) *Population Dynamics Module* (*PDM*), (2) *Housing Dynamics Module (HDM)*, and (3) *Empirical Module (EM)*. The individual modules are organized at different geographic and demographic scales, which allow us to incorporate varying levels of details appropriate for each module. For example, migration flow is modeled at the regional level whereas the housing market is modeled at the city level. Each module is envisaged as a tightly coupled system to allow for the exchange of information. However, we first develop each module separately in order to verify and validate results from individual modules.

ABM is used to build the *HDM*, which simulates behaviors of important actors for the housing market. Slums develop from the bottom-up as a result of such behaviors rather than from the more traditional top-down normative constraints of classical models [25]. Our framework is designed to explore the links between individual behaviors and aggregate outcomes. For example, it can be used to show how different household behaviors in the housing market can lead to the emergence of different slum patterns. It can also be used to explore how different urban policies that influence such behavior may lead to outcomes that were not anticipated originally. For detailed description of agents, their behavioral rules and simulation experiments, we refer interested readers to Patel et al. [51].

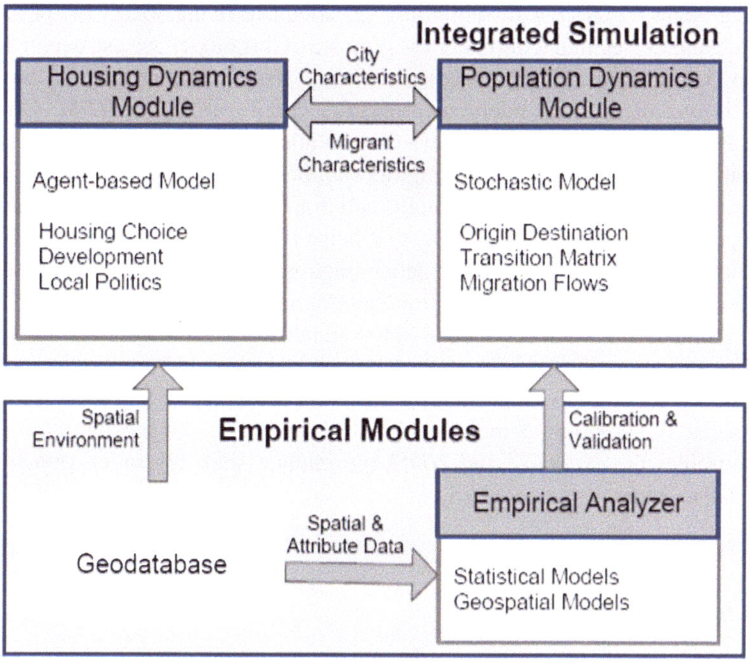

Fig. 1 Integrated simulation framework (Source: [50])

A GIS is used for the *EM*, which has two sub-components, the *Geodatabase* and the *Empirical Analyzer*. The *Geodatabase* is used to store and represent the spatial environment of a city. As slums develop as a result of humans interacting with their spatial environment, this framework is designed to explore links between spatial configuration of a city and emergence of slum patterns. For example, it can be used to show how different land parcel sizes within a city could lead to different slum patterns, as will be shown later. This explicit representation of the spatial environment allows us to conduct spatially explicit policy experiments that are not possible in abstract models such as Barros [26] and Patel et al. [51].

Finally, a Discrete Event Simulation (DES) is used to build the *PDM*, which simulates population growth for the modeled city. As slums develop as a result of the gap between demand created by this population growth and existing supply of housing within a city, this module is envisaged to capture the links between regional population dynamics and slum formation within a city. For example, it can be used to show how varying migration rates can lead to varying slum patterns within a city. This module is under development and we plan to integrate it in the next phase of the model.

The very nature of the framework requires the identification of slums at the individual household level, and hence we can leverage the widely accepted UN-Habitat [52] definition of slums which states that a household is a slum household if they lack any one or more of these five basic housing elements: (1) access to safe drinking water, (2) access to sanitation, (3) adequate living space, (4) permanent structure and (5) secured tenure. For the purpose of this model, we use overcrowding criteria (i.e. adequate living space) to identify slums since density on any land parcel is one of the key emergent properties coming from the *HDM*. Once the slums are identified, our model shows the evolution of key output parameters such as number and size of slums, housing density, housing price to income ratio, etc., over space and time in the form of maps and graphs. The *Empirical Analyzer*, once developed, will provide a set of tools that helps in calculating these key indicators from simulation outputs within our geosimulation framework. We also present the empirical analysis of the slum location patterns in Ahmedabad (see section "Case Study: Ahmedabad") that could be used to validate the geosimulation model results. In following subsections, we describe *HDM* and *EM* before we present the empirical analysis of the slum location patterns in Ahmedabad (see section "Case Study: Ahmedabad"). In section "Simulation Results", we present the geosimulation model that combines *Geodatabase* and *HDM* to simulate slum formation dynamics in Ahmedabad, India.

Housing Dynamics Module

The *HDM* simulates household residential choice behavior in a spatially explicit housing market. The type of simulation used for this purpose is an ABM, which allows us to capture the heterogeneity of decision-making with respect to location choice behavior of households. The *HDM* models the micro-processes of housing choice behavior at the household level and captures the emergent macro-phenomenon of formation and expansion of slums at city level. The main agents in this module are households, developers and politicians. A prototype model for the *HDM* can be seen in Patel et al. [51]. The spatial environment of this model includes housing units, land parcels that contain these housing units, and finally, electoral wards that contain these land parcels. As the simulation progresses, the *HDM* receive a set of newly formed households along with their characteristics in each time period, the main agents driving housing demand in the model. Two other types of agents, local politicians and developers are also modeled in this module (as they are important actors for the housing supply).

Agent behaviors were informed by survey-based studies as advocated by Robinson et al. [53] and other extensive reviews of various modeling efforts (e.g. [26, 48, 49]). The assumptions underlining the behavioral rules, such as preferences to live closer to Central Business District (CBD) or live within budget constraint, are also supported by the theories developed in residential choice literature (e.g. [54–57]). Moreover, behavioral rules were supplemented by calibrating micro-process parameters based on their fit to macro outcomes [53].

Empirical Module

The land parcels (i.e. individual housing sites) along with several attributes (e.g. housing price, size etc.) form the spatial environment in our conceptual framework, which is stored in the *Geodatabase* component of the *EM* as shown in Fig. 2. Electoral ward boundaries are also superimposed to form an appropriate spatial environment for politician agents. GIS is used to consolidate ward level attributes such as percentage slum population in each ward, required for micro-processes for politician agents to determine their behavior. The spatial environment evolves as the simulation progresses (e.g. percentage slum in a particular ward will change as a result of household agents' residential location choice). Aspatial parameters such as economic growth of the city, level of informality of economy, etc., are also an important part of this module that is supplied exogenously and affects change in household characteristics during the simulation such as income levels.

The simulation outputs are then compared with the empirical analyses on various parameters such as the slum incidence rate, slum locations, size distribution and densities for calibrating and validating the geosimulation model. It also helps to identify uncertainty and errors in terms of input data, parameterization, model outputs, etc., along with sensitivity testing [58].

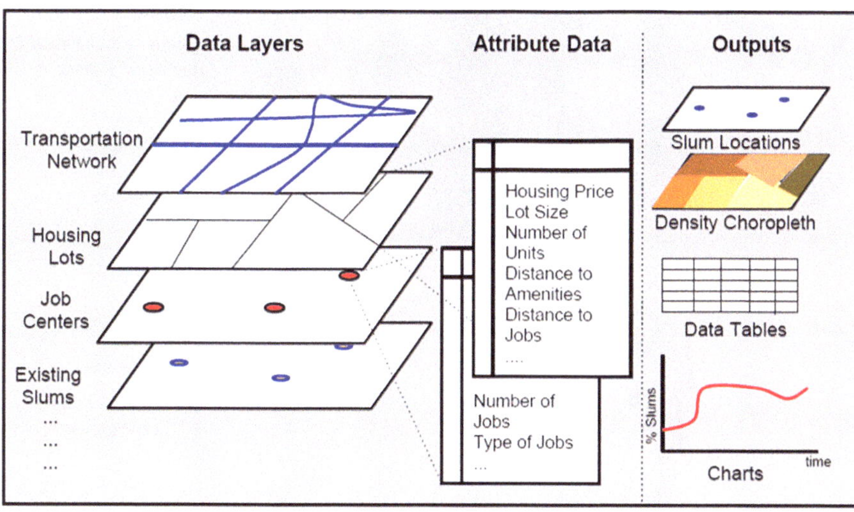

Fig. 2 Geodatabase with data layers and outputs (Source: [50])

Case Study: Ahmedabad

Ahmedabad is the sixth-largest city of India with a population of 3.5 million [59], of which 1.5 million people (41%) live in approximately 1668 slums spread across the city as shown in Fig. 3. We have chosen Ahmedabad as our case study for several reasons. First, the second tier cities in India such as Ahmedabad are at the forefront of the urbanization process and hence more relevant for policymakers (Authors' interviews with policymakers in India, 2011). Second, Ahmedabad has the required data on slums, which allows us to validate our model. Third, the moderate size of Ahmedabad is computationally feasible to simulate on a desktop computer.

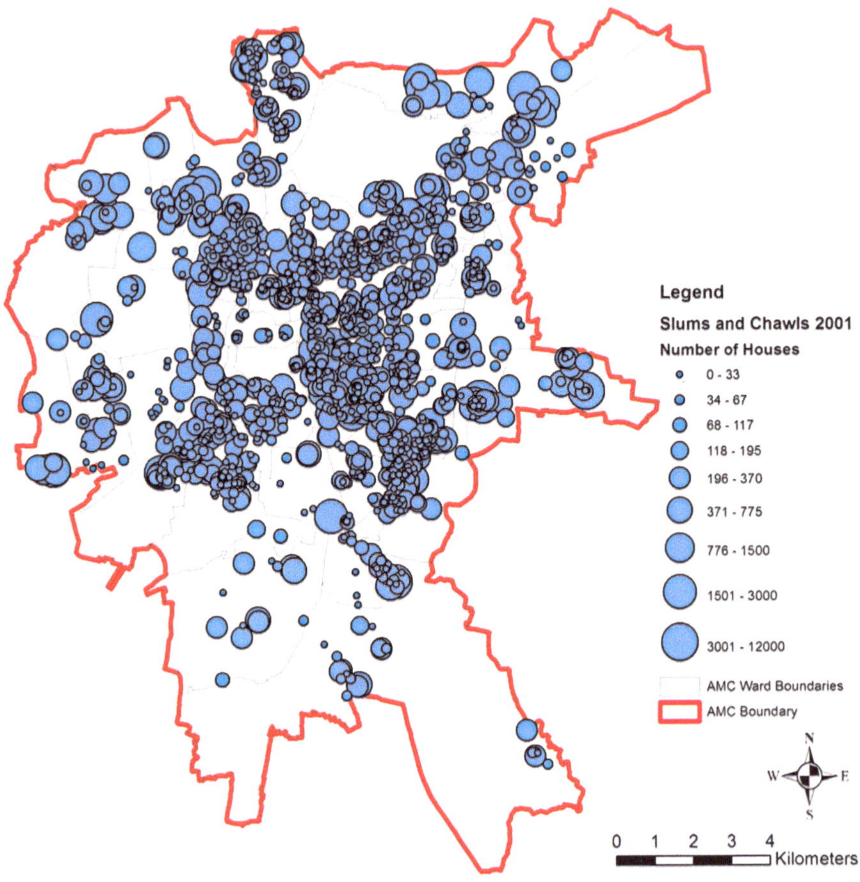

Fig. 3 Slum locations and slum sizes in Ahmedabad, 2001 (Source: [50])

Slums Data

The Ahmedabad Municipal Corporation (AMC) collected slum location data in 2001 with the help of field NGOs named Saath and Sewa. According to this data, there were 710 slums and 958 *chawls* (a formal but dilapidated and overcrowded slum-like housing) in Ahmedabad in 2001. The survey focused on the status of housing and basic infrastructure in these slums and *chawls*. The information collected included: the number of houses, types of houses, locations, the geographic area of the land parcels, ownership, and the status of basic infrastructure such as water, sanitation, street lights, etc.

Slums Mapping

The data contained several variables indicating slum locations. The set of variables that indicated a location were Survey Number, Town Planning (TP) Scheme Number, and Final Plot (FP) Number. Survey Numbers and FP Numbers in combination with TP Scheme Numbers were used to uniquely identify a land parcel, which were then used to locate individual slums and *chawls* on a base map. There were several records that either had missing Survey Number or FP Number or both. For those cases, the actual address and the ward numbers were used to identify the location. On limited occasions, the authors applied their personal knowledge of the city's geography to identify the locations. Addresses often indicated the proximity to other slums or the known landmarks. Using a combination of these methods, 641 of 710 slums and 896 of 958 *chawls* were located on the base map as shown in Fig. 3. The remaining 131 slums and *chawls* could not be located due to insufficient locational information.

Descriptive and Spatial Analysis

A descriptive analysis was then conducted to understand the spatial patterns of slums and *chawls* in Ahmedabad. Slums and *chawls* were not differentiated for this purpose and are referred as slums in the rest of this chapter. Slums vary in size as defined by the number of houses. The largest slum in Ahmedabad had more than 10,000 houses and the smallest slum had only three houses.

As shown in Fig. 4, there is a large number of smaller slums (with less than 250 houses) and only a small number of large slums (with more than 250 houses). The mean slum size has 197 houses (Standard deviation of 553 houses). The variation in the size of the slums relates to several factors including availability of vacant land at that location, age of the slum, accessibility to other amenities, etc. Moreover, the slum size distribution provides a stylized understanding of the slum formation

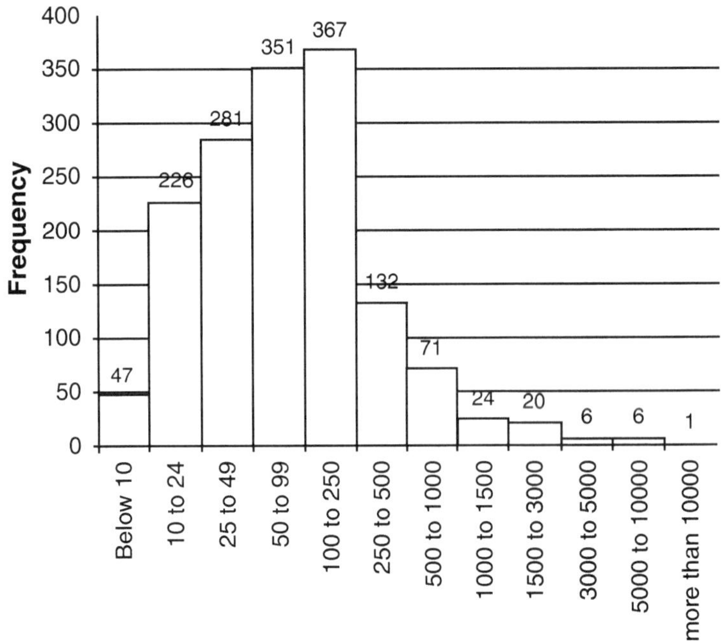

Fig. 4 Slum size distribution by number of houses in Ahmedabad, 2001 (Source: [45])

pattern, i.e. there are a large number of smaller slums and a small number of large slums. It therefore, provides us with a basis for qualitative validation during the development stage of the model.

A commonly used summary statistics to understand the spatial pattern of any point data is the centroid. The centroid is a measure of central tendency for spatial pattern similar to the statistical mean. In Ahmedabad, the centroid of all slum locations was within 0.3 km of the centroid of the city itself. This indicates that the locations of slums follow the spatial growth pattern of the city. In other words, the spread of slums is balanced in all directions from the city center. Another useful statistic is the standard distance. The standard distance provides a single summary measure of feature distribution around their center similar to the way a standard deviation measures the distribution of data values around the statistical mean. The standard distance for distribution of slum locations in Ahmedabad was found 4.7 km. About 1001 or 65% of slums were within 4.7 km (one standard distance) from the centroid indicating that the slum locations are concentrated around the city center of Ahmedabad.

Spatial Clustering

Slum locations were analyzed to explore if slums were randomly located, clustered or dispersed within Ahmedabad. A spatial statistic called the Nearest Neighbor Index (NNI) was calculated for revealing underlying pattern of slum locations. The NNI is expressed as the ratio of the observed mean distance between the nearest neighbors to the expected mean distance for a hypothetical random distribution. The pattern is considered clustered when the index is less than 1 and dispersed when greater than 1.

The NNI was calculated using Spatial Statistics tools of ESRI's ArcGIS 10 and it was found to be 0.46, indicating clustering of slums in Ahmedabad. The value of the Z-score (−40.3) suggests that this clustered pattern is not a result of random chance ($p < 0.01$). This suggests that there are some underlying spatial processes at work. We hypothesize that such processes are related to the locational preferences of households and the spatial configuration of housing market, which is accounted for in our conceptual simulation framework.

Once it was identified that there exists a spatial clustering of slums in Ahmedabad, the Kernel density estimation method was used to reveal the underlying clustering of slums. Kernel density calculates the density of point features around each output raster cell. A smoothly curved surface is fitted over each point. The kernel density is then presented in the form of a choropleth map of densities.

Figure 5 shows the areas where the slums are clustered as well as areas where the slums are sparse. As it is evident, there are several hot spots of slums in Ahmedabad. Particularly, at several locations in the Eastern part of the city, the kernel density value is high compared to the locations in the western part of the city. There are few clusters in the Northern region but Northeastern and Southeastern regions have a sparse density of slums.

Simulation Results

To show how our conceptual geosimulation framework could be used for a real world city to study slum formations, specifically using the *HDM* and the *EM*, we now present a policy scenario analysis. The model was tested with several initial conditions. In the first hypothetical scenario, the model was initiated with populating the "Walled City" without any pre-existing slums. As the simulation progressed, the emergence of slums was observed, purely as a result of human-environment interaction. This experiment partly explains how slums came into existence in a city over time. As seen in Fig. 6, the simulation first showed formation of new slums within Walled City in the starting few years and eventually slums were dispersed to peripheries. Such a pattern is similar to the empirically observed pattern in the city of Ahmedabad as shown in previous section.

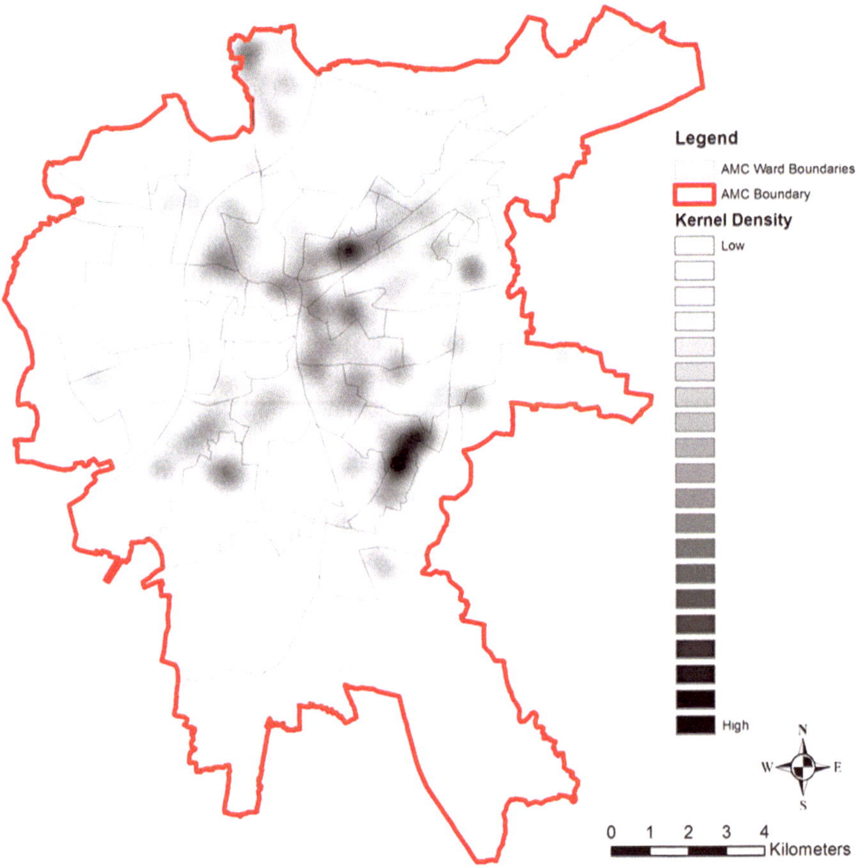

Fig. 5 Clustering of slum locations in Ahmedabad, 2001(Source: [50])

In order to verify the model behavior, we conducted an urban growth pattern experiment. In this experiment, we assessed if a city grows to its current extent and increases in density over time. The city was populated with half the wards located in the center as the initial condition and then the model was run for 30 years. It was observed that the model reaches to the spatial extent of the city in 30 years as shown in Fig. 7, after which it attains population growth with increased density. This pattern is very similar to the empirically observed spatial growth of the city of Ahmedabad.

Once the model behavior was understood, some policy experiments were conducted in order to assess effectiveness of our model as a policy support tool. In one such experiment, the values of number of housing units allowed per unit area were varied. Ahmedabad's Development Plan [60] imposes development control regulations, which stipulate Floor Space Index (FSI, also known as Floor Area Ratio). FSI controls how much built area can be added per unit of land area. The

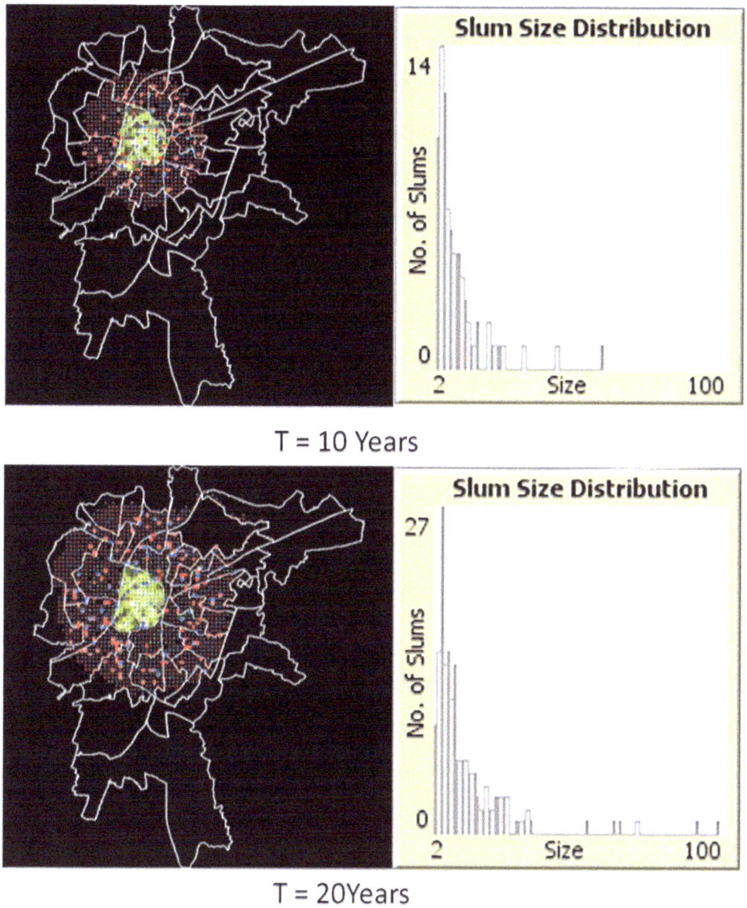

Fig. 6 Hypothetical scenario with "Walled City"

residential density is thus determined by this policy measure. While this policy does not restrict the supply of housing in totality, i.e. housing market can produce the same number of units over larger area to meet with the demand, it does restrict the supply of housing at particular locations. The values tested for Ahmadabad were the existing housing stock (base scenario), twice the current conditions and finally five times current conditions in scenario 3.

As shown in Table 1, the increase in FSI induces lower slum population (55% in base scenario to 45% in higher FSI scenario). While slums still form, they form at fewer locations. They tend to be larger and denser as highlighted in Table 1. As we increase FSI to five times from the base scenario, slum density almost doubles from 47,000 persons per sq. km to 88,000 persons per sq. km. This trend might be indicative of the fact that when the formal development takes place at a higher density (as reflected in higher FSI), slum density also increases. The area under

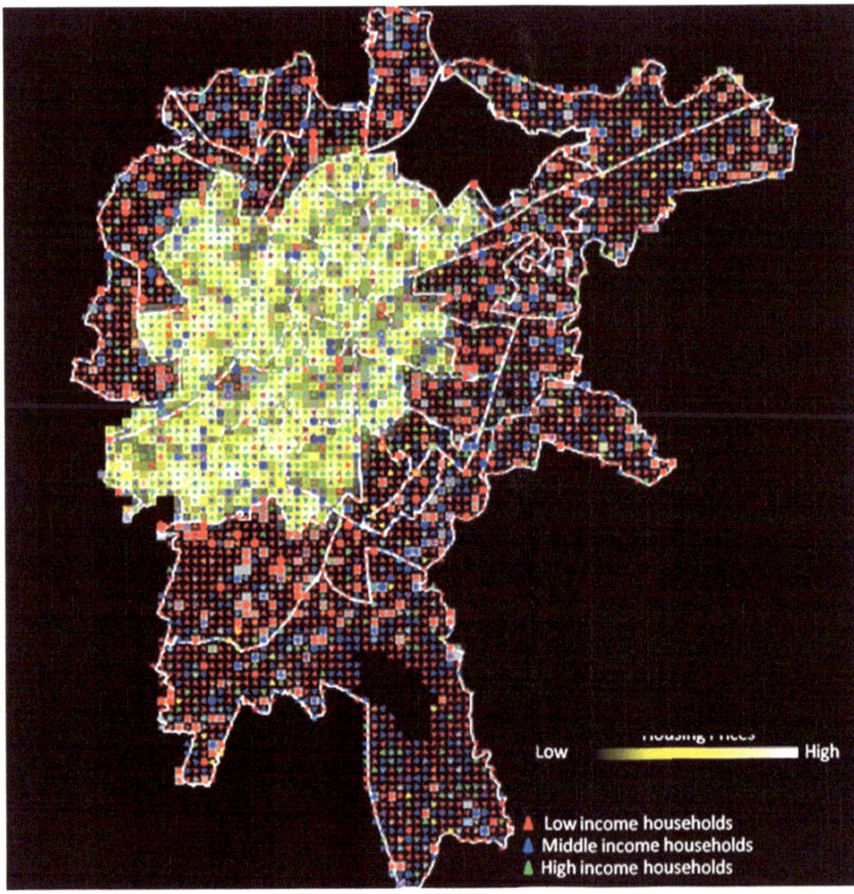

Fig. 7 Spatial sprawl experiment

Table 1 Impacts of floor space index on simulation outcomes

	100 units per site (base scenario)	200 units per site	500 units per site
Percent slums	55%	50.2%	45%
Smallest slum (population)	200	300	600
Largest slum (population)	27,600	28,900	32,000
Number of slums	401	220	171
Area under slums (percent of total area of the city)	14.5%	8.5%	6.4%
Slum Density (Persons per sq. km)	47,400	76,543	88,889

slum has also observed a decline. It is possible that slums follow the pattern of formal development and hence if formal development takes place in a compact area, slums tend to be compact as well. In other words, slums will follow the spatial development patterns of the formal city. However, empirical analysis of slums in cities with varying levels of compactness is required to further validate this hypothesis.

Discussion and Future Research Directions

This chapter presented a conceptual geosimulation model that explores slum formation. After presenting the initial framework, we demonstrated the application of the model to the city of Ahmedabad. The model presented here is an effort to develop a theoretical framework, which is capable of generating slum patterns similar to that observed in the real world. Ideally, if it is to be considered as a good urban planning support tool, it should be able to predict slum locations with precision. However, such a system inherently requires data on several aspects that are often not available in developing countries. For example, our model could produce better results if housing price data was available for multiple years. This would allow for several assumptions to be replaced with actual data that can help in calibrating and validating this model further. Data availability is improving in developing countries (e.g. massive data collection under RAY in India) and it is hoped that our model will benefit from improved data.

While the model presented here is simple, it does generate real world patterns of slum formation over space and time. There are several limitations to our model but many of these are seen in nearly all simulation modeling endeavors [61]. For example, to move towards a truly "comprehensive" understanding is challenging as there are many variables and factors such as political will, economic dynamics, and "unforeseen forces" that cannot be fully modeled and captured in a modeling and simulation exercise [58].

It should also be noted that detailed calibration and validation of this model are not carried out due to the lack of consistent data for multiple time-periods. In addition, the goal of this chapter was to conduct exploratory research and test the feasibility of the modeling framework. Although, Census 2011 has already been carried out in India, detailed ward level data is being compiled but not yet available at the time of writing this chapter. It is hoped that the 2011 census and slum mapping under RAY will provide comparable data, which can be used to calibrate and validate the model. In the future, the model could be initiated with slum locations in 2001. The simulation could then be carried out for the period between 2001 and 2011 with population growth and migration estimates from the Census 2011 data. The simulated outputs could then be compared to the observed slum locations in 2011 (new slum location data has been already collected under RAY in 2011 but it is uncertain when it will be publicly available).

Furthermore, model development, calibration and validation could also benefit from extending the spatial analysis. For example, the data for the year in which each slum came into existence could provide temporal evolution of the slum pattern observed in year 2001. This analysis could be further enriched if the year of construction for each housing unit within those slums is also obtained. Such data could show when each slum came into existence, whether those slums increased or decreased in size and population and so on. Temporal variable could provide a dynamic view of evolution of slum systems that could then be used to validate dynamic patterns generated within simulation.

Nonetheless, our chapter sets a way for developing a science of slum formation. We integrate spatial GIS data with ABM to overcome the limitation of individual methods. We also use empirical analysis of slum location patterns to verify and validate our model. However, understanding human behavior remains a challenge to ABM generally [62], but more so with respect to slums. For example, there has been much research on residential decision-making in the developed world but not in developing counties and in particularly in slums. We hope that our work will precede collections of such new type of behavioral data. Data limitations in developing countries also call for new methods for sharing data. For example, detailed base maps are very rarely available for slums. However, there is a growing interest in using Volunteered Geographic Information (VGI) to map slums. For instance, the Map Kibera project (http://mapkibera.org) and Transparent Chennai project (http://www.transparentchennai.com/). Such efforts would give us the ability to have a more detailed spatial footprint of slums. All in all, we hope to contribute towards formation of a dedicated research agenda that develops a science of slums both in terms of geosimulation modeling and empirical spatial analysis of slums.

Acknowledgment This research was supported by a grant from the Geography and Spatial Sciences Program of the National Science Foundation (NSF-BCS-1225851).

References

1. UN-Habitat (2010) State of the cities 2010–11. UN-Habitat, Nairobi, Kenya. Available at http://www.unhabitat.org/content.asp?cid=8891&catid=643&typeid=46&subMenuId=0&AllContent=1. Accessed on May 20th, 2012
2. UN-Habitat (2003) The challenge of slums. United Nations Human Settlements Programme, Sterling
3. Anzorena J, Bolnick J, Boonyabancha S, Cabannes Y, Hardoy A, Hasan A, Levy C, Mitlin D, Murphy D, Patel S, Saborido M (1998) Reducing poverty: some lessons from experience. Environ Urban 10(1):167–186
4. Pugh C (2001) The theory and practice of housing sector development for developing countries, 1950-99. Hous Stud 16(4):399–423
5. United Nations (2000) United Nations Millennium Declaration. General Assembly Resolution No. 2 of Session 55. Available at http://www.undemocracy.com/A-RES-55-2.pdf. Accessed on May 20th, 2012
6. Government of India (2005) Jawaharlal Nehru National Urban Renewal Mission. Ministry of Urban Employment and Poverty, Alleviation and Ministry of Urban Development, New Delhi
7. Times of India (2009) UPA's target—a slum free India in 5 years, Times of India (5th June). Available at http://bit.ly/1XguIRY. Accessed on January 7th, 2012
8. MHUPA (2011) Status Note on Rajiv Awas Yojana, Ministry of Housing and Urban Poverty Alleviation, Government of India, New Delhi, India. Available at http://mhupa.gov.in/W_new/NOTE_RAJIV_AWAS_YOJANA.pdf. Accessed on January 15th, 2012

9. Mayo SK, Gross DJ (1987) Sites and services-and subsidies: the economics of low-cost housing in developing countries. The World Bank Econ Rev 1(2):301–335
10. Pugh C (1989) The World Bank and Urban Shelter in Bombay. Habitat Int 13(3):23–49
11. Pimple M, John L (2002) Security of tenure: Mumbai's experience. In: Durand-Lasserve A, Royston A (eds) Holding their ground: secure land tenure for the urban poor in developing countries. Earthscan, London, pp 75–85
12. Cities Alliance (1999), Cities without slums action plan. Cities Alliance, Washington, DC. Available at http://www.citiesalliance.org/sites/citiesalliance.org/files/CA_Docs/brln_ap.pdf. Accessed on September 12th, 2012
13. Chauhan U, Lal N (1999) Public-private partnerships for urban poor in Ahmedabad: a slum project. Econ Polit Wkly 34(10–11):636–642
14. Mukhija V (2001) Enabling slum redevelopment in Mumbai: policy paradox in practice. Hous Stud 18(4):213–222
15. Burra S (2005) Towards a pro-poor framework for slum upgrading in Mumbai, India. Environ Urban 17(1):67–88
16. Gulyani S, Bassett EM (2007) Retrieving the baby from the bathwater: slum upgrading in Sub-Saharan Africa. Environ Plan C 25(4):486–515
17. Gulyani S, Bassctt EM (2008) Rcvisiting ... rctricving thc baby from thc bathwatcr: slum upgrading in Sub-Saharan Africa. Environ Plan C 26(5):858–860
18. Mahadevia D, Naranayan H (1999) Shangaing Mumbai—politics of evictions and resistance in slum settlements: Working Paper 7. Center for Development Alternatives, Ahemedabad
19. Bamberger M, Gonzalez-Polio E, Sae-Hau U (1982) Evaluation of sites and services projects: the evidence from El Salvador, World Bank Staff Working Paper No. 549, Washington, DC
20. Takeuchi, A., Cropper, M. and Bento, A. (2006) The welfare effects of slum improvement programs: the case of Mumbai, World Bank Policy Research Working Paper Number 3852, Washington, DC. Available at http://papers.ssrn.com/sol3/papers.cfm?abstract_id=922978.
21. Viratkapan V, Perera R (2006) Slum relocation projects in Bangkok: what has contributed to their success or failure? Habitat Int 30(1):157–174
22. Abbott J (2002) A method-based planning framework for informal settlement upgrading. Habitat Int 26(3):317–333
23. Sietchiping, R. (2004), A geographic information systems and cellular automata-based model of informal settlement growth, Ph.D. thesis, School of Anthropology, Geography and Environmental Studies, The University of Melbourne, Melbourne. Available at http://repository.unimelb.edu.au/10187/1036.
24. Wilson AG (2000) Complex spatial systems: the modeling foundations of urban and regional analysis. Pearson Education, Harlow
25. Batty M (2005) Cities and complexity: understanding cities with cellular automata, agent-based models, and fractals. MIT, Cambridge
26. Barros J (2012) Exploring urban dynamics in Latin American cities using an agent-based simulation approach. In: Heppenstall A, Crooks AT, See LM, Batty M (eds) Agent-based models of geographical systems. Springer, New York, pp 571–590
27. Young G, Flacke J (2010) Agent-based model of the growth of an informal settlement in Dar es Salaam, Tanzania: An Empirically Informed Concept. In: Swayne D, Yang W, Voinov A, Rizzoli A, Filatova T (eds) Proceedings of the 2010 International congress on environmental modelling and software: modeling for environment's sake. Ottawa, Canada
28. Batty M (1976) Urban modelling: algorithms, calibrations, predictions. Cambridge University Press, Cambridge
29. White R, Engelen G (1993) Cellular automata and fractal urban form: a cellular modelling approach to the evolution of urban land use patterns. Environ Plan A 25(8):1175–1199
30. Benenson I, Torrens P (2004) Geosimulation: automata-based modeling of urban phenomena. Wiley, Chichester
31. Jacobs J (1961) The death and life of great American cities. Vintage Books, New York
32. Maguire DJ (2005) Towards a GIS platform for spatial analysis and modelling. In: Maguire DJ, Batty M, Goodchild MF (eds) GIS, spatial analysis and modelling. ESRI, Redlands, pp 19–39

33. Wegener M (2000) Spatial models and GIS. In: Fotheringham AS, Wegener M (eds) Spatial models and GIS: New potential and new models. Taylor & Francis, London, pp 3–20
34. Brail RK, Klosterman RE (eds) (2001) Planning support systems: integrating geographic information systems, models and visualisation tools. ESRI, Redlands
35. Engelen G, White R, Nijs T (2003) Environment explorer: spatial support system for the integrated assessment of socio-economic and environmental policies in the Netherlands. Integr Assess 4(2):97–105
36. Maguire DJ, Batty M, Goodchild M, F. (eds) (2005) GIS, spatial analysis and modelling. ESRI, Redlands
37. Clarke KC, Gaydos LJ (1998) Loose-coupling a cellular automaton model and GIS: long-term urban growth predictions for San Francisco and Baltimore. Int J Geogr Inf Sci 12(7):699–714
38. Clarke KC, Gazulis N, Dietzel CK, Goldstein NC (2006) A decade of SLEUTHing: lessons learned from applications of a cellular automaton land use change model. In: Fisher P (ed) Classics from IJGIS: twenty years of the International Journal of Geographical Information Science and Systems. Taylor & Francis, Boca Raton, pp 413–426
39. Landis J (1994) The Californian urban futures model: a new generation of metropolitan simulation models. Environ Plan B 21(4):399–420
40. Landis J, Zhang M (1998) The second generation of the California urban futures model. Part 1: model logic and theory. Environ Plan B 25(5):657–666
41. Landis JD, Monzon JP, Reilly M, Cogan C (1998) Development and pilot application of the California Urban and Biodiversity Analysis (CURBA) Model, University of California at Berkeley, Institute of Urban and Regional Development, Monograph 98-01. Available at http:/ / escholarship.org/uc/item/5713p6g6
42. Crooks AT, Heppenstall A (2012) Introduction to agent-based modelling. In: Heppenstall A, Crooks AT, See LM, Batty M (eds) Agent-based models of geographical systems. Springer, New York, pp 85–108
43. Torrens PM (2006) Simulating sprawl. Ann Assoc Am Geogr 96(2):248–275
44. Xie Y, Batty M, Zhao K (2007) Simulating emergent urban form: Desakota in China. Ann Assoc Am Geogr 97(3):477–495
45. Torrens PM (2002) Cellular automata and multi-agent systems as planning support tools. In: Geertman S, Stillwell J (eds) Planning support systems in practice. Springer, London, pp 205–222
46. Wise S, Crooks AT (2012) Agent based modelling and GIS for community resource management: acequia-based agriculture. Comput Environ Urban Syst 36(6):562–572
47. Robinson DT, Brown D (2009) Evaluating the effects of land-use development policies on ex-urban forest cover: an integrated agent-based GIS approach. Int J Geogr Inf Sci 23(9): 1211–1232
48. Augustijn-Beckers E, Flacke J, Retsios B (2011) Simulating informal settlement growth in Dar es salaam, Tanzania: an agent-based housing model. Comput Environ Urban Syst 35(2): 93–103
49. Vincent OO (2009) Exploring spatial growth pattern of informal settlements through agent-based simulation, MS [in Geographical Information Management & Applications (GIMA). Utrecht University, Delft University of Technology, Wageningen University and the International Institute for Geo-Information Science and Earth Observation]). Wageningen, Netherlands. Available at http://www.msc-gima.nl/index.php/modelling.
50. Patel A, Koizumi N, Crooks A (2012) Simulating spatio-temporal dynamics of slum formation in Ahmedabad, India, 6th urban research and knowledge symposium on cities of tomorrow: framing the future at Barcelona, Spain, October 8–10, 2012
51. Patel A, Crooks AT, Koizumi N (2012) Slumulation: an agent-based modeling approach to slum formations. J Artif Societies Social Simul 15(4). Available at http://jasss.soc.surrey.ac.uk/ 15/4/2.html
52. UN-Habitat (2006) State of the world's cities 2006/7, UN-Habitat, Nairobi, Kenya. Available at http://www.unhabitat.org/pmss/listItemDetails.aspx?publicationID=2101. Accessed on May 20th, 2012

53. Robinson DT, Brown D, Parker DC, Schreinemachers P, Janssen MA, Huigen M, Wittmer H, Gotts N, Promburom P, Irwin E, Berger T, Gatzweiler F, Barnaud C (2007) Comparison of empirical methods for building agent-based models in land use science. J Land Use Sci 2(1):31–55
54. Alonso W (1964) Location and land use: toward a general theory of land rent. Harvard University Press, Cambridge
55. Benenson I, Omer I, Hatna E (2003) Agent-based modelling of householders migration behaviour and its consequences. In: Billari FC, Prskawetz A (eds) Agent-based computational demography: using simulation to improve our understanding of demographic behaviour. Physica, New York, pp 97–115
56. Florida R (2002) The rise of the creative class: and how its transforming work, leisure, community and everyday life. Basic Books, New York
57. Soja EW (2000) Postmetropolis: critical studies of cities and regions. Blackwell, Malden
58. Evans AJ (2012) Uncertainty and error. In: Heppenstall A, Crooks AT, See LM, Batty M (eds) Agent-based models of geographical systems. Springer, New York, pp 309–346
59. Census of India (2001) Census Data Products available from Office of the Registrar General and Census Commissioner, New Delhi, India. http://censusindia.gov.in/ Accessed on January 11th, 2012
60. AUDA (1997) Revised draft development plan of AUDA—2011 AD. Ahmedabad Urban Development Authority, Ahmedabad
61. Crooks AT, Castle CJE, Batty M (2008) Key challenges in agent-based modelling for geo-spatial simulation. Comput Environ Urban Syst 32(6):417–430
62. Kennedy W (2012) Modelling human behaviour in agent-based models. In: Heppenstall A, Crooks AT, See LM, Batty M (eds) Agent-based models of geographical systems. Springer, New York, pp 167–180

Incorporating Urban Spatial Structure in Agent-Based Urban Simulations

Haoying Wang

Introduction

In recent decades, increasingly sophisticated models have been proposed to understand the complex and interdependent social and physical factors involved in the development of sustainable urban systems. A recent OECD (Organization for Economic Co-operation and Development) report points out that, despite recent advances in computational capacities, methodological difficulties still prevent the development of efficient and user-friendly urban modeling tools [1]. This gives a fair overview of our current status on urban system research. The challenges we face in modeling urban systems become more and more structural as we keep improving computation technology. Over the second half of the twentieth century, the research approach on urban systems has gradually transformed from traditionally physical design-focused to a framework with more attention on social and economic processes (e.g. [2–5]). The transition features at least three new standing pillars of urban modeling: behavior component, spatial interaction, system dynamics. Incorporating these modeling aspects into urban system models calls for a more integrated structural understanding of urban systems. Such a structural understanding of urban systems is not only necessary to the management of city operations, but also fundamental to public policy-making. It requires modeling of urban systems to take into account all physical/geographic, social, economic components, and their interrelationships in a simplified but informative way.

To understand the structure of urban systems, a starting point is the behavioral motivation of the city: why do people live in cities? Gutkind [6] answers the question as following: an unfulfilled longing for the amenities and distractions of city life

H. Wang (✉)
New Mexico Institute of Mining and Technology, Socorro, NM 87801, USA
e-mail: halking.econ@gmail.com

J.-C. Thill, S. Dragicevic (eds.), *GeoComputational Analysis and Modeling of Regional Systems*, Advances in Geographic Information Science,
DOI 10.1007/978-3-319-59511-5_9

has driven people living in cities, where men have developed more differentiated habits and needs within proximity. Recently, Glaeser [7] answers the question with more enthusiasm: people come to cities in search for something better. Cities are proximity, density, and closeness at a much larger scale. The human nature of desire for connections and proximity, searching for more and better choices, has driven people to work and live in cities. How this human nature is organized into a landscape where different people, culture, and sectors are integrated is the key to understand the structure of urban systems. It is easy to understand why the rich and the middle-class choose to live in the cities and bear higher living cost, for example, but why the poor? One observation is that, the urbanization of poverty may be explained by the better access to public transportation in cities [8]. In a nutshell, it is the diversified preferences that drive urban residents' choices, and the urban system exists as an aggregate representation of all individual choices.

From a modeling perspective, the difficulty also lies in between calibrating household behavior and aggregate representation. The complexity in understanding household behavior comes from the fact that urban residents face an even larger choice set, while that is also what is fascinating about cities. Location choice is the major decision of each household in the city, because many of the amenities and distractions of city life are likely associated with location. Location is also the basic component of any urban systems. All developed locations and open space left in between constitute a continuum of urban landscape, which is the framework urban systems build upon.

Given location choices and income constraints, households decide on the amount of consumption through a series of decision making mechanisms. The current simulation approach to urban systems seeks to link together different sub-systems and markets through modular architecture with substantial geographical details and level of behavioral realism (e.g. [9, 10]). The advantage of such approach is that it is good at simulating short run evolution of urban systems at an extremely disaggregate level. The disadvantage, on the other hand, is that such multi-agent system is vulnerable to structure change, especially in the long run. In other words, current urban simulation models tend to perform well in descriptive explanation, but lack strong ability of prediction (e.g. in land use change and ecosystem applications). In today's regional economy, not only is the gradual-adjustment type of public policy process important, but also the public policy for long term regional development planning. To address both types of policy needs, it is critical to develop urban simulation tools which are robust to structure change and capable of predicting urban evolution in relatively long run.

In this chapter, we focus on the role of urban spatial structure in urban simulations, and how it helps to strengthen the current simulation approach to urban systems. As discussed in Crooks et al. [11], one of the key challenges to ABM simulation in geocomputation is to what extent the model is rooted in independent theory. Basing on household behavior, urban spatial structure theory integrates economy system, land use, and transportation system together, which provides an easy to implement framework for simulating urban systems. Methodologically, urban spatial structure theory also provides a trackable way to measure performance and efficiency of urban systems, which is valuable to public policy-making. Section "Components of Agent-Based Urban Simulation" discusses basic components of agent-based urban simulation. In sections "Incorporating Urban Spatial Structure", "Transportation and Congestion: An Application", and "ABM Simulation: Land Development and Congestion," a monocentric city simulation model of transportation cost and congestion effects is developed to illustrate how urban spatial structure models can be integrated with ABM simulation. Section "Concluding Remarks" concludes the chapter with discussion on policy implications and future research.

Components of Agent-Based Urban Simulation

Urban simulation models are constructed to address operational needs in planning and policy-making for increasingly complex urban systems. Many well-known urban simulation models have two basic functionalities: land use and transportation. From a modeling perspective, an agent-based simulated urban system should have three categories of building components: households, spatial interaction, and landscape. In this section, we discuss the role of each category in urban simulation. These components are also essential constituents of urban spatial structure models.

Household Behavior

The human behavioral component is the building block of many observed social and economic phenomenon. The first basic behavioral component of a urban system is household—each household acts as a node in the network of urban systems. The decision and choice made by one household can directly or indirectly affect the behavior of other households across the city. In the economics approach, the (rational) behavior of a household is usually summarized into a mathematical form—utility function. The utility function approach provides a simplified way to represent the inter-relationship among all available choices. In the context of urban modeling, these choices are usually being categorized into housing consumption, non-housing consumption, and transportation consumption. Given household income as a binding constraint, the balance among three consumption categories can be realized through household location choice. Thus, optimal location choice is equivalent to utility maximization for a rational household.

Household behavior can be generalized to all sorts of agents—business owner, land developer, social planner, and etc. Similarly, the utility function can also be generalized to generic objective function—profit function, social welfare function, and etc. To completely describe the behavior of these agents, a decision making mechanism has to be established basing on the objective function. In most of mathematical social science fields, optimization theory is employed to develop decision making mechanism for agents. The idea is to maximize (or minimize) the objective by optimizing the combination of all available choices. Note that, for many of the urban system problems, a socially optimal decision is not necessarily optimal to all agents because of the resource constraints. On the other hand, if all agents make individual optimal decisions, the aggregate social outcome is also not necessarily optimal to the society (e.g. [12]). Such inconsistency between aggregate modeling and aggregated individual modeling has been a major challenge to analytical approach to urban systems. The advantage of agent-based simulation approach to urban systems, however, is to explore the aggregate social outcome (i.e. emerging properties) from a disaggregate level which the analytical equilibrium approach often fails to do [13].

Spatial Interaction

Individual households as agents are not isolated from each other. Urban households live within very close proximity, thus mutual interactions are an indispensable part of urban life. Households interact with each other through two important mechanisms: social network and market. Many systems take the form of networks, and all non-economic and some economic components of urban systems are connected through social networks. An important property of social network is the so-called small world effect or neighborhood effect, which means the network effects tend to be localized [14]. In urban spatial context, many of the spillover effects are associated with social networks, which is something to take into account in urban modeling and public policy-making. Bramoullé and Kranton [15], for example, show that individuals who have active social neighbors have high benefits from public goods with only little effort. The social network effects can influence household location choice, even though other factors are also important. Ettema et al. [16] suggest that social interactions between households and between individuals potentially have an influence on household location, mobility and activity choices. Wang [17] shows that neighborhood spillover effects through housing markets can affect the whole land development process in an urban area.

The connection through markets is more measurable, at least from the economic perspective. Households may compete with each other on the market—for instance, the labor market, where over-supply is often the case. Households may also corporate with each other through the market—for instance, form a labor union or business alliance, so that everyone can benefit from collective bargaining. In ABM simulation, a common approach of modeling market mechanism is to allow trade between agents [18]. Trade between agents is more like an atomic market, which is quite different from the macro market that all households can potentially participate. Coordination between atomic markets like trade mechanism and the macro market still remains a challenge to ABM simulation. In urban simulation, how to integrate different macro markets (e.g., housing market and labor market) into one simulation framework is a more pressing challenge, because to inform policy-making an understanding of the linkage among different markets is critical.

In short, because of the existence of social networks and markets, the aggregate social outcome in urban systems is no longer a simple adding-up of individual decisions. An urban simulation model which fails to consider the consequence of these interaction mechanisms may produce biased results.

Landscape

Landscape is the physical foundation of urban simulation. All scientific modeling requires some level of abstraction or simplification of reality and observed phenomenon. In urban modeling, the spatial configuration of agent activities matters. The conceptualization of urban landscape varies across different disciplines. For example, ecologists pay more attention on the structure of impervious surface and its impacts on ecosystem processes (e.g. [19]). Economists are more interested in the residential pattern and the spatial distribution of economic activities (e.g. [20]). These alternative perspectives on urban landscape are not independent from each other as they seem to be. The structure of impervious surface, for instance, is just a physical description of road system and residential development.

Landscape can be generated from image or GIS data of original urban layout using visualization techniques [21]. This approach is often used in scenario-based case studies. Another approach is to design landscape geometrically following certain pattern of urban configuration, and the transportation system is usually integrated as part of landscape. In a two-dimension urban simulation, landscape can be modeled in grid or circular form. Circular form is usually used to model monocentric urban structure, where each ring can be defined as a model unit. Grid form is more generalized, and it can be used in both monocentric and non-monocentric urban modeling. The smaller the circular rings and grid cells, the more realistic is the simulation. However, there is always a trade-off between computation time and level of details in time, space, and agents that a simulation model can represent. In practice, the choice of landscape form depends on the purpose of simulation and computation power available.

Incorporating Urban Spatial Structure

In this chapter, an ABM simulation on transportation cost and congestion effects is developed to illustrate the role of urban spatial structure in agent-based urban simulation. Due to space limit, the model is confined to the monocentric city model only. Different components of the simulation model will be discussed. The theory of urban spatial structure has inspired many analytical and empirical insights about urban systems, which should be integrated in urban simulations [11]. In general, analytical models like urban spatial structure models provide more tractable step-by-step procedures for simulation than heuristic models do.

The simulation framework is designed based on an urban spatial structure model which integrates household behavior, market interactions, and urban landscape. The idea is to illustrate how we can learn more about urban system dynamics through emergent properties by incorporating urban spatial structure with the ABM simulation approach. The basic setup for the urban spatial structure is following. The city consists of a continuum of households, living across the urban area. The homogeneous urban land is divided into many areas (residential and non-residential), not necessarily equally, each of which has fixed boundary. Households within each region have three major consumption categories: housing/land (residential, industrial, commercial, and etc.), non-housing, and transportation. In the model, we focus on transportation and congestion effects.

Transportation and Congestion[1]: An Application

The interaction between residential land use and transportation land use, which may be generalized to the interaction between land use and infrastructure, can result in potential negative externalities. Among which, congestion is the biggest concern in urban development policy. As Wheaton [22] points out, if urban land is allocated to the highest paying use (e.g. as in the Herbert-Stevens model [23]), aggregated land rent is maximized only if there is no externalities. In many of the conventional urban development models, especially spatial equilibrium models, transportation cost is given exogenously and with no congestion effects. In part, this is because congestion cost depends directly on the choice of travel/commuting routes. Modeling travel pattern even with low degree of realism poses challenges to the framework of spatial equilibrium models. On a two-dimension plane, roads and streets can go any direction, modeling travel pattern and congestion essentially becomes a high-dimension problem. Therefore, in either analytical modeling or simulation modeling, certain simplifications have to be made upon the structure of travel patterns. The advantage of simulation approach is that, it allows more details and flexibility in model implementation. In this section, an analytical urban spatial structure model with congestion is introduced, which can be solved as a closed-city optimal control problem. Basing on the analytical model, a dynamic simulation is designed to illustrate how urban simulation can be used to inform policy-making.

[1]Following Solow [27], congestion cost is defined as the cost of travel per person per mile at any point, which depends on two factors: the number of travelers using that part of the transportation system, and the amount of land allocated to transportation use at that point

Static Approach with a Closed-City

Given a circular monocentric city where N consumers commute inwards either to the central business district (CBD, the central labor market), or another region (the local labor market) between home and the CBD. The commuting distance (t),[2] if ignoring the local labor market, can be measured by the ray from the CBD to home. The city is a closed environment, with border at distance B. Following Solow [24] and Wheaton [22], an intermediate variable is created to reflect the potential travel demand, $n(t)$, equal to the number of households residing beyond distance t. In Solow [24] and Wheaton [22], this variable represents the number of commuters passing region t on their way to work in the CBD.[3] In this chapter's framework, residents may choose to work locally, thus the actual travel demand in region t can be less than $n(t)$. Intuitively, the marginal cost of travel in region t is expected to be a positive function of travel demand, and a negative function of transportation capacity in region t.

Travel demand and transportation capacity in region t can be defined as following. Let s denotes a region between region t and the CBD, i.e., $0 \leq s \leq t$, and $\alpha_{t,s}$ be the proportion of residents who live in region t and choose to work in region s. Assuming that all regions are discrete, and $s = 0$ represents the CBD region, then by definition $\sum_{s=0}^{t} \alpha_{t,s} = 1$.[4] The travel demand at region s, $D(s)$, can be expressed as:

$$D(s) = \sum_{i=0}^{s} \sum_{t=i}^{B-1} \left(n(t) - n\left(t + 1\right)\right) \alpha_{t,i} \tag{1}$$

If all residents choose to work either in their residing region t, or in the CBD region, then the travel demand can be simplified to:

$$D(t) = \sum_{j=t}^{B-1} \left(n(j) - n\left(j + 1\right)\right) \alpha_{j,0} \tag{2}$$

In continuous case, $D(t)$ can be written as:

$$D(t) = -\int_{t}^{B} n'(z)\alpha_z dz \tag{3}$$

where α_z is the proportion of residents who live in region z and choose to work in the CBD region, which can be a constant or a function of distance z.

---

[2]In this chapter, t is used as a discrete integer to denote both commuting distance and regions to simplify notation. This implies that all regions have the same width, but different areas

[3]For convenience, in the circular monocentric city model the distance to the CBD, t, is often being used to index land use region as well. In this case, a land use region is a ring around the CBD

[4]An implicit assumption here is that, residents living in region t do not choose to work in regions beyond t. Residents who work in regions beyond t are better off by choosing to live in their working region, because the congestion cost increases as it gets closer to the CBD

An implicit boundary condition here is $n(B) = 0$, due to the closed-city assumption. Transportation capacity is denoted as the fraction of the land allocated to roads and streets in region t, $\nu(t)$. Following Wheaton [22], the urban land development planning can be formulated as an optimal control problem, in continuous case:

$$\begin{cases} \underset{X(t),Q(t)B}{Max} & \int_0^B \left[\frac{Y-T(t)-X(t)}{Q(t)}\right] 2\pi t\,(1-\nu(t))\,dt + \left[A-\pi B^2\right] \\ & R_a + \beta\left[U(X,Q) - U_0\right] \\ & T'(t) = c\left(\frac{D(t)}{2\pi t\nu(t)}\right) \\ \text{Subject to :} & \\ & n'(t) = -\frac{2\pi t(1-\nu(t))}{Q(t)} \end{cases} \tag{4}$$

with two boundary conditions:

$$\begin{cases} T(0) = 0 \\ n(B) = 0 \end{cases} \tag{5}$$

Y, $T(t)$, $X(t)$, and $Q(t)$ are household income, transportation cost, non-land consumption (the numeraire, price is standardized to 1), and land consumption, respectively. A is the total land area available, and R_a is the opportunity rent of urban land (e.g. agricultural land rent). $U(X, Q)$ is the household utility function. $c(\cdot)$ is the marginal transportation cost, which is a function of the ratio of travel demand to transportation capacity $\frac{D(t)}{2\pi t\nu(t)}$. $c'(\cdot)$ and $c''(\cdot)$ are usually assumed to be positive (e.g. [25]). The maximization problem in Eq. (4) can be solved following optimal control theory (see [22]).

Dynamic Approach with an Open-City

The static approach to urban land development planning in Eq. (4) ignores the urban evolution process. In reality, the urban evolution proceeds as a gradual process and takes decades to adjust [26]. Instead, the urban authority can choose to plan development stage by stage, i.e., planning and developing one region each time period rather than the whole urban area at once. The gradual development process, in many important aspects, is in analogy to the concept of regional economic evolution. At different stages of development, changes of economic and institutional environment can lead to updated perspective and goal on urban development planning. Therefore, a dynamic disequilibrium approach provides a better way to frame the planning problem, which is also one of the main advantages of simulation approach to urban modeling [10].

In the circular monocentric city, the development process goes naturally from the CBD to outside suburban area ring by ring. The land use and economic landscape may show path dependence, but new development can be treated as another planning problem conditional on previous development. Without loss of generality, index each ring region by natural numbers ($t = 1, 2, 3, \ldots,$ with the CBD being region 0), and for any region t the optimization problem becomes:

$$\underset{X(t),Q(t)}{Max}\left[\frac{Y - T(t) - X(t)}{Q(t)}\right] 2\pi t\,(1 - v(t)) + \beta\,[U\,(X, Q) - U_0] \qquad (6)$$

If t represents the newly developed edge region, then the change of transportation cost ΔT only depends on the travel demand originated from region t and the transportation capacity of region t. From the first constraint in Eq. (4), given that the distance horizon is discrete (and $\Delta t = 1$), we have:

$$\Delta T = T(t) - T\,(t-1) \approx T'(t)\Delta t = T'(t) = c\left(\frac{n(t)\alpha_t}{2\pi t v(t)}\right) \qquad (7)$$

However, the transportation cost $T(t)$ (not $\Delta T(t)$) is not solely determined by conditions in region t, instead it shows path dependence:

$$T(t) \approx T\,(t-1) + c\left(\frac{n(t)\alpha_t}{2\pi t v(t)}\right) \qquad (8)$$

By the recurrence relation, with boundary conditions $T(0) = 0$ and $n(t+1) = 0$, Eq. (8) can be written as:

$$T(t) \approx c\left(\frac{n(t)\alpha_t}{2\pi t v(t)}\right) + c\left(\frac{n(t)\alpha_t + [n\,(t-1) - n(t)]\,\alpha_{t-1}}{2\pi\,(t-1)\,v\,(t-1)}\right) + \ldots$$
$$= \sum_{i=1}^{t} c\left(\frac{\sum_{s=i}^{t} [n(s) - n\,(s+1)]\,\alpha_s}{2\pi i v(i)}\right)$$

Following Solow [27], choose an exponential form for $c(\cdot)$, $c\left(\frac{n(t)\alpha_t}{2\pi t v(t)}\right) = k\left(\frac{n(t)\alpha_t}{2\pi t v(t)}\right)^m$, thus

$$T(t) \approx \sum_{i=1}^{t} k\left(\frac{\sum_{s=i}^{t} [n(s) - n\,(s+1)]\,\alpha_s}{2\pi i v(i)}\right)^m \qquad (9)$$

where k and m are positive constant parameters. Note that, if α_s is constant for all regions, i.e., $\alpha_s = \alpha$, then $\sum_{s=i}^{t} [n(s) - n\,(s+1)]\,\alpha_s = n(i)\alpha$. $n(i)$ is the population residing beyond distance i, and $n(i)\alpha$ is the portion of that population who work in the CBD region. In this case, Eq. (9) can be simplified into:

$$T(t) \approx \sum_{i=1}^{t} k\left(\frac{n(i)\alpha}{2\pi i v(i)}\right)^m \qquad (10)$$

Given transportation cost $T(t)$ computed according to either Eq. (9) or Eq. (10), and other parameters, the optimization problem in Eq. (6) can be solved from the following first order necessary conditions:

$$\begin{cases} \frac{U_Q}{U_X} = \frac{Y - T(t) - X(t)}{Q(t)} \\ U_0 = U\left(X(t), Q(t)\right) \end{cases} \tag{11}$$

In the closed-city model, U_0 can be determined endogenously. In the open-city model, U_0 is usually set as exogenous [22]. In Eq. (6) and Eq. (11), Y and U_0 are exogenous parameters. Y can be considered as the average income level in a given region (e.g. census tract). U_0 can be interpreted as the minimum living standard or quality of life in the region given the income level Y. The idea is that Y and U_0 are not two independently determined parameters. The two parameters can also be interpreted at individual household level.

The optimum conditions in Eq. (11) are similar to those of spatial equilibrium models, at least in the mathematical form. The essential difference is that the transportation cost now depends on the travel demand and transportation capacity from all previous stages of development. Put another way, the transportation cost for residents in the newly developed region now reflects the congestion effects created when they pass through all previously developed regions on the way to the CBD. Since traffic congestion is a mutual effect, from a social planner's perspective, therefore, the extra transportation cost imposed on residents located in previously developed regions by the congestion effects also needs to be taken into account.

ABM Simulation: Land Development and Congestion

The trade-off between transportation capacity and congestion does not disappear as long as there exists land scarcity. Traffic congestion is a price that urban residents have to pay for taking the advantage of concentration of amenities and economic activities by living in cities. Congestion, as the result of many individual trip decisions, driving habit, and transportation mode choices, is a complicated phenomenon to model. As Lindsey and Verhoef [28] point out, there is no single best way to model traffic patterns and congestion. For the purpose of modeling land use and transportation planning, it is adequate to capture only the main stationary relationship. In this section, an ABM simulation is implemented based on the monocentric urban spatial structure.

The goal of ABM simulation is to explore emergent properties out of a complex and open-ended system. In the context of urban modeling, ABM complements spatial equilibrium based models in both behavioral foundation (or micro-diversity as in Crooks et al. [11]) and system dynamics. There are three main components in an ABM simulation: stochastic component, decision making mechanism, aggregated representation. Stochastic components are the input to the model, which drives the process dynamics. Decision making mechanism, usually built upon a set of rationality and behavioral assumptions and optimization theory, is a simplified description of the individual behavior. Aggregated representation is more about output analysis. The results of ABM simulation are often not as neat as those of analytical equilibrium models. Therefore, certain level of aggregation (e.g. graphical and statistical analysis) is necessary to interpret the results and their implications. One of the concerns on ABM simulation practice is that the theoretical implications of many simulation models often remain implicit and hidden behind the mask of ad hoc assumptions about model structure and system process [11]. Therefore, it is imperative to clarify and lay out these major components in ABM simulations.

In the ABM simulation developed below, the stochasticity comes mainly from the population (consists of agents) growth process and household (agent) income variations. The decision making mechanism is designed basing on the open-city model developed in section "Transportation and Congestion: An Application". For computational purpose, some aspects of the structural model may be simplified.

Simulation Setup

In this ABM example on congestion cost, the landscape for model development is a monocentric circular city which consists of a CBD region in the center and residential regions surrounding the CBD. To study the urban land development dynamics, the simulation starts from a city with zero population and none residential development at the beginning. The development process of the city includes two sub-processes: population growth and new land development, which are also where the potential stochastic components come into play. Instead of modeling a birth-and-death process, the simulation only focuses on the net population growing process, which is assumed to follow a stochastic arrival process. The income level of each agent (i.e. household) is drawn from a statistical distribution that defines the range of income across the city.

Another important input to the simulation is the amount of land devoted to residential development and transportation capacity in each region. Transportation capacity can be considered as a local public good, which also has spillover effects (by reducing congestion effects) to households from other regions. If the provision of transportation capacity is funded through property tax (by taxing housing expenditure), then there exists an optimal level of ν—the proportion of land devoted to transportation capacity. The determination of a socially optimal ν is not straightforward, because ν apparently depends on the total (taxable) housing expenditure. At the same time, each individual household's housing expenditure depends on the transportation cost and therefore ν. To simplify, changes of ν can be considered as exogenous policy shocks. In this simulation, ν is assumed to follow an exogenous distribution with respect to distance t (to be discussed later).

The core element of the decision making mechanism is the household utility function. For each household, the disposable income is allocated to three different expenditures: housing, non-housing, and transportation. Given a constant α—the proportion of households working in the CBD, basing on Eq. (10) transportation cost is same for all households within a given region. Therefore, there are only two decision variables left in each household's consumer problem. The household decisions on housing and non-housing consumptions are made basing on the optimal conditions in Eq. (11)[5]. In this simulation a Cobbs-Douglas utility structure is specified [29], that is

$$U(X, Q) = \gamma_0 X^{\gamma_X} Q^{\gamma_Q} \tag{12}$$

where γ_0, γ_X, and γ_Q are constant parameters. While each household is differentiated by its income level, households may also be differentiated through the relative preference on housing and non-housing consumption, i.e., choosing different γ_X and γ_Q. Upon solving the household optimization problem, for each residential region t, all individual optimal housing consumptions constitute the aggregated housing consumption. The total land available for residential development in region t is given by $2\pi t(1 - \nu(t))$. Denoting the aggregated housing consumption in region t as $\sum_{i=1}^{h} Q_i^*(t)$, with h being as the total number of households in the region, $Q_i^*(t)$ the optimal housing consumption, then a measure for residential development density (ρ) in the region can be defined as:

$$\rho_t = \frac{\sum_{i=1}^{h} Q_i^*(t)}{2\pi t(1 - \nu(t))} \tag{13}$$

The residential development density, that measures the tension between residential land demand and supply, is an emergent property in this simulation. More specifically, the ABM simulation can help to illustrate the dynamic relationship between transportation cost and land development density, as well as the development density distribution with respect to distance. The change of development density from the scenario with congestion effects to the scenario without congestion effects is also interesting to explore.

Another interesting emergent property in this simulation is the housing price dynamics. As being implicit in Eqs. (6) and (11), the (unit) housing price in the analytical model is endogenously determined. The housing price solved through Eq. (11) is the individual willingness to pay for housing of each household. The existence of such heterogeneity of housing prices within a region can be explained by the residential sorting process [30]. Within a given region, households who are willing to pay more for a unit of housing are more likely to reside at location with better amenities or public services. Through the sorting process, household location choices within the region therefore reflect their willingness to pay. In this simulation, the micro sorting process is not explicitly modeled. There are two ways to look at housing price: (1) the cross-sectional housing price distribution within each region;

[5]Even though the optimal conditions here are derived to maximize land rent from a representative household (or social planner)'s perspective, they are equivalent to those first order necessary conditions of a household expenditure minimization problem. The minimized expenditure equals exactly to the household disposable income net of transportation cost which depends only on distance t. The reason for this result is that housing price is endogenous in the model, which exhausts any disposable income net of non-housing and transportation expenditure

(2) the distribution of housing price with respect to distance. In spatial equilibrium models, there exists a basic trade-off relationship between transportation cost and housing price, so that households are indifferent between locations across the city. In the proposed ABM simulation model, given the existence of congestion effects, how the relationship would change becomes a policy relevant question. So does the relationship between housing price and development density.

Parameterization

In this simulation, both the dynamic nature of the system and the open-ended environment of the model define a terminating simulation—the urban system is unlikely to reach a steady state. Given the setup of the model, two criteria can be chosen to terminate the simulation: (1) the simulation terminates after average housing price reaches certain level, for example, its opportunity cost—agricultural rent; (2) the simulation terminates after the city expands beyond a given boundary (e.g. $t \leq 10$). Depending on the goal of simulation, either criterion can be a reasonable choice.

The values of all key parameters in the simulation are set based on the scenario of large U.S. metropolitan areas. According to the U.S. Bureau of Labor Statistics Consumer Expenditure Survey in 2011, for instance, urban households on average spent $50,348; and $17,226 of which was on housing consumption, $2586 of which was on transportation. Given this empirical evidence on income allocation, the parameters γ_X and γ_Q are set to 0.5 and 0.3, respectively[6]. According to Arnold and Gibbons's [31] analysis of urban impervious surface coverage, about $5-10\%$ of suburban land, $20-30\%$ of urban land, and $40-60\%$ of commercial center land is devoted to roads and parking. Therefore, the proportion of land devoted to transportation capacity, $\nu(t)$, is specified as a decreasing function of distance t in the range of $5-40\%$. Similar to Wheaton [22], the parameter β in the transportation cost is set to 1.1, which is a very conservative specification on congestion effects. Further sensitivity analysis can be performed to explore the impact of parameter choices. All simulation parameters are summarized in Table 1.

Simulation Results

The simulations are programmed in MATLAB and implemented on a 64-bit Windows 7 operating system, with a 3.40 GHz Intel Core i7–2600 processor and 12.0 GB RAM. For a simulation (with graphing) with both congestion scenario and

[6]The parameter 0.5 and 0.3 are chosen based on relative income allocation. 0.5: 0.3 $\approx$ non-housing consumption net of transportation cost: housing consumption

Table 1 Parameters in the ABM simulation

Variable	Value	Definition
T	10	Number of residential regions
POP_t	Triangular (2,5,4)[a]	Net population growing process (in 10,000)
γ_0	1	Utility function parameter
γ_X	0.5	Utility function parameter
γ_Q	0.3	Utility function parameter
k	0.01	Transportation cost function parameter
m	1.1	Transportation cost function parameter
α	0.5	Proportion of households working in the CBD
$\nu(t)$	$\left(40 - \frac{35(t-1)}{T-1}\right)\%$	Land devoted to transportation capacity
Y	Uniform [30,000,100,000]	Household income level
U_0	$U_0 = Y/2$	Household desired utility level

[a] All generated numbers are rounded to integers

no congestion scenario[7], the simulation CPU time ranges from 100 to 130 seconds. Given the range of the city, the CPU time increases with the number of agents (population size). There are two major endogenously determined variables in this simulation: housing price and transportation cost. The two variables are also highly policy-relevant.

The housing price distributions (kernel density estimation with Epanechnikov kernel and optimal bandwidth) are presented in Fig. 1. Due to space limitation, four regions (1, 4, 7, 10) are included only. All housing prices are standardized (divided by the maximum price and multiplied by 100) so that the maximum price equals to 100. Note that the graphs only show the relative distribution of housing prices within each region, which varies from region to region. As the distance to the CBD increases, moving from the CBD to suburban area, the price distribution becomes less skewed. That is, housing price is more uniformly distributed across households in suburban area. One possible explanation for this phenomenon is that, given household income follows a uniform distribution, in the suburban area household income level has more impact on the willingness to pay (individual housing price) for housing consumption.

Another way to look at housing price is through the aggregate housing price level in each region. Figure 2 shows the relationship between aggregate housing price and distance to the CBD. The dashed line (red) indicates an approximately negative linear relationship between housing price and distance to the CBD under no congestion scenario (see footnote 7). Under no congestion scenario, the marginal price change with respect to distance to the CBD is constant. With congestion effects

[7]In the congestion scenario, the transportation cost is calculated according to Eq. (10). In the no congestion scenario, the transportation cost per unit distance is set equal to the transportation cost at $t = 10$ under congestion scenario divided by 10. In other words, at region $t = 10$, the total transportation costs in both scenarios are the same (see Fig. 3)

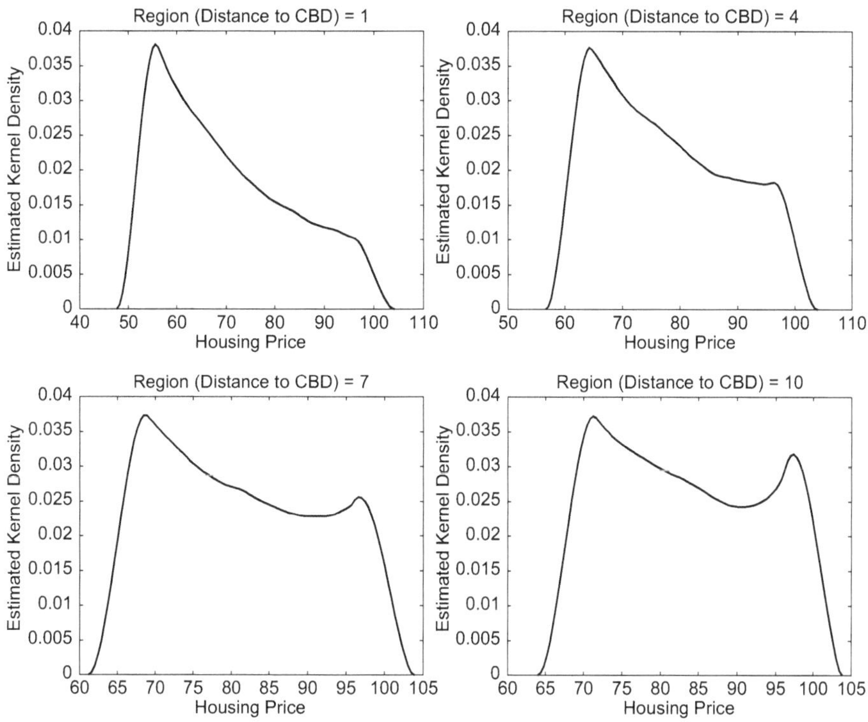

Fig. 1 Housing price distribution at different regions. Note: All housing prices are standardized with maximum price equals to 100. The density curves are kernel estimation with Epanechnikov kernel and optimal bandwidth

considered, housing price decreases quickly first, and then slows down as moving further from the CBD. Under the congestion scenario, the housing price level change reflects both a distance effect and a congestion effect. Both effects lower the housing price. The distance effect reflects the fact that, the further moving from the CBD, the higher the transportation cost and therefore the lower the housing price. The congestion effect, on the other hand, has a diminishing effect. In the regions near to the CBD, congestion tends to be more severe thus dominates the distance effect. This can be seen from the part where the solid (blue) line is under the dashed line (red) in Fig. 2. In the regions far from the CBD, the congestion effect is reduced and the distance effect becomes dominant.

The change of transportation cost works in the opposite direction to the change of housing price. According to the spatial equilibrium principle, if something is attractive in one location, then we should expect to see something unattractive offsetting it in the same location [32]. In this model, housing price and transportation cost offset each other. In Fig. 3, the dashed line (red) shows the transportation cost without congestion effects, where total transportation cost is in a direct relationship with distance to the CBD. The marginal transportation cost in this case is constant

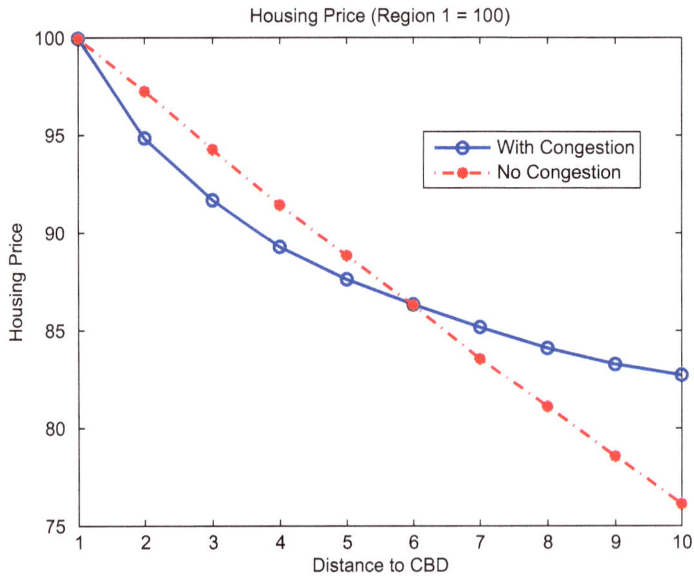

Fig. 2 Average housing price and distance to the CBD. Note: The aggregate housing prices are standardized with the price in region 1 equals to 100

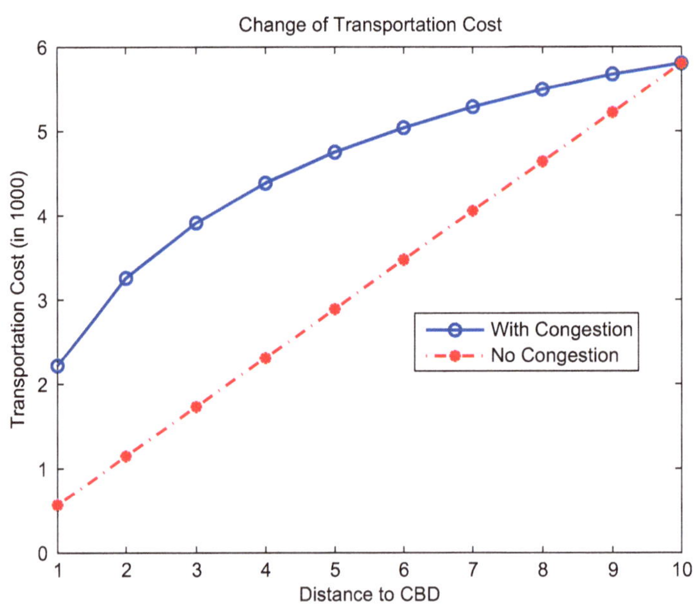

Fig. 3 Transportation cost and distance to the CBD

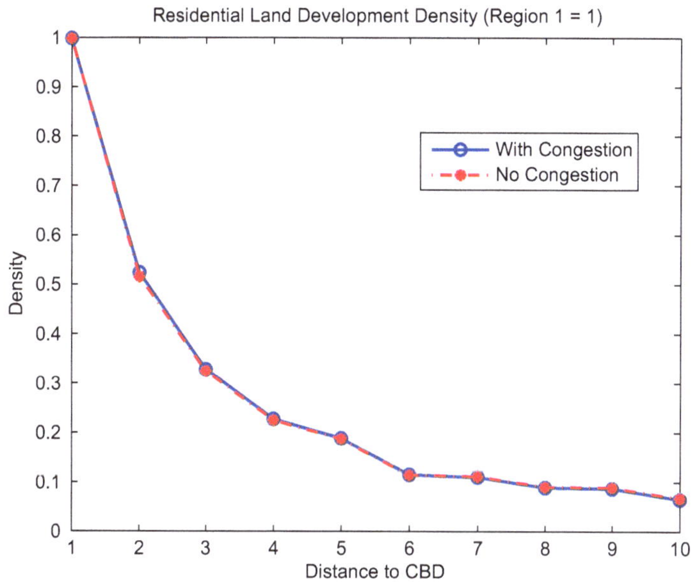

Fig. 4 Residential land development density and distance to the CBD. Note: The density measure is standardized with density in region 1 equals to 1

(see footnote 7). When there are congestion effects associated with travel, the total transportation cost becomes higher as expected. The diminishing trend of marginal transportation cost in this case reflects the fact that congestion effect is reduced as households reside further from the CBD. One other result to note is that, the difference between transportation costs under two different scenarios is not maximized at region 1. The maximum difference is reached around region 3.

As discussed in section "Simulation Setup," another way to explore the simulation outcome aggregately is to look at residential development density in each region. In this model, both the housing demand side (population) and housing supply side (amount of land in each residential region) are exogenously determined. Given that these two factors directly determines the pressure on land development, thus the land development density is likely to follow an exogenously determined pattern as well. In other words, the existence of congestion effects should not have a strong impact on land development density across all regions. The results presented in Fig. 4 confirm this conclusion. In Fig. 4, the development density measure is standardized (divided by the maximum density) with density in region 1 equals to 1. The development density under two different scenarios is almost overlapping with each other, even though there indeed exists small differences (see Table 2). Note that the congestion effects are also a function of distance and the size of the city (e.g. the radius of urban area in reality), which becomes important especially in an open-city model.

Table 2 Residential land development density and distance to the CBD

Region	1	2	3	4	5	6	7	8	9	10
With congestion	1.0000	0.5255	0.3295	0.2292	0.1905	0.1167	0.1120	0.0912	0.0890	0.0668
No congestion	1.0000	0.5178	0.3267	0.2270	0.1893	0.1164	0.1133	0.0925	0.0913	0.0694

Note: all numbers reported are corresponding to Fig. 4

Discussion

The advantage of the ABM simulation approach to urban systems is that it has a solid behavioral foundation of individual decisions. Depending on the context of modeling, the simulation procedure still needs guidance on model structure from analytical approach. In the simulation model presented above, we have incorporated urban spatial structure models and spatial equilibrium theory into simulation. The strength of these independent theories is that they provide simplified and structural ways to understand a complex system. Built upon which, simulation models can become a powerful tool in facilitating structural understanding of urban systems while with adequate level of spatial details.

The current model still hinges on the classic monocentric urban spatial structure with homogeneous landscape. The limitations of such models could be relaxed in at least two ways. First, the literature has long been paying attention on the development of non-monocentric models. The difficulty with developing non-monocentric urban spatial structure is mainly on the analytical treatment of spatial dimensions. This could be a bottleneck in integrating the analytical approach and the ABM approach, but it also points to a fruitful future research direction. Another development in the literature that could help to refine the modeling of urban system dynamics is the residential sorting process. The entire urban area may never reach an equilibrium. At a smaller scale, however, households can sort across different locations (e.g. within a community) and reach a local equilibrium. This requires urban simulation to take into account the existence of microstructures within the urban system.

Sensitivity analysis, which many existing ABM simulation models fail to emphasize, is an important part of aggregate representation. In some sense, sensitivity analysis is as important as parameter calibration. In every simulation model, certain parameters have to be exogenously given or calibrated. The sensitivity of simulation results with respect to the choice of exogenous parameters is necessary knowledge for understanding the results. In the model presented above, parameter m—a transportation cost parameter—is an important parameter to the model [22]. Figs. 5 and 6 show how the change of m (from 1.0 (Fig. 5) to 1.2 (Fig. 6), the default value in the model is 1.1) influences the main results of simulation.

Combining Figs. 2, 3, 4, 5, and 6, as parameter m changes, we can see that the main patterns of housing price, transportation cost, and development density

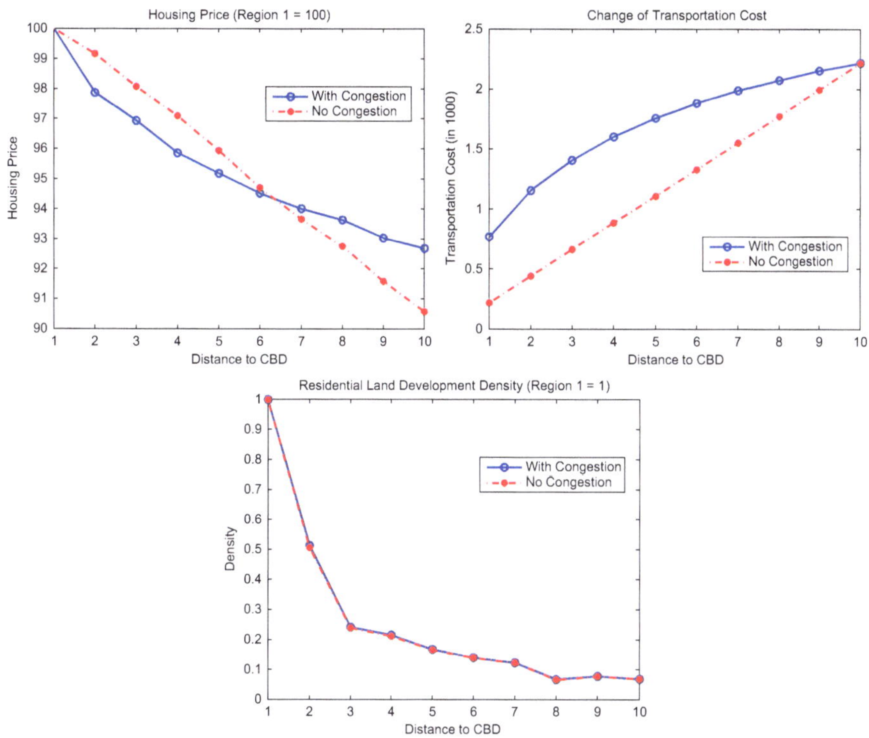

Fig. 5 Sensitivity analysis on transportation cost parameter m ($m = 1.0$)

between congestion scenario and no congestion scenario still hold. The noticeable changes in the results are mostly from the magnitude of specific measures. Therefore, as far as the specification on transportation cost function is concerned, the simulation results are robust. Similarly, sensitivity analysis on other key parameters (e.g. proportion of households working in the CBD) can be performed.

Relating to the model in this chapter, the commuting cost in the city also depends on the travel route choice. In this chapter's simulation model, the transportation system consists of symmetric ray-style routes and all households choose the shortest route to commute. An alternative scenario would be allowing households to choose among different travel routes. Unless the urban configuration is asymmetric and heterogeneous, then there is only negligible difference between the two scenarios. On the other hand, allowing for travel mode choice could lead to substantial difference in the outcomes, because different travel modes directly imply different levels of transportation cost given other factors.

Though the simulation model is only for illustration purpose, we can still learn some policy implications from the outcomes. The first policy-relevant result is the underestimate of transportation cost (in classic spatial equilibrium models) and

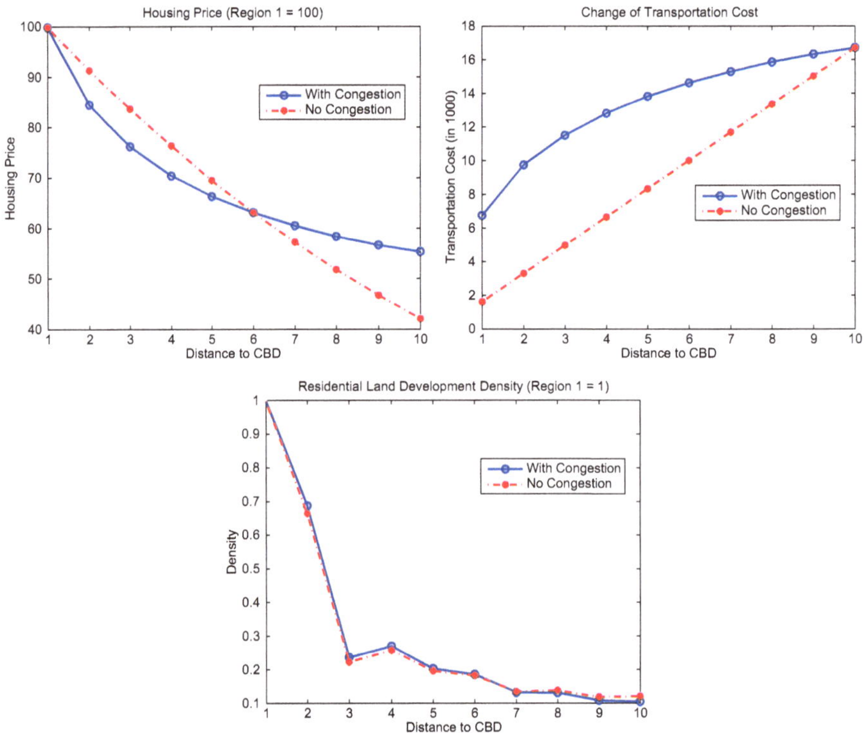

Fig. 6 Sensitivity analysis on transportation cost parameter m ($m = 1.2$)

the nonlinearity of transportation cost, as shown in Fig. 3. The underestimate of transportation cost is due to the ignorance of congestion cost. The ABM simulation helps to inform the nonlinearity of transportation cost, which is valuable for designing and evaluating public transportation system. A land use policy-relevant result is that land development density is insensitive to the existence of congestion costs (Fig. 4). This on the other hand implies that land development density depends more on overall urban spatial structure and demographics. Therefore, both economic planning and land use planning have important impacts on land use density.

Concluding Remarks

The modern city is an arrangement between its residents and local governments from both an institutional and a financial perspective. Seeking for efficient public policy and proper government intervention is essential to the sustainability of such an arrangement. Because of the mobility and heterogeneity of the population, it is often difficult to keep track of all individual household location and consumption decisions. On the other hand, public policy tends to provide general prescription for diversified individual preferences. How to aggregate the individual preferences into a form that policy makers can practice on is a critical task of urban modeling. The ABM simulation approach proposes a way to visualize urban systems so that well-founded social and economic implications can be derived to inform public policy-making. Though the multi-agent systems introduce solid behavioral foundation to urban modeling, the current urban simulation methodology still needs emphasis on structural understanding of urban systems. White and Engelen [33] raise two major concerns on high-resolution simulation models of urban and regional systems, for example, regarding the evaluation of simulation results and model predictability. One solution to address these issues is to incorporate urban spatial structure theory into urban simulations, which is the main theme of this chapter.

In this chapter, the linkage between major components of urban simulation and urban spatial structure models are discussed. Upon which, an ABM simulation model of urban land development is proposed with focus on transportation cost and congestion effects, to illustrate the role of urban spatial structure in urban simulation. The goal of the chapter is twofold. The first goal is to stress the importance of analytical modeling as the skeleton of urban modeling, even with the simulation approach. A modular architecture of urban simulation is not necessarily informative regarding results evaluation and model predictability. A further goal is to emphasize how urban spatial structure models can help to integrate household behavior, individual decision making, and aggregate model representation together. The simulation example provided in the chapter, though only for illustration purpose, gives at least some sense on how the combination of analytical modeling and ABM simulation can be an efficient and informative approach to urban modeling.

Still, there are many challenges ahead in urban modeling. For example, the development of theories on social interactions, networks, and matching mechanisms has substantially pushed the limit of our knowledge on human behavior and system dynamics. How to incorporate these new research into urban modeling is both a theoretical question and an empirical matter. Another under-researched area of urban simulation is the model calibration, which plays a critical step towards good model predictability. Similarly, calibrating model specification and parameters is also both a theoretical issue and an empirical issue. All these challenges and therefore potential future research directions will certainly have profound impacts on urban and regional modeling.

Lastly, the focus of the chapter is to suggest how we could use well-established urban spatial structure models in economics and urban studies to strengthen current agent-based urban simulation studies. The chapter does not intend to criticize current urban spatial models. Instead, the chapter argues that we should incorporate them to improve current agent-based urban simulation practices.

References

1. OECD Global Science Forum (2011) Effective modelling of urban systems to address the challenges of climate change and sustainability. http://www.oecd.org/sti/sci-tech/49352636.pdf. Accessed 28 Sep 2016
2. Clarke M, Wilson AG (1983) The dynamics of urban spatial structure: progress and problems. J Reg Sci 23:1–18
3. Fujita M, Krugman P, Venables AJ (1999) The spatial economy: cities, regions, and international trade. MIT Press, Cambridge MA
4. Jacobs J (1969) The economy of cities. Random House, New York
5. Putman SH (1984) Integrated urban models: policy analysis of transportation and land use. Pion Ltd, London
6. Gutkind EA (1962) The twilight of cities. The Free Press of Glencoe, New York
7. Glaeser EL (2011) Triumph of the city. The Penguin Press, New York
8. Glaeser EL, Kahn ME, Rappaportand J (2008) Why do the poor live in cities? The role of public transportation. J Urban Econ 63:1–24
9. Ettema D, de Jong K, Timmermans H, Bakema A (2007) PUMA: multi-agent modelling of urban systems. In: Koomen E, Stillwell J, Bakema A, Scholten HJ (eds) Modelling land-use change. Springer, Netherlands, pp 237–258
10. Waddell P, Borning A, Noth M et al (2003) Microsimulation of urban development and location choices: design and implementation of UrbanSim. Netw Spat Econ 3:43–67
11. Crooks A, Castle C, Batty M (2008) Key challenges in agent-based modelling for geo-spatial simulation. Comput Environ Urban Syst 32:417–430
12. Hardin G (1968) The tragedy of the commons. Science 162:1243–1248
13. Portugali J (2000) Self-organization and the city. Springer
14. Milgram S (1967) The small world problem. Psychol Today 2:60–67
15. Bramoullé Y, Kranton R (2007) Public goods in networks. J Econ Theory 135:478–494
16. Ettema D, Arentze T, Timmermans H (2011) Social influences on household location, mobility and activity choice in integrated micro-simulation models. Transp Res A Policy Pract 45:283–295
17. Wang H (2016) A simulation model of home improvement with neighborhood spillover. Comput Environ Urban Syst 57:36–47
18. Epstein JM, Axtell R (1996) Growing artificial societies: social science from the bottom up. Brookings Institution, Washington, DC
19. Paul MJ, Meyer JL (2001) Streams in the urban landscape. Annu Rev Ecol Syst 32:333–365
20. Henderson V, Mitra A (1996) The new urban landscape: developers and edge cities. Reg Sci Urban Econ 26:613–643
21. Vanegas CA, Aliaga DG, Benes B, Waddell P (2009) Visualization of simulated urban spaces: inferring parameterized generation of streets, parcels, and aerial imagery. IEEE Trans Vis Comput Graph 15:424–435
22. Wheaton WC (1998) Land use and density in cities with congestion. J Urban Econ 43:258–272
23. Herbert JD, Stevens BH (1960) A model for the distribution of residential activity in urban areas. J Reg Sci 2:21–36
24. Solow RM (1973) Congestion cost and the use of land for streets. Bell J Econ Manag Sci 4:602–618
25. Keeler TE, Small KA (1977) Optimal peak-load pricing, investment, and service levels on urban expressways. J Polit Econ 85:1–25
26. Black D, Henderson V (2003) Urban evolution in the USA. J Econ Geogr 3:343–372
27. Solow RM (1972) Congestion, density and the use of land in transportation. Swed J Econ 74:161–173
28. Lindsey R, Verhoef E (2000) Congestion modeling. In: Hensher DA, Button KJ (eds) Handbook of transport modeling. Elsevier, Pergamon, Amsterdam, pp 353–374
29. Cobb CW, Douglas PH (1928) A theory of production. Am Econ Rev 18(1):139–165

30. Bayer P, Timmins C (2005) On the equilibrium properties of locational sorting models. J Urban Econ 57:462–477
31. Arnold CL Jr, Gibbons CJ (1996) Impervious surface coverage: the emergence of a key environmental indicator. J Am Plann Assoc 62:243–258
32. Glaeser EL (2007) The economics approach to cities. National Bureau of Economic Research Working Paper, No. 13696
33. White R, Engelen G (2000) High-resolution integrated modelling of the spatial dynamics of urban and regional systems. Comput Environ Urban Syst 24:383–400

The ILUTE Demographic Microsimulation Model for the Greater Toronto-Hamilton Area: Current Operational Status and Historical Validation

Franco Chingcuanco and Eric J. Miller

Introduction

This chapter reports on the Integrated Land Use, Transportation, Environment (ILUTE) Demographic Updating Module (I-DUM) which updates the residential population demographics of the ILUTE model system. ILUTE is an agent-based microsimulation model that dynamically evolves the urban spatial form, economic structure, demographics and travel behavior over time for the Greater Toronto-Hamilton Area (GTHA). It has been designed to be a credible, policy-sensitive decision support tool for transportation and land use planning [1–4].

In particular, the chapter provides a comprehensive update on I-DUM's operational status, as well as presents some historical validation tests. I-DUM has recently undergone significant development and has reached a state of maturity where a 100% synthetic GTHA population of persons, families and households is being tested against a 20-year historical (1986–2006) period.

Having operational and validated demographic microsimulation models is important to integrated urban models in that they:

1. Provide population levels and their attributes, required by behavioral models (e.g., residential mobility and location choice, automobile ownership, activity/travel, etc.) [5].

F. Chingcuanco
Oliver Wyman, 120 Bremner Boulevard, Toronto, ON, Canada M5J 0A8
e-mail: franco.chingcuanco@oliverwyman.com

E.J. Miller (✉)
Department of Civil Engineering, University of Toronto, 35 St. George St., Toronto, ON, Canada M5S 1A4
e-mail: miller@ecf.utoronto.ca

J.-C. Thill, S. Dragicevic (eds.), *GeoComputational Analysis and Modeling of Regional Systems*, Advances in Geographic Information Science,
DOI 10.1007/978-3-319-59511-5_10

2. Supply inputs to work and school commuting models.
3. Endogenously maintain the representativeness of model agents and their attributes as the simulation progresses (e.g., households moving in and out of the study area) [5].
4. Accommodate the dependency between short-term (e.g., start/finish school) and long-term (e.g., residential mobility) household decisions throughout different life-cycle stages [6].
5. Support the implementation of activity-based models (e.g., TASHA [7]) that are required to address the full range of transportation policies facing twenty-first century cities [8].
6. Facilitate extending integrated urban models to include other processes of interest (e.g., urban energy use [9]) by serving as inputs that drive these models.

The more disaggregate nature of a microsimulation is also highly desirable in order to enhance behavioral fidelity and reduce aggregation bias [5]. It can easily be argued that the relatively limited impact that disaggregate mode choice models have had on travel demand modeling can be rooted in the difficulty of projecting the required population socio-demographic attributes [10].

The rest of this paper is organized as follows. Section "Literature Review" briefly reviews demographic microsimulation. Readers familiar with this topic should skip ahead to section "The ILUTE Model System", which gives an overview of the ILUTE model system. Afterwards, section "Overview of the ILUTE Demographic Updating Module" gives a high-level description of I-DUM. In particular, the section describes its design and implementation, the data sources used, the demographic attributes generated and maintained throughout the simulation, and the demographic processes modeled. Section "Descriptions of Individual I-DUM Processes" then gives a detailed description of each of the I-DUM processes being modeled. Afterwards, section "Simulation Results" presents and discusses the results from the full population 20-year validation runs and touches on the model's computational performance. Finally, a conclusion follows as well as an outline of future research directions for the I-DUM.

Literature Review

Microsimulation is a general method to exercise a disaggregate model over time [5]. It is used to analyze complex and/or dynamic systems with many elements that interact with each other. For this type of system, a closed-form analytical expression is often not available due to the complex nature of its processes. In this case, computer-based simulations offer the best alternative to make intelligent predictions by evolving the system through time.

In the context of integrated urban modeling, microsimulation derives from applied econometrics where it was used to apply quantitative methods onmicrodata

[11]. Due to its disaggregate nature, microsimulation can better capture the complex interactions between policy and social-economic life [12]. It can also determine the distributional impacts of policy measures [13].

Demographic Microsimulation Mechanics

A number of demographic microsimulation models have been built in order to analyze issues such as retirement, population projection, labor supply, and other matters related to household life-cycle changes. Comprehensive reviews of existing models can be found in Morand et al. [14] and Ravulaparthy and Goulias [15], who collectively examine sixteen models built for different regions. For most of these models, the demographic events represented can be categorized as: population changes (in- and out-migration, birth and death); household formation (marriage/cohabitation, divorce/separation, children leaving homes); and the education, health and work status of the population.

Microsimulation models typically have one of two starting points: a cross-section of the population or a birth cohort. In both cases, the initial step is to define the agents (e.g., households) where a starting point could be a snapshot of the population of interest, such as disaggregate records from a census [16]. However, such data are often not available due to privacy and cost concerns. One way around this is to use different sources of publicly available aggregate data to synthesize a base population.

There are two main approaches in synthesizing a population: combinatorial optimization and synthetic reconstruction [16]. Once a synthetic population has been created, the microsimulation engine acts on the agents in the simulation. The occurrences of demographic events (ageing, marriage, etc.) are evaluated for each member, and their attributes (age, marital status, etc.) are updated once these events have been identified. The goal is to maintain the representativeness of the base sample throughout the simulation.

Demographic Microsimulation Typology

Microsimulation models can differ in the way they execute events over time (continuous vs. discrete) and how they manage relationships among population members (open vs. closed models) [14]. For continuous time models, the durations of all possible state transition events are generated for each member of the population. The first event to occur is executed, and this procedure is repeated using the first event as the starting point. In contrast, discrete time models treat time periods one after the other, "stepping through" time in the classic sense. These models execute all possible state transition events that are realized at every time step. While continuous time models may have some theoretical advantages, they are often more complex to implement and less transparent than their discrete time counterparts.

In a closed demographic model, the simulation usually starts with a sample of the population, which includes links between population members (e.g., family ties). Members can enter/exit the population through birth/death and in-/out-migration events. Throughout the simulation, the relationships among the members are tracked and the changes are propagated throughout their social networks. For instance, if agents X and Y get married, new links are formed between them. Both agents are full population members being simulated. In contrast, open models do not maintain associations. Using the same example, if agent X gets married, a new spouse will be attached to agent X as an attribute. The new spouse is not a full population member being explicitly simulated, but only exists to properly model agent X's marriage and life path.

ILUTE Demographic Microsimulation

I-DUM is a closed and discrete time demographic model. Being closed, social networks are maintained throughout the simulation, which can be useful for modeling social travel behavior [17]. In addition, the spatial distribution of these social networks (e.g., where one's parents live) arguably also serve as "anchor points" that characterize household residential search behavior [18]. With respect to its treatment of time, ILUTE uses a modified discrete time approach that supports multiple temporal scales. This allows models with different time periods to be integrated into the model system in a simple and transparent manner. The next section first gives a brief overview of ILUTE before I-DUM and the validation results are discussed.

The ILUTE Model System

ILUTE is an object- and agent-based microsimulation model of an urban system, where the system state is evolved from an initial base case to some future end state in discrete time steps. The system state is defined in terms of the individual persons, households, dwelling units, firms, etc. (the agents) that collectively define the urban region being modeled. The attributes of these agents are evolved by simulating their behavior (changes in residential location, labor force activity, etc.) over time. Figure 1 summarizes key elements of the current implementation.

As shown in this figure, key processes modeled within ILUTE include the following:

- A 100% synthetic population of persons, families, households and dwelling units for each census tract in the study area has been constructed from 1986 Census data using a modified iterative proportional fitting (IPF) procedure [19].

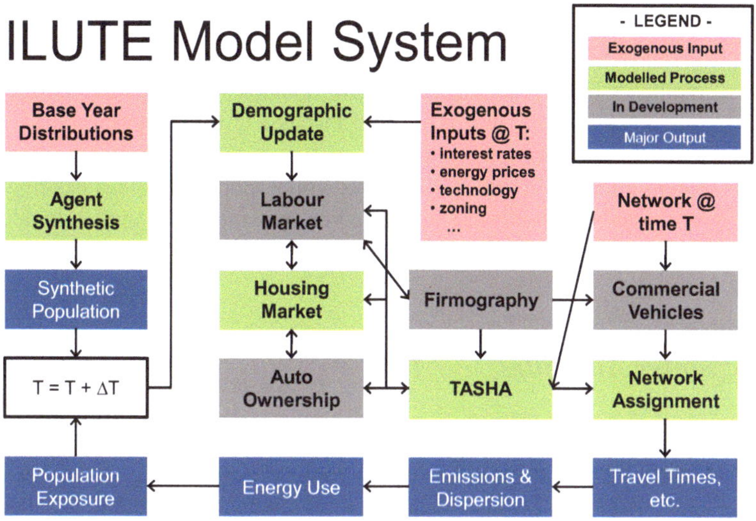

Fig. 1 The ILUTE model system and key processes modeled

- Resident population demographics are updated each time step, which is the focus of this chapter.
- The labor market component evolves the labor force in terms of: entry/exit of persons to/from the labor market; mobility of workers from one job to another; allocation of job seekers to available jobs; and the determination of worker salaries/wages. Preliminary models have been developed in [20, 21].
- The housing market component evolves residential locations over time [22, 23]. It includes the endogenous supply of housing by type and location [24], as well as the endogenous determination of sales prices and rents [25]. Initial results are presented in [26].
- Household automobile ownership is dynamically evolved using models of household vehicle transactions and vehicle type/vintage choice [27, 28].
- Once household demographics, labor market characteristics, residential location and auto ownership levels have been determined, the activity/travel patterns for each person within each household for a typical weekday are estimated using the agent-based microsimulation model TASHA (Travel/Activity Scheduler for Household Agents) developed by the ILUTE team [7].
- There has also been work done on developing environmental [29, 30] and energy [9] modeling components within ILUTE.
- It is the intent to implement some form of firmographic model within ILUTE. This has not yet been accomplished, and so for current historical model system testing purposes, observed employment levels by occupation and industry for each census track in the study area are exogenous inputs to the simulation. Preliminary firmographic frameworks are presented in [21, 31]. Another long-term project is to implement a microsimulation-based commercial vehicle movements model in ILUTE [32].

Overview of the ILUTE Demographic Updating Module

Given a synthetic base population, I-DUM updates these agents' attributes at each time step. New agents are introduced through birth and in-migration, while agents exit through death and out-migration events. Unions between agents are formed through a marriage market, while a divorce model dissolves existing ones. Transitions to new households are also triggered by a move-out model. In addition, each person's driver's license ownership and education level are managed.

Demographic Attributes

Population members in ILUTE are represented by household, family, and person agents. Households are defined as one or more persons living within the same dwelling unit. They can consist of any combination of individuals and families. Families are defined either as husband-wife couples with or without children, or single parents living with children. Links between these members are explicitly maintained throughout the simulation, which allows family relationships to be tracked over time. Note that all families and individuals must belong to a household.

Table 1 lists the demographic attributes of the person class. All these attributes are maintained and/or modeled for all agents across the entire simulation. Persons have an exclusive association with a family or a household. Hence, when either the FamilyId or HouseholdId is non-zero, the other by definition is zero. Agents maintain family relationships through identifiers (e.g., SpouseId). Sex, MaritalStatus and EducationLevel are enumeration types, which are defined data types of named constants. There is also a flag to signify driver's license ownership.

Both family and household classes have member lists: households have a list of families and individuals and families have a list of members. Like person agents, families maintain associations with their households through a household ID. Similarly, households have unique dwelling IDs, which imply a one-to-one mapping between households and dwelling units.

Table 1 Person class demographic attributes in defined in I-DUM

Attribute	Type	Attribute	Type	Attribute	Type
MyID	int	ExSpouseIdList	List<int>	EducationLevel	none
HouseholdId	int	ChildIdList	List<int>		kindergarden
FamilyId	int	SiblingIdList	List<int>		elementary
MotherId	int	DriversLicense	bool		highschool
FatherId	int	MaritalStatus	single		college
SpouseId	int		married		undergrad
Age	short		divorced		graduate
Sex	male/female		widowed		

I-DUM Processes

I-DUM is executed in yearly time-steps. A bottom-up approach is employed in which the demographic evolution emerges through the sequential updating of each person. The whole module can be broken down into a sequence of three main parts. First, demographic events are evaluated for each agent in the simulation. The process takes a list of agents, uses their attributes to compute transition probabilities (e.g., age), evaluates these events (e.g., death), and adds the agents to respective lists (e.g., list of deceased agents). After all the possible state transitions have been determined for the entire population, all the realized events are processed to reflect their changes. For instance, the family relationships of deceased agents are managed (e.g., the spouse is widowed). A cleanup process is executed to delete or convert invalid families and households after they have been updated.

Table 2 lists the demographic processes modeled as well as the factors that drive them. Depending on data availability, these models range from simple empirical probabilities (birth) to more advanced methods such as hazard (divorce) and logit (education) models. They can also either be static or dynamic. The letters under the "Data Code" column match the data sources found in Table 3, which describe the geographic levels of the data. Model outcomes are conditioned on the agent's current state. For instance, the likelihood of a birth event is a function of a female's age, marital status and current year of the simulation. Each of these models is described in further detail in the next section.

Table 2 Demographic processes modeled in I-DUM and a summary of factors that drive their transition probabilities

Process	Factors	Temporal	Type	Data code
Birth	Age; marital status	Dynamic	Rate-based	A, B, C, G, I
Death	Age; marital status; gender	Dynamic	Rate-based	A, B, C, G, I
Marriage	Age; marital status; gender	Dynamic	Rate-based and Logit	D, E, H
Divorce	Ages; marital status; years of birth	Static	Hazard	J
Move out	School/job changes; gender	Static	Hazard	K
Driver's license	Age; gender; geographic location	Dynamic	Rate-based	L
Education level		Dynamic	Logit	Under development
Out-migration		Dynamic	Rate-based	F, G
In-migration		Dynamic	Rate-based	F, G

Table 3 Data sources for the I-DUM models by level of aggregation

Data code	Data source and description	Sources
A	Public Use Microdata Files, by census metropolitan area (1986, 1991, 1996, 2001, 2006)	[33]
B	Estimates of population by sex and age group, by census division (1986–2002) [051–0016]	[34]
C	Estimates of population by sex and age group, by census division (1996–2006) [051–0052]	[34]
D	Marriages by marital status and age of groom and bride, Canada (2000–2002) [101–1005]	[34]
E	Estimates of population by marital status, age group and sex, Canada, provinces and territories (1986–2006) [051–0010]	[34]
F	Total population, census divisions and census metropolitan areas (1986–2006) [051–0034]	[34]
G	Components of population growth, by census division (1986–2006) [051–0035]	[34]
H	Estimates of births, deaths and marriages, Canada, provinces and territories (1986–2006) [053–0001]	[34]
I	Ontario births and deaths registry, by municipality (1986, 1991, 1996)	[34]
J	General Social Survey on the family, Canada (1995)	[35]
K	General Social Survey on family transitions, Canada (2006)	[35]
L	Transportation Tomorrow Survey, by wards (1986, 1991, 1996, 2001, 2006)	[36]

In addition to parallelization concerns, I-DUM has also been designed for modularity. This allows components to be easily replaced. For instance, a hazard divorce model was recently implemented in place of an older rate-based one, with very minor code modifications.

Data Sources

A list of the data sources used by the I-DUM models are found in Table 3. The "Data Code" column matches that of Table 2 to map the respective data sources to the I-DUM processes they drive. Except for the Ontario birth and death registries (item I), all the data are publicly available. Data sources B–H are available through the Canadian socioeconomic information management system, which is a database maintained by Statistics Canada. Sources A, J and K are housed under the Computing in the Humanities and Social Sciences (CHASS) data center while source L is provided by the Data Management Group (DMG). Both CHASS and DMG are University of Toronto data centers.

Data with varying spatial and temporal levels are used, and the best available proxy data are employed when needed. Many of the empirical probabilities employed combine various data sources to get a comprehensive cross-section across the required socio-demographic dimensions (e.g., age groups, gender, and marital status) and time periods.

Descriptions of Individual I-DUM Processes

This section gives detailed descriptions of the individual models used that drive I-DUM. It also contains estimation results as well as rate calculations that explain how the agents in ILUTE make demographic decisions, and how the changes from these events are propagated throughout the simulation. I-DUM manages demographic relationships and seeks to maintain reasonable population, family and household counts throughout the simulation. These variables serve as important inputs to other ILUTE components. For instance, the new households that result from marriages, births and divorces are key drivers to ILUTE's residential housing market.

I-DUM is initialized with a set of agents/objects which are synthesized from base year Census (and perhaps other) data. A 100% population of persons, families, households and dwelling units for each census tract in the study area has been constructed for 1986 using a modified IPF procedure [19] that:

- Simultaneously generates these four objects in a fully consistent manner.
- Permits a large number of object attributes to be included in the synthesis.
- Is computationally efficient.
- Makes full use of multiple multivariate tables of observed data.
- Is extendable to include additional elements (e.g., household auto ownership, which is not yet included in the synthesis procedure).
- For model testing purposes, either the full 100% population can be used, or a smaller subset, randomly drawn from the full population, can be used to speed up run times, with all other model elements and processes (e.g., building supply, etc.) being appropriately scaled.

The education model is not discussed in this chapter but will be presented in a separate work along with the ILUTE labor force model.

Marriage

Marriages in ILUTE are broken into three main steps: (1) a marriage event occurs in which potential marriage candidates join a marriage market; (2) a marriage market is executed where potential grooms and brides are paired off; and (3) the family relationships and attributes of the new couple are processed (e.g. setting husband-wife relationships, transferring existing children, etc.). Marriage events, which trigger an individual to join the marriage market, are driven by rates. These rates are calculated through empirical probabilities for population cross-sections across 13 age groups, three marital statuses (single, divorced, widowed), 20 time periods (1987–2006), and gender.

After a marriage event, the individuals join a marriage market in which they are paired with other potential brides and grooms. The matching is executed under a utility maximization framework. A potential bride or groom is randomly chosen from the marriage pool. A choice set is generated for this candidate by drawing agents of the opposite gender from the pool. The candidate's utilities for these matches are calculated based on [37]. These utilities are based on the potential couple's incomes, education levels, and the male/female ratios in their respective geographic areas. These utilities are converted to choice probabilities via a multinomial logit formulation, and a match is made through simulation. A logit formulation is used in order to introduce some stochasticity in the matching. The marriage market is discussed further in [2].

After the marriage market is cleared, the family relationships of the new couple are updated. Depending on the situations of the individual newlyweds, this could include forming new households or merging existing ones, as well as transferring any children over. Newlyweds with new households enter the housing market. Note that at the start of the ILUTE simulation, marriage durations for the base population are estimated from census data using a regression model. This is critical for the operation of the divorce module, which is discussed in the subsection below.

Note that the marriage module intends to include common law unions between males and females. The authors chose to continue denoting these events as a "marriage" to follow convention (e.g., "marriage market"), as well as to be consistent with prior ILUTE work. The authors also caution that there is some inconsistency with this intention and the data, as the marriage process data (e.g., for matching males and females) only account for officially recognized marriages. More importantly, the current marriage module can only handle heterosexual unions. Same-sex unions are not explicitly modeled, but are implicitly accounted for in the non-marital household formation process briefly described in subsection "Moving Out".

Divorce

The ILUTE divorce process evaluates whether a divorce event occurs for existing husband-wife couples in the simulation. The agents' attributes (e.g. marital status) and family relationships are also updated. A spouse is moved out and a new household is created for this agent. Custody is also handled according to Ontario aggregate rates (59% for mother single custody and 33% for joint custody). A proportional hazards regression model (Table 4) was estimated using the 1995 and 2001 General Social Surveys on the family (source J in Table 3) to model divorce decisions. Due to a lack of data, the divorce model does not include a temporal component, i.e., the same regression is applied for all divorces across the 20-year simulation.

Table 4 Proportional-hazards regression results for the divorce model

Variable	Coef.	Exp (Coef.)	S.E.	t-Stat	Pr(>|Z|)
hPreviousDivorce	0.675	1.964	0.108	6.243	0.000
wPreviousDivorce	0.571	1.770	0.124	4.610	0.000
hAgeSquaredFrom25	−0.001	0.999	0.001	−2.898	0.004
wAgeSquaredFrom25	0.002	1.002	0.001	2.983	0.003
withIn5Yrs	−0.119	0.888	0.073	−1.622	0.105
marriedAfter1960s	0.664	1.943	0.109	6.120	0.000
wMarriedBefore1950s	−0.473	0.623	0.325	−1.454	0.146
hMarriedBefore1950s	−0.595	0.551	0.323	−1.842	0.065
hBornBefore1945	−0.232	0.793	0.096	−2.404	0.016
wBornBefore1945	−0.391	0.676	0.108	−3.632	0.000
hBornAfter1959	0.129	1.138	0.146	0.885	0.376
wBornAfter1959	0.270	1.310	0.119	2.265	0.023
Number of observations	25,262				
Number of events	5012				
Likelihood ratio test	9341	on 14 df	$p = 0$		
Wald test	11,467	on 14 df	$p = 0$		
Score (Logrank) test	29,660	on 14 df	$p = 0$		

Birth

The birth process handles all birth related events in ILUTE, including evaluating the birth event, updating the attributes of the mother, creating the new born baby, and managing family relationships (e.g. adding parent-child links, creation of a new family, etc.). The birth rates are calculated through empirical probabilities for population cross-sections across seven age groups, four marital statuses (single, married, divorced, widowed) and 20 time periods (1987–2006). If the new mother is married or already has children, then the new baby is simply added to the mother's existing family and household. Otherwise, a new family and household are created, and the agents enter the housing market.

Death

The death process handles all death related events in ILUTE, including evaluating the death event, removing the deceased from the simulation, and managing family relationships (e.g. making the spouse a widow, making the children orphans or finding new guardians, exiting the housing market if active, etc.). The death rates are calculated through empirical probabilities for population cross-sections across 24 age groups, four marital statuses (single, married divorced, widowed), 20 time periods (1987–2006) and gender. When the household head agent of a non-individual household dies, a new agent is designated.

Table 5 Proportional-hazards regression results for the move out model

Variable	Coef.	Exp (Coef.)	S.E.	t-Stat	Pr(>|z|)
LiveParents15	−0.127	0.880	0.002	−54.750	<2e-16
School	1.620	5.053	0.002	817.020	<2e-16
Job	1.259	3.520	0.002	505.970	<2e-16
Male	−0.141	0.869	0.002	−80.400	<2e-16
Number of observations	1497				
Number of events	906				
Likelihood ratio test	748,022	on 4 df	p = 0		
Wald test	721,136	on 4 df	p = 0		
Score (Logrank) test	834,458	on 4 df	p = 0		

Moving Out

A move out process is used to transition young adults into moving out from their families into their own households. New households are created for transitioning agents and they enter the housing market. A proportional hazards regression model was estimated using the 2006 General Social Surveys on family transitions (Source K in Table 3). Table 5 displays the estimation results. Similar to divorce, the move out model does not include a temporal component. A complementary household formation process is used to create and maintain non-family households with more than one individual (e.g., student roommates, friends sharing an apartment, etc.).

Driver's License

The driver's license process has two functions: grant drivers' licenses to eligible candidates; and revoke these licenses when drivers get too old to drive. The Transportation Tomorrow Surveys (Source L in Table 3) was primarily used to calculate the driver's license acquisition and revocation rates, which are taken for cross-sections across three levels of aggregation of the 46 TTS planning districts in the GTHA, 80 valid ages (16–95), 20 time periods (1987–2006) and gender.

Out-Migration

The out-migration process manages all out-migration related events in ILUTE. Out-migration numbers for the GTHA census divisions (Toronto, Durham, Peel, York, Halton and Hamilton) were taken from Statistics Canada. These values were divided by the corresponding GTHA census division populations to obtain the out-migration rates for six census divisions and 20 years. Out-migration events are handled in the same manner as death, though out-migrating household heads have the decision to out-migrate their entire families with them. At present, there is a 75% chance of this happening, and if this event is true, the family members are simply added to the out-migration persons list.

In-Migration

The in-migration process manages all in-migration related events in ILUTE. Unlike out-migration, in-migration does not require calculating in-migration rates. Instead, actual in-migration numbers are used to synthesize in-migrant agents for each year. The attributes of the in-migrating agents (e.g., age, gender, household status, etc.) are determined from the data. Note that these in-migration numbers are scaled down by a factor that corresponds to ILUTE's base population size, and these factors were calibrated to get the observed total population numbers per year.

When new agents are immigrated in, their corresponding families and households are also built. A process that builds familial relationships across a batch of agents, which is also used by ILUTE's population synthesizer, is executed. There may be some advantages of synthesizing in-migrant agents directly from data distributions instead of randomly drawing from the observed data. This alternative is intended to be explored.

Simulation Results

This section presents a 20-year (1986–2006) simulation run for a fully synthesized population against historical data for the GTHA. The simulation starts with over 6.5 million agents (4.1 million persons, 1.1 million families, 1.4 million households), and the overall number of agents grow past 10 million after a 20-year run. On a computer with an i7-2600 processor (3.4 GHz, 4 cores) with 16 GB of RAM running on a 64-bit Windows 7 operating system, the simulation takes just under 10 min to complete, including 2.5 min to load a base population and form the initial relationships among the agents.

While the figures below aggregate the simulation outputs for the entire region, each simulation process follows the geographic level of detail afforded by the data, as defined in Tables 2 and 3. Furthermore, note that the empirical rates to drive these models are known ex-post, as the objective of this entire section is to illustrate the performance of running the full I-DUM. That is, the focus is to demonstrate a valid model system, and less on building accurate individual models (e.g., in-migration forecasts).

Figures 2 and 3 compare the 1986 and 2006 age distribution of males and females in ILUTE with historical data. Each of the four sets of bar graphs sum to 100%. For the most part, the simulation produces the correct age distributions by gender

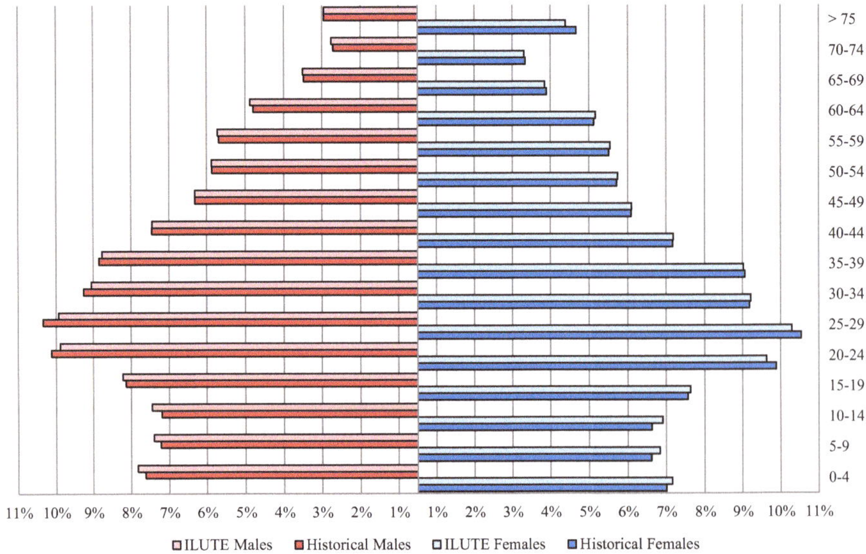

Fig. 2 1986 ILUTE vs. historical age distributions for males and females

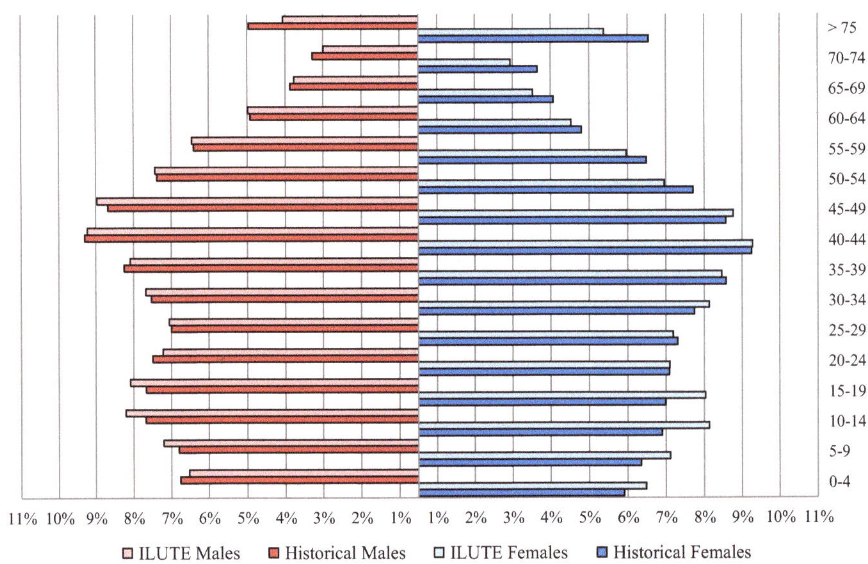

Fig. 3 2006 ILUTE vs. historical age distributions for males and females

after 20 years. Although the simulation under predicts females greater than 75 and over predicts 10–19 year olds, the errors are relatively small (in the order of 1% absolute error per age group). Figures 4 and 5 add another dimension by plotting the 2006 distribution of males and females by age and marital status for ILUTE

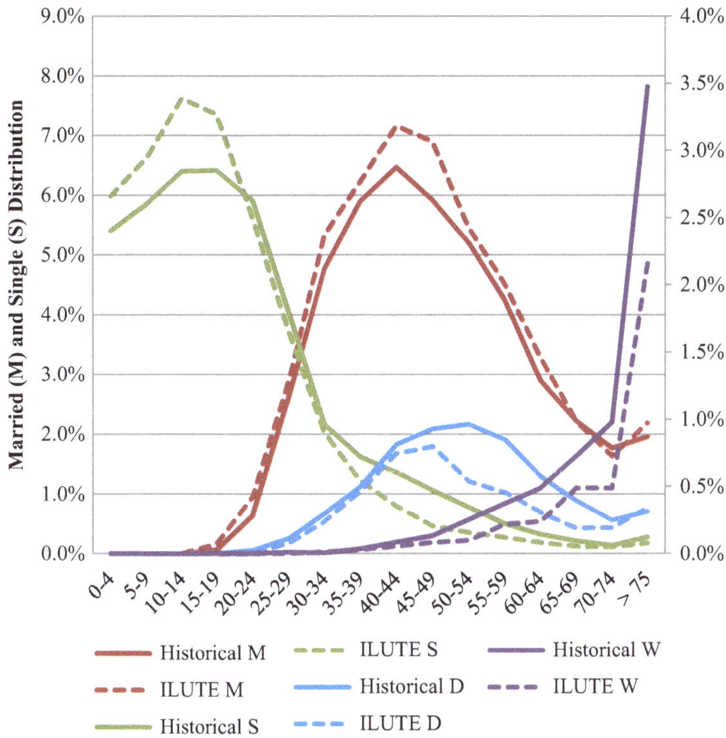

Fig. 4 2006 ILUTE vs. historical female population by marital status and age

and their corresponding historical values. The areas under each set of marital status curves sum to 100%. Note the presence of two axes (married and singles on the left, widowed and divorced on the right) and the scale difference between the male and female widowed and divorced axes. Again, the distributions are generally tracked quite well after the 20-year simulation. The under and over predictions of females illustrated in Fig. 3 are revealed in Fig. 4 to correspond to widowed and single agents.

Besides maintaining the proper marital status and age distributions for person agents, I-DUM also seeks to preserve the correct distribution of household types. Table 6 presents simulated vs. historical household type distributions for 4 years (2006 data are not available). ILUTE tends to produce too many single individuals and too few single families as the simulation progresses. This discrepancy can be attributed to multiple factors, including: birth and marriage rates being too low, divorce and move out rates being too high, and the out-migration model's insensitivity to socio-demographic factors. The overproduction of female widows (Fig. 4) can also be related to this issue.

Figure 6 plots the birth, death and out-migration rates (left axis) as well as the absolute population levels (right axis) for ILUTE and the corresponding historical

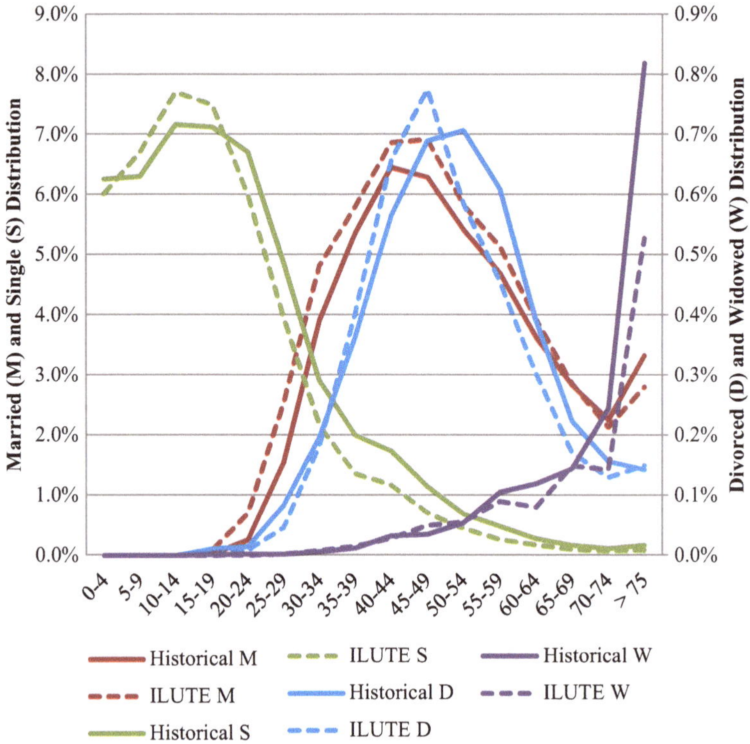

Fig. 5 2006 ILUTE vs. historical male population by marital status and age

Table 6 ILUTE vs. historical household type distributions

		Single individual (%)	Multiple individual (%)	Single family (%)	Single family and individuals (%)	Multiple family (%)
Census	1986	20.8	2.8	74.0	2.2	0.1
	1991	21.4	3.7	71.6	3.1	0.2
	1996	22.0	3.0	72.6	2.2	0.2
	2001	22.2	2.9	72.6	2.1	0.2
ILUTE	1986	21.1	3.3	74.1	1.0	0.5
	1991	23.3	2.8	71.8	1.8	0.4
	1996	25.3	2.4	70.3	1.7	0.3
	2001	27.3	2.2	68.7	1.5	0.3

benchmarks. The birth and death rates seem to perform quite well (with a slight under prediction of deaths), but the out-migration rates start off a bit too high. While the model corrects itself as the simulation progresses, this may be due to the population levels increasing faster than they should have. That is, a larger denominator results in lower out-migration rates. Absolute population levels are also plotted on the same figure for comparison. While ILUTE starts with about

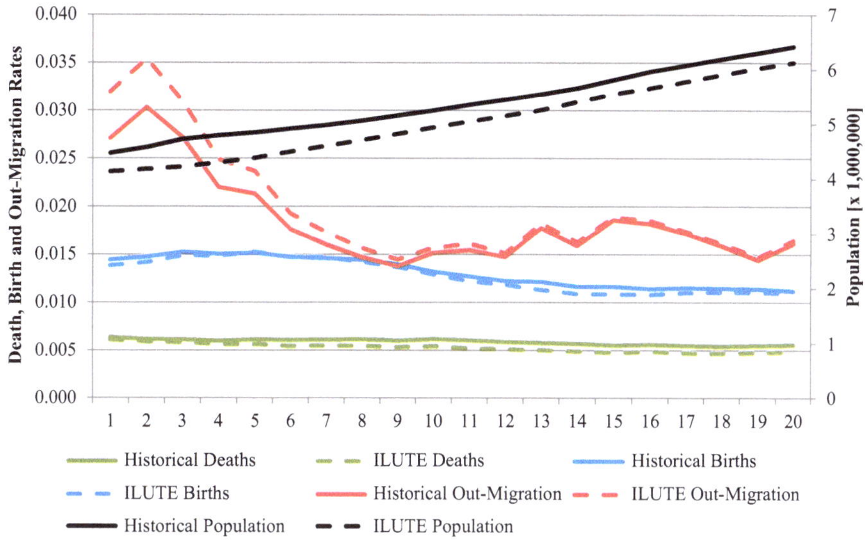

Fig. 6 ILUTE vs. historical birth, death and out-migration rates and total population levels

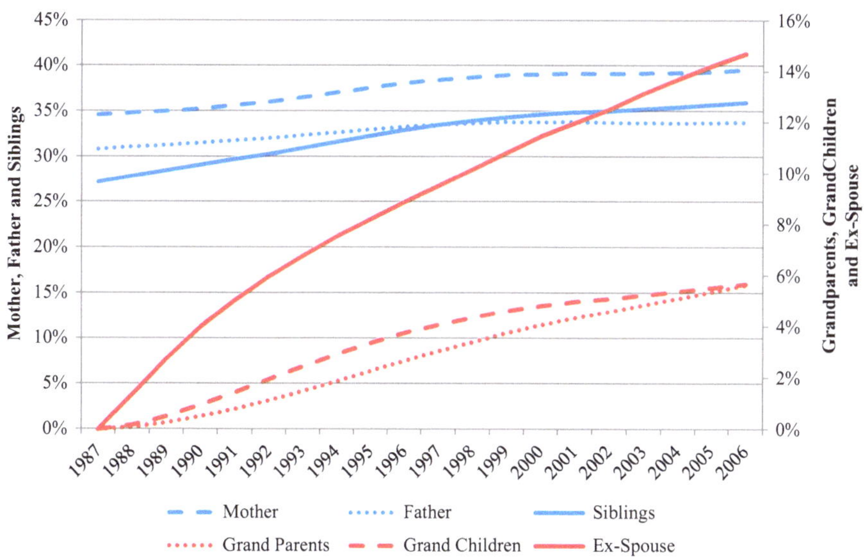

Fig. 7 Percent of ILUTE population with a particular relationship over time

300,000 less persons in 1986, the rate of growth seems to match the observed values quite well. The delta in the base population numbers is an issue with the population synthesis and is currently being investigated.

Following this, Figs. 7 and 8 demonstrate how social networks are built and maintained throughout the simulation. At the very start of the simulation, only

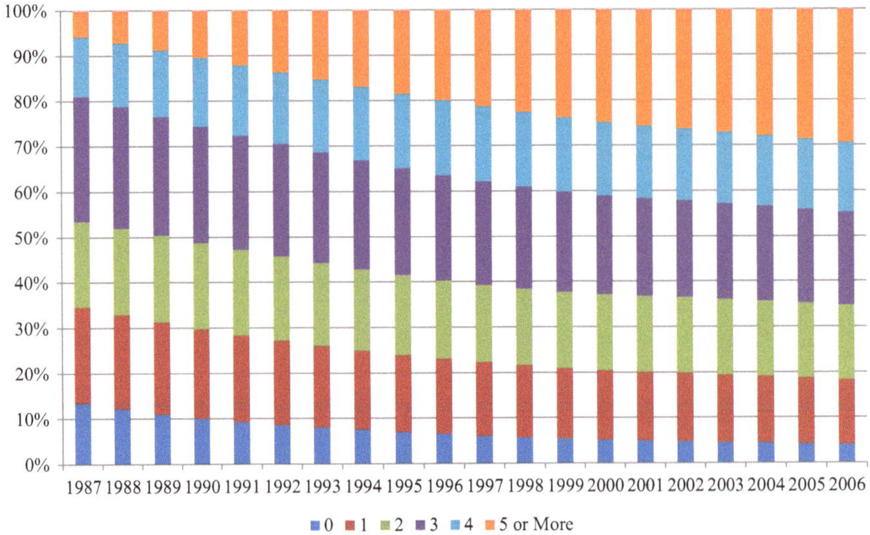

Fig. 8 Percent of population with social connections over time

synthesized families have relationships among each other (e.g., parent and child association). As the simulation progresses, agents start to build secondary associations (e.g., grandparent-grandchildren links) through intermediate agents (i.e., the parent). Histories are recorded as shown by the growing ex-spouse list. Note that the percentage of agents that have relationships plateau out due to agents exiting the simulation. Figure 8 depicts the population's growing social network connectivity throughout the 20 years. As mentioned earlier, these social connections can help predict the spatial choices of people (e.g., residential location choice, destination choice, etc.) and is beneficial to be maintained.

Figure 9 compares the distribution of divorces in ILUTE from 1986 to 2006 to historical values, which illustrates the utility of tracking agent histories across the simulation. The marriage date of each agent couple is maintained, and this is used in evaluating divorce decisions. While the plot demonstrates a well performing divorce model, it also reaffirms the performance of the marriage model. For example, since the divorce model uses a hazard function with age-related covariates, agents would have to get married at the right age and find partners with the appropriate age differences to get the correct divorce distributions shown. Preliminary results of I-DUM's marriage market can be found in [1, 2].

Discussion and Future Directions

This paper presents the operational status of the ILUTE Demographic Updating Module (I-DUM). The performance of I-DUM is then compared against historical observations across multiple dimensions. In general, I-DUM exhibits a strong

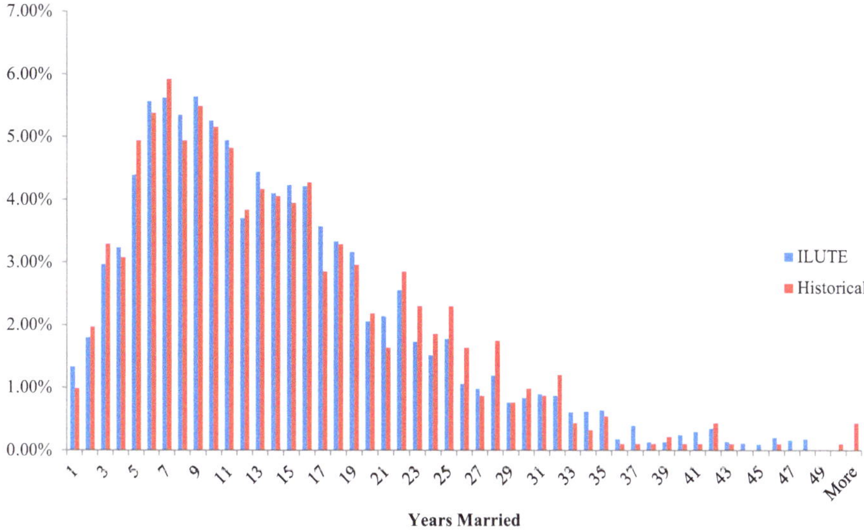

Fig. 9 1986–2006 ILUTE vs. historical divorces by years married

performance, and the authors have confidence that it can maintain the validity of inputs to the other behavioral models in ILUTE.

Note that multiple simulation runs have also been conducted to explore the uncertainty of the outputs, which are important for validating microsimulation models. The results (not shown here) are distributed very tightly around the single-run outputs presented in this paper, which is reasonable given that relatively simple demographic models are used throughout. This also suggests that clear demographic patterns could emerge across millions of simulated agents, despite their heterogeneity.

As discussed previously, ex-post values are used in building the individual models. A focus going forward is to conduct demographic forecasting exercises. While finding new independent sources of data would be helpful, it is also possible to estimate the models for half the simulation period (1986–1996) and evaluate its performance going forward (1997–2006). Some components of I-DUM are still under development (education and driver's license), and this is also the focus of current research. Future research steps include integrating I-DUM with models of labor force participation and automobile ownership, which require operational and validated education and driver's license sub-models.

Acknowledgments The authors would like to thank Bilal Farooq, Adam Rosenfield, Gurbani Paintal, James Vaughan and David Wang for their contributions. This work was funded by a Canada NSERC Discovery Grant.

References

1. Miller EJ, Farooq B, Chingcuanco F, Wang D (2011) Historical validation of an integrated transport – land use model system. Transp Res Rec: J Transp Res Board 2255:91–99
2. Farooq B, Miller EJ, Chingcuanco F, Giroux-Cook M (2013) Microsimulation framework for urban price-taker markets. J Transp Land Use 6(1):41–51
3. Salvini PA, Miller EJ (2005) ILUTE: an operational prototype of a comprehensive microsimulation model of urban systems. Netw Spat Econ 5(2):217–234
4. Miller EJ, Hunt JD, Abraham JE, Salvini PA (2004) Microsimulating urban systems. Comput Environ Urban Syst 28:9–44
5. Miller EJ (2003) Microsimulation. In: Goulias K (ed) Transportation systems planning: methods and applications. CRC Press, Boca Raton
6. Eluru N, Pinjari AR, Guo JY, Sener IN, Srinivasan S, Copperman R, Bhat CR (2008) Population updating system structures and models embedded in the comprehensive econometric Microsimulator for Urban Systems. Transp Res Rec: J Transp Res Board 2076:171–182
7. Miller EJ, Roorda MJ (1831) A prototype model of household activity/travel scheduling. Transp Res Rec: J Transp Res Board 2003:114–121
8. Goulias K, Kitamura R (1992) Travel demand forecasting with dynamic microsimulation. Transp Res Rec 1357:8–17
9. Chingcuanco F, Miller EJ (2012) A microsimulation model of urban energy use: modelling residential space heating demand in ILUTE. Comput Environ Urban Syst 36(2):186–194
10. Miller EJ, Kriger DS, Hunt JD (1998) Integrated urban models for simulation of transit and land use policies: Final Report. TCRP Web Document 9
11. Orcutt GH (1957) A new type of socio-economic system. Rev Econ Stat 39(2):116–123
12. Mitton L, Sutherland H, Weeks M (2000) Introduction. In: Mitton L, Sutherland H, Weeks M (eds) Microsimulation modelling for policy analysis. Cambridge University Press, Cambridge
13. Joachim M (1994) Microsimulation—a survey of methods and applications for analyzing economic and social policy. FFB discussion paper no. 9, Universität Lüneburg, Germany
14. Morand E, Toulemon L, Pennec S, Baggio R, Billari F (2010) Demographic modelling: the state of the art. SustainCity Working Paper 2.1a, Paris
15. Ravulaparthy S, Goulias K (2011) Forecasting with dynamic microsimulation: design, implementation, and demonstration. University of California Transportation Center. UCTC-FR-2011-07, Santa Barbara
16. Müller K, Axhausen KW (2010) Population synthesis for microsimulation: state of the art. Presented at the swiss transport research conference, Ascona
17. Carrasco JA, Miller EJ (2008) How far and with whom do people socialize? Empirical evidence about distance between social network members. Transp Res Rec: J Transp Res Board 2076:114–122
18. Huff OJ (1986) Geographic regularities in residential search behaviour. Ann Assoc Am Geogr 76(2):208–227
19. Pritchard DR, Miller EJ (2012) Advances in population synthesis: fitting many attributes per agent and fitting to household and person margins simultaneously. Transportation 39(3): 685–704
20. Hain M (2010) Labour market model of the greater Toronto and Hamilton area for integration within the integrated land use, transportation, environment modelling system. Master's thesis, University of Toronto
21. Harmon A (2013) A microsimulated industrial and occupation-based labour market model for use in the integrated land use, transportation, environment (ILUTE) modelling system. Master's thesis, University of Toronto
22. Habib MA, Miller EJ (2008) Influence of transportation access and market dynamics on property values: multilevel Spatio-temporal models of housing price. Transp Res Rec: J Transp Res Board 2076:188–191

23. Habib MA, Miller EJ (2009) Reference-dependent residential location choice model within a relocation context. Transp Res Rec: J Transp Res Board 2133:56–63
24. Haider M, Miller EJ (2004) Modeling location choices of housing builders in the greater Toronto area, Canada. Transp Res Rec: J Transp Res Board 1898:148–156
25. Farooq B, Miller EJ (2012) Towards integrated land use and transportation: a dynamic disequilibrium based microsimulation framework for built space markets. Transp Res A Policy Pract 46(7):1030–1053
26. Rosenfield A, Chingcuanco F, Miller EJ (2013) Agent-based housing market microsimulation for integrated land use, transportation, environment model system. Presented at the 2nd international workshop on agent-based mobility, traffic and transportation models, methodologies and applications (ABMTRANS'13), Halifax
27. Mohammadian A, Miller EJ (2003) Dynamic modeling of household automobile transactions. Transp Res Rec: J Transp Res Board 1831:98–105
28. Duivestein J (2013) Household vehicle fleet decision-making for an integrated land use, transportation and environment model. Master's thesis, University of Toronto
29. Hatzopoulou M, Miller EJ, Santos B (2007) Integrating vehicle emission modeling with activity-based travel demand modeling: a case study of the greater Toronto area (GTA). Transp Res Rec: J Transp Res Board 2011:29–39
30. Hao J, Hatzopoulou M, Miller EJ (2010) Integrating an activity-based travel demand model with dynamic traffic assignment and emission models: an implementation in the greater Toronto area. Transp Res Rec: J Transp Res Board 2176:1–13
31. Mostafa T, Roorda MJ (2013) A framework and analysis of firm evolution processes. Presented at the METRANS international urban freight conference (I-NUF) long beach
32. Roorda MJ, Cavalcante RA, McCabe S, Kwan H (2010) A conceptual framework for agent-based modelling of logistics services. Transp Res E: Logist Transp Rev 46(1):18–31
33. Statistics Canada. Public use microdata files. http://www5.statcan.gc.ca/bsolc/olc-cel/olc-cel?catno=11-625-XWE&1ang=eng. Accessed 28 June 2012
34. Statistics Canada. Canadian socio-economic information management system. http://www5.statcan.gc.ca/cansim/home-accueil?lang=eng. Accessed 28 June 2012
35. Statistics Canada. The general social survey. http://www5.statcan.gc.ca/bsolc/olc-cel/olc-cel?catno=89F0115X&CHROPG=1&lang=eng. Accessed 28 June 2012
36. Data Management Group. Transportation tomorrow survey. http://www.dmg.utoronto.ca/transportationtomorrowsurvey/index.html. Accessed 28 June 2012
37. Choo E, Seitz S (2013) The collective marriage matching model: identification, estimation and testing. In: Choo E, Shum M (eds) Structural econometric models, Advances in econometrics, vol 31. Emerald Group, Bingley, pp 291–336

Part III
Heuristics, Data Mining, & Machine Learning

Machine Learning and Landslide Assessment in a GIS Environment

Miloš Marjanović, Branislav Bajat, Biljana Abolmasov, and Miloš Kovačević

Introduction

The synergy of Geographic Information System (GIS) and various computational methods has stimulated regional spatial modeling in the last couple of decades. Various research fields, ranging from fundamental and applied environmental disciplines to natural hazards assessment, have benefited from such trends. Regional planning and decision-making have also indirectly become easier while the public is becoming more involved and better informed on the topic. This is reflected by an increase of interest in spatial modeling, amongst different scientific communities. Landslide assessment is one of many contemporary topics in environmental research that can benefit from advances in spatial modeling techniques.

Google's Insight for Search (covering 2004 – present) indicates a considerable ascent in an interest for keywords such as "landslide," "debris flow," "landslide hazard," and "susceptibility." The interest seems particularly high in areas affected by these hazards, especially after catastrophic landslide events that are closely followed by local and international media coverage. According to the Google News services, there have been nearly 90,000 landslide casualties worldwide in the past decade, which is more than 5% of the global natural hazard toll [1].

Growth of scientific interest in landslide topics is manifested by increasing number of academic publishing trends. Various scientific teams have shown interest in landslide-related topics since the late 1980s till present. In this period, research activities have grown exponentially resulting in 150–200 scientific articles per

M. Marjanović • B. Abolmasov
Faculty of Mining and Geology, University of Belgrade, Belgrade, Serbia

B. Bajat (✉) • M. Kovačević
Faculty of Civil Engineering, University of Belgrade, Belgrade, Serbia
e-mail: bajat@grf.bg.ac.rs

J.-C. Thill, S. Dragicevic (eds.), *GeoComputational Analysis and Modeling of Regional Systems*, Advances in Geographic Information Science,
DOI 10.1007/978-3-319-59511-5_11

year [2]. Research is mostly conducted in areas commonly affected by landslides, such as Circum Pacific Region, mountainous regions in the Alps, the Himalaya, and other volcanic and seismic areas worldwide. However, relatively few countries including Italy, USA, Canada, UK, China, France, Japan and Spain report the majority of landslide research [1–4]. Landslides are one of the most wide-spread and complex natural phenomena and therefore require a multidisciplinary approach. It is expected that the number of studies and research in this area will continue to grow [2, 5].

Some of these global trends have been our principal research motifs, and will be briefly discussed hereinafter. Particular interests are shown for regional studies due to their applicability on one hand and scientific contribution on the other. The modeling of natural hazards such as landslides poses a challenge to researchers and local planners since they are often of non-linear nature. Machine Learning (ML) techniques are growing in popularity in environmental science research because they can be integrated with GIS to solve such non-linear and nonparametric problems and do not require specific distributions or other constraints over input variables. Integration with GIS enabled ML regression and supervised classification tasks which are now essential parts of landslide susceptibility zoning and predictive semi-automated landslide mapping.

Related Work

Early works in GIS-based landslide susceptibility assessment appeared in the 1970s when GIS software and computer hardware components became more readily available [3]. Pioneering attempts involved simple solutions including heuristic and simple statistical non-predictive models. As technology progressed, these early approaches rapidly moved towards the implementation of more sophisticated mathematical and statistical models. Data that existed only in analogue form became available in digital form so that various mathematical and statistical computations became possible via GIS. These GIS computations have been recently enriched by numerous ML techniques that are particularly useful for addressing non-linear classification problems and include Decision Trees (DT), Artificial Neural Networks (ANN), Support Vector Machines (SVM) and Logistic Regression (LR).

Machine Learning in Landslide Assessment

Decision Tree (DT) algorithms are often denoted as classification data mining algorithms that reveal complex relations between data elements, i.e. inputs and outputs, unlike conventional black-box models that conceal the relation between inputs and outputs [6]. DTs expose how solution is conceptualized by using tree-like structure, which starts from the most important inputs and stems onward to the least important. Apart from hierarchy of inputs, DTs provide insight into threshold values of inputs that are critical for achieving classification. For instance, given the slope and altitude as input variables, a DT can reveal that slope is more important and that values steeper than e.g. 5° and higher than 300 m indicate potential landslides. One of the few attempts to implement Decision Tree algorithms within a landslide susceptibility framework was a study in South Korea [6] that applies this technique to data-mining of the national database of engineered slopes. The objective was to identify the most important attributes that affect slope stability. Another case study from Japan [7] handled a similar problem. Both studies revealed some important relations between causal factors and landslides, but yielded models of landslide susceptibility with accuracy of only 70%. DTs are more applicable in the Expert Systems design [8] because they give an insight into the particular conditions that may be related with the actual landslide occurrences and are likely to provide reliable landslide susceptibility maps. Unfortunately, they can be deficient in predicting, i.e. mapping new landslide occurrences. One comparative study [9] asserts that Decision Tree techniques cope with overoptimistic assessments due to model overfit even with extensive input data pre-processing. In such cases, other ML techniques, including LR and ANN are more reliable.

One of the most popular and most broadly used techniques in landslide assessment is multi-layered feed-forward ANN Multi-Layered Perceptron (neurons are processed from one layer to another), usually in combination with a back-propagation learning algorithm [10]. A description of pioneering work and a few comparative studies are worth mentioning here. Initially, problems of multi-dimensionality and non-linearity in input data were solved by using ANN black-box algorithms for prediction of a system's behavior rather than using a complex and never completely defined deterministic modeling. This approach was first experimented in the case studies in South Korea [11] where an ANN procedure, trained over a landslide susceptibility map and based on likelihood ratio technique, obtained fairly precise models. Therein, overfitting had been addressed as a serious drawback and the usage of an independent testing area (that was not included in the training stage) was suggested as a precautionary measure. As in most of other published studies [12–16], the advantages of the ANN application include independency from particular data distribution, mixing of ordinal and nominal data types, and the power to generalize i.e. to spatially/temporally predict new landslide occurrences. However, at the same time the drawbacks were recognized in GIS integration issues including time-consuming data preparation, demanding fitting/optimization of the ANN parameters, and associated optimization problems of the back-propagation learning algorithm, durable training and evaluation period. In comparison to other techniques (logistic regression, cluster analysis, fuzzy approach) applied by the same authors [12–16], ANNs can be characterized as one of the most successful techniques in landslide susceptibility assessment.

Practice of SVM in geo-spatial modeling is a more recent development. Pioneering work in application to landslide susceptibility [17, 18] include comparison of single-class vs. two-class (binary) SVM in the Hong Kong area. The authors demonstrate how the latter provided better conditions for algorithm training and testing. Another study [19] modeled only one type of landslide phenomena, i.e. debris flows, by comparing SVM and fuzzy approach. SVM approach outperformed the fuzzy method in their study and it was considered appropriate and more convenient for this kind of assessment in the area of interest (Yunnan Province, China). There have been only a few comparative approaches over individual case studies. The first concerned a case study from the Ecuadorian Andes [9] and applied Logistic Regression, Decision Trees and SVM. The author emphasized the necessity of thorough input data preparation and pointed to the overoptimistic accuracy of ML techniques, which turned out to be less efficient than Logistic Regression models. More recent comparative research [20] has given a good insight into perspectives on landslide assessment methodologies. Various modeling methods have been compared, showing that several methods including ANN and SVM can be very accurate. As a result, a host of different case studies have been encouraged by these findings. Experimenting with SVM and comparing it to the other ML or conventional methods has thus been furthered [21–26].

Logistic Regression has a longer tradition in natural hazard assessment. It was proven successful in numerous case studies [27], but has lately been broadly challenged by other ML techniques. Logistic Regression is typically discussed in comparative case studies [9, 28, 20], however there are several contributions depicting Logistic Regression in greater detail [29, 30]. Their findings confidently promote the logistic method as very reliable and very convenient in landslide assessment framework. In addition, a very interesting approach has been proposed in a Southern Norway case study [31] wherein Geographically Weighted Regression [32] variants have been applied with Global regression models (Logistic Regression and Spatial Regression). They revealed that Geographical Weighting, i.e. incorporating spatial correlation structure in regression, refines global regression models and enhances predicting performance.

Current State of the Machine Learning Implementation in GIS

Although promising, ML implementation has encountered numerous issues in environmental scenarios [33]. GIS has offered the possibility of a hyper-production of various terrain attributes at unprecedented resolutions which has caused data overload for many current hardware capacities. For instance, it is quite common to work on areas of 100 km^2 or larger in the regional landslide assessment framework with fair 10 m resolution, requiring excessively large datasets made up of millions of pixels. Each pixel is an instance with allocated spatial coordinates and additional coordinates, which can be represented by geological, geomorphological, environmental or other information usually called terrain attribute or conditioning factor. Recent GIS-driven developments in Geomorphometry introduced dozens of new parameters that describe or quantify the terrain surface or its hydrological features. Similarly, various synthetic/statistical parameters became available through Geostatistics and GIS so that the number of attached coordinates can easily reach tens of dozens. One way to resolve the "big data" issue in many fields has been through the use of cloud computing, a perspective field of computer science. It is possible to host a computing task on actual and virtual machines or on web-hosted computer clusters. At present, the integration of cloud computing with GIS software used in environmental research requires some programming skills and sophisticated hardware infrastructure.

Integration of ML in GIS is yet another issue. There are several examples of tight coupling of various ML techniques in GIS platforms. For instance, there are default modules for Multiple Regression, DTs, and ANNs (Multi-Layered Perceptron, SVM and Self Organizing Maps) in IDRISI Taiga and ENVI 4+. However, these modules are more adequate for Remote Sensing and Image Analysis tasks which limits the type and amount of the input data (multi-channeled images with 8-bit depth of a single channel are usually required). To our experience, these modules have also shown instability and are difficult to use with larger datasets. Some default modules for Multiple Regression are also available in typical commercial (ArcGIS) and open-source (SAGA GIS, Q-GIS) GIS platforms. Other ML modules are only available as add-ins such as SpatialDataModellerTools and SVMTools for ArcGIS 10+. They contain either Radial Basis Function ANN modules or SVM modules, but our impression is that the analysis is limited in terms of data type and size and it seems to be more Image Analysis oriented, i.e. adapted for specific image types of according Satellite Sensors. On the other hand, there are comprehensive solutions such as R Development Core Team [34] and MatLab that offer several different packages for ML implementation (kernlab, e1071), but require some programming skills. R has an additional advantage since it is coupled with SAGA GIS open-source platform. Some other standalone solutions such as MachineLearningOffice[1] [35] or Weka 3+ [36] are probably the most convincing solutions. They all suffer from the same issues in coupling with GIS, however each of them have strong points and weaknesses uniquely appreciated by different users, making further argument on this topic relative.

Modeling Principles

Landslide assessment stands for a structured gathering of the available information, processing/modeling with that information, and forming a judgment about it in a transient workflow [37, 38]. This workflow unfolds through stages of preprocessing, implementation or modeling, and post-processing, wherein modeling plays the crucial role. The principal ideas of landslide investigations and the general assessment of landslides revolve around several postulates [3, 5]:

[1]This open-source software can only be found on a supplementary CD of the book Kanevsky et al. 2009, which gives practical examples for using the software.

- Slope failures do not occur randomly or by chance, but rather as a result of the interplay of different conditions that are governed by different physical processes and laws;
- Landslides leave more-or-less distinct footprints (upon activation or after reasonable period of inactivity) that could be mapped in the field or remotely;
- Similar types of landslide movement may result in similar landslide footprints;
- Principle of historical recurrence of landslides implies that the landslides are likely to reoccur on the same location, once activated in the past;
- Principle of Uniformitarianism (past and present are keys for the future) implies that the slope failures are more likely to occur under conditions that have led to instability in the past or in other environmentally similar locations. Therefore, the knowledge gained on landslides can be generalized and expanded to other areas where similar conditions apply;
- Implicitly, conditions that are not taken into account in the model do not affect it drastically, and those conditions that are taken in consideration do not change systematically over space and time (time/space invariant).

By relying on these postulates, it is possible to propose two types of models:

1. Landslide susceptibility models, which represent zoning of spatial probability of landslide occurrence on a relative scale, i.e. from 0 to 1 or from low to high (in this context these are rather re-interpretative than predictive probabilistic models);
2. Predictive models of landslides, which map new potential landslides (in this context these are discrete predictive models).

Machine Learning Implementation Via Classification Task

Landslide susceptibility and prediction models can be identified as ML based classification tasks that should map landslide and non-landslide instances. The classification task could be automated which leads to the supervised learning procedure. The procedure assumes that an expert is presented with a possibly small representative region (training area) with all the necessary data (terrain attributes and landslide classes). An ML algorithm subsequently uses the available training data to learn the mapping between the values of various terrain attributes[2] that are acquired for the particular area of interest and the classes. After learning the mapping rule, the algorithm applies it over the rest of the area and gives an automated prognosis of the spatial distribution of landslides or their susceptibility zones.

[2]Input 2D raster datasets are organized in the way that each grid element (pixel) represents a data instance with attached spatial reference and supplementary thematic information (geological, morphometric, environmental and their derivatives).

The corresponding learning problem could be formulated in the following way. Let $P = \{\mathbf{x}|\mathbf{x} \in R^n\}$ be the set of all possible pixels extracted from the raster representation of a given area. Each pixel is represented as an n-dimensional real vector $\mathbf{x}$, where coordinate x_i represents the value of the ith terrain attribute associated with the pixel $\mathbf{x}(\mathbf{x} = <x_1, x_2, \ldots, x_n>)$. Further, let $Z = \{z_1, z_2, \ldots z_l\}$ be the set of l disjunctive, predefined landslide classes (a multi-class case, where $l > 2$). A function $f_c:P \rightarrow Z$ is called a classification if for each $\mathbf{x} \in P$ it holds that $f_c(\mathbf{x}) = z_j$ whenever a pixel $\mathbf{x}$ belongs to the landslide class z_j. In practice, for a given area one has a limited training set of m labeled pairs $(\mathbf{x}, z)_j, j = 1, \ldots, m$ where $\mathbf{x} \in R^n$ and $z \in Z$. The machine learning approach tries to find a function f_c' which is a good approximation of a real unknown function f_c using only the examples from the training set and a specific learning method.

A very brief explanation of some of the most common classification techniques is going to be given here, while more detailed explanations of particular techniques are easily found elsewhere [35, 39, 40].

Decision Tree (Fig. 1a) is a tree-like graph in which nodes represent testing units and branches represent outcomes of the tests. In each node a test is performed over an associated attribute value of the instance that enters the node (here, instances are terrain pixels in the form of $\mathbf{x} = <x_1, x_2, \ldots, x_n>$). Instances flow through the tree until the terminal nodes (leafs) are reached, and in which classification is performed by assigning a class label z that corresponds to a particular leaf.

During the training phase, one uses the examples to build the tree in which each leaf collects the instances of the same class or at least majority of them belong to the same class. The main task is to select best discriminative attributes starting from the root node of the tree. Given a node n and the set of input instances S_{in}, one tries to select the attribute A that best separates S_{in} into subsets $S_{out}(A = v)$. Chosen attribute A must not been previously used in parent nodes. Each $S_{out}(A = v)$ contains instances with the same values v or at least the majority of them are the same, so that each $S_{out}(A = v)$ can be assigned one of the z classes. The notion of "best separation" could be measured using the entropies of resulted subsets concerning the class labels z of its members. Ideally, each subset would contain all instances of the same z class (entropy is zero; branches are terminated with leaf nodes). *Gain Ratio* measure which is based on the entropy is commonly used to test candidate attributes for the given node n (higher the value of GR, better the attribute). There are a lot of possible trees that classify all training examples correctly, but learning procedures prefer less complex (more general) solutions.

The classification mapping f_c' is reached by converting tree paths into sets of equivalent rules (usually simplified/pruned) since one could interpret each class z as a disjunction of conjunctions of constraints over attribute values down the tree path.

ANN are parallel processing devices that mimic the behavior of biological neurons operating in a living brain. Multilayered Perceptron (MLP) is a type of ANN in which all neurons are organized into at least two processing levels: output layer and minimum one hidden layer (Fig. 1b). Inputs to the network represent attribute values (x_i from vector $\mathbf{x}$) and each input is connected to all neurons from the first hidden layer. While neurons in the same layer are not connected, inter

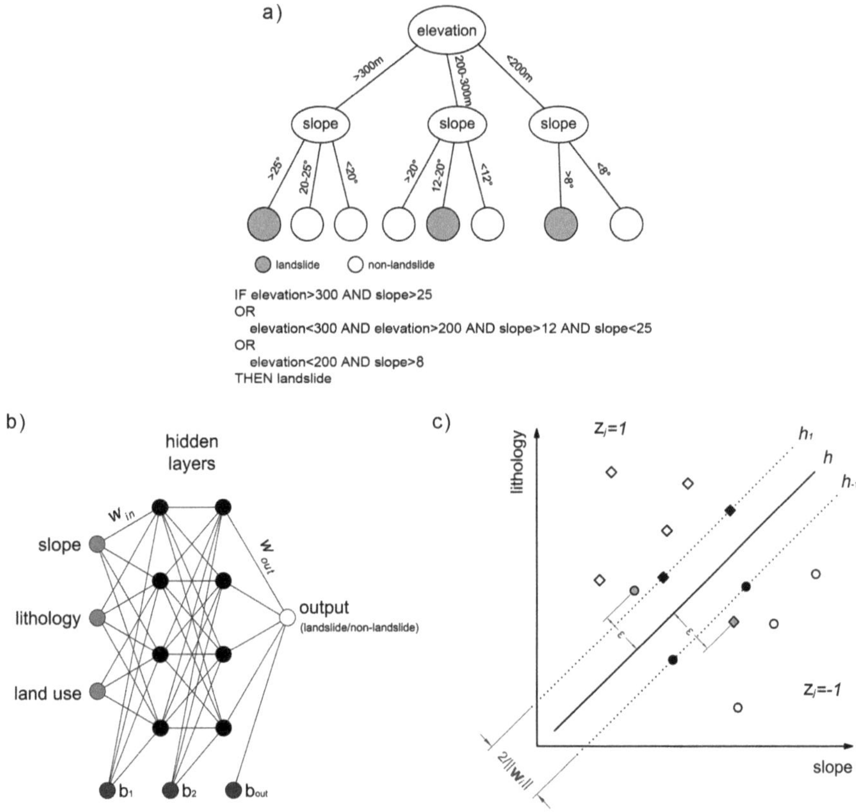

Fig. 1 Machine learning classifiers: (**a**) Example of decision tree; (**b**) Example of MLP; (**c**) Binary SVM classification example (h - hyper plane, $h_{1,-1}$ - margins, circles are landslides and squares are non-landslides, bold instances are Support Vectors)

level connections are organized in a following way: from each neuron in a non-output layer there exists a connection to each neuron in the next layer. Connections are attributed by real weights that model, to some extent, their significance to the outcome of the network. A neuron represents a processing unit that calculates the weighted sum of its inputs and transforms it to the output value using some predefined sigmoid-shaped transfer function (e.g. logistic). For the purpose of a binary classification (Fig. 1b) the output layer consists of a single neuron which outputs zero or one depending on the class z of a landslide instance $\mathbf{x}$.

In the training phase, a network is presented with the training examples $(\mathbf{x}, z)_j$, $j = 1, \ldots, m$ and the weights on connections are updated in each iteration in order to decrease the difference between the outputted value $f_{\text{network}}(\mathbf{x}_j)$ and the desired value z_j. MLP is commonly trained using a back-propagation algorithm which performs a gradient descent over the surface of the error function $f_e = f_e(\mathbf{w})$. The initial vector of connection weights ($\mathbf{w}$) is chosen randomly and f_e could be a mean squared error function or a cross-entropy function. MLP is very effective classifier, but when compared to decision trees it lacks in human readability and interpretability of the produced model.

SVM algorithm is a binary classifier (Fig. 1c) that constructs a classification boundary by using a simple linear function as a separation hyperplane between landslide and non-landslide instances in the attributes space ($f_c' = \mathrm{sgn}(\mathbf{wx} + b)$). Many possible separation hyperplanes exist, but the idea of the learning process is to find the one in the middle of the widest margin between the examples of the two classes. Further, it allows some examples to lie on the wrong side of the hyperplane (gray examples from Fig. 1c) while widening the margin. The trade-off between the margin width (=2/norm($\mathbf{w}$)) and the number of incorrectly classified examples is controlled by a real parameter C. The solution for the weight vector $\mathbf{w}$ is a linear combination of some training points called support vectors and the classification function transforms into:

$$f_c'(\mathbf{x}) = \mathrm{sgn} \sum_{i=1}^{m} \alpha_i z_i \left(\mathbf{x} \cdot \mathbf{x}_i\right) + b \tag{1}$$

Real coefficients α_i are found in the training phase. Since the real problems produce linearly non-separable classes, SVM learning assumes mapping of an original attribute space into a high dimensional *feature* space where points become linearly separable (φ:$\mathbf{x} \rightarrow \varphi(\mathbf{x})$). After mapping, the classification function becomes:

$$f_c'(\mathbf{x}) = \mathrm{sgn} \sum_{i=1}^{m} \alpha_i z_i \left(\varphi\left(\mathbf{x}\right) \cdot \varphi\left(\mathbf{x}_i\right)\right) + b \tag{2}$$

However, one does not have to know the mapping function $\varphi(\mathbf{x})$, and the dot product from (Eq. [2]) could be calculated using a special family of functions called *kernels* in the attribute space: $k(\mathbf{x}, \mathbf{x}_i) = \varphi(\mathbf{x})\cdot\varphi(\mathbf{x}_i)$. Kernels enable SVM to learn complex separating surfaces in the attribute space which transform to linear hyperplanes in the corresponding feature space. Gaussian Radial Basis Function exp.$(-\gamma||\mathbf{x}$-$\mathbf{x}_i||)^2$ is one of the most common kernel function. It is controlled by the kernel width parameter γ. Like MLP, SVM models cannot be interpreted in a human readable manner.

Sampling Strategy

Selecting the training area is a very delicate procedure and requires particular strategies. The optimal approach is to build a sufficiently accurate model with a smaller number of training examples, leading to a reduced engagement of the expert, less demand on the hardware, and provides for quicker modeling. This is usually done by sampling 50–70% of all instances randomly throughout the area, for training. The remaining part is reserved for testing the model. On the other hand, there are predictive models that require a more meaningful sampling, thereby entailing that the training data are spatially constrained by manual intervention of an expert (Fig. 3b). The training area sampling is particularly important because of the risk of overfitting. In overfitting, scenario models that are usually too complex tend to decrease in training misclassification errors at the expense of increasing the testing error. Thus, one is forced to trade-off the model's complexity for its fitting; i.e. the model's variance against its bias. However, it is also possible to take precautionary measures to suppress overfitting such that the aforementioned trade-offs do not affect the modeling choice overly. There are a number of ways reported to partly reduce overfitting in models [35, 39].

One solution is to train the algorithm through the k-fold Cross-Validation (CV). It is based on repetitive learning and validation over the training split. In k-fold CV, k stands for the number of partitions of the training split and therefore also represents the number of iterations. In the first run, one partition is taken for validation while $k-1$ partitions are merged together for learning. A different partition takes the validation role for each subsequent iteration, while the remaining $k-1$ partitions take over the learning role until all k iterations are finished. In turn, the procedure yields a result for one configuration/combination of the algorithm parameters. The CV needs to be repeated for each parameter configuration if one seeks the optimal parameter combination to give the best generalization. It is therefore preferable that the algorithm does not have too many parameters to optimize.

Another method is to generate training and testing splits that have balanced class sizes, and equivalent proportions of instances in the split (i.e. #landslide = #non-landslide). The latter is not always feasible in spatial modeling due to the usual abundance of one class and scarcity of another or several other classes (as in the case of landslide assessment, where non-landslide class is usually much larger). This is especially pronounced if the adopted training/testing sampling strategy is spatially constrained by an additional expert's criterion, as mentioned before.

It is also possible to be more exclusive about the input data (terrain attributes) so that the sampling noise can be removed at its source. For instance, all auto-correlated terrain attributes can be removed because they are bringing redundant information into the model. Further, an attribute selection based on *Information Gain* [39] for example, might be performed. The modeling can then be performed iteratively so that one terrain attribute with the lowest rank is removed from the input dataset (leave-last-out) after each iteration. The attribute removal is meaningful and justified if the error threshold decreases or remains the same while the number of inputs decreases.

Performance Evaluation

Finally, it is necessary to evaluate the model objectively so that it can be critically assessed and compared with other models. There are numerous evaluation parameters that are based on a contingency table, accounting for false positives and negatives. Receiver Operating Characteristics (ROC) [41] represents one such evaluation metric that depicts relative trade-offs between benefits and costs, i.e. evaluating true positive rate (tp_{rate} or hit rate) and false positive rate (fp_{rate} or false alarm rate). These are the coordinates of a 2D plot defined as a ROC space. The ROC curves are functions in that space, given that their contingency table parameters True Positive Rate and False Positive Rate match ROC space coordinates for a given probability threshold [42]. Analyzing ROC curves offer additional benefits. Firstly, there is commonly used Area Under Curve (*AUC*) parameter for evaluation, but there is also a possibility to describe the performance of the model qualitatively based on the curve shape and placement in the ROC space. For instance, the model with random performance will match the diagonal of the ROC plot area whereas conservative and liberal models will have skewed curves toward the lower-left sector and upper right sector of the ROC plot, respectively. Such qualitative descriptions help in choosing among models with similar *AUC* values.

Apart from considering *AUC* and the qualitative characteristics of the curve, it is also important to adapt to the particular modeling framework. In landslide assessment for instance, conservative models are preferred as they are on the safe side. Accordingly, the trade-off between false negatives and false positives is obvious as the tolerance to false negatives must be minimal and false positives are even desired. The false negative rate (fn_{rate}) is therefore another important evaluation parameter that needs to be considered [43].

Promising attempts towards even more specific and more objective evaluation schemes have been reported [44, 45], but these approaches still need to be tested in the landslide assessment framework.

Practical Example: Halenkovice Case Study

A practical example of a landslide assessment in the Halenkovice study area in Czech Republic (Fig. 2) is herein provided in order to illustrate how GIS and ML are used to address the presented problem. Both landslide susceptibility and predictive landslide mapping types of the models are going to be discussed. The objective was to challenge ML techniques, SVM in particular, to see how they perform in landslide assessment and to experiment with different scale/resolution of training data.

The area is situated near the Halenkovice Plateau in the Outer Western Carpathians in SE Moravia (Czech Republic) and extends over roughly 60 km^2. It is composed of Mesozoic and Tertiary flysch rock formations and it is segmented locally by Paleogene basins and grabens with typical marine and locally lacustrine evolution. Predominant rock types of the flysch formations are stratified sandstones, alternating with conglomerates. These are inter-layered by thin segments of clay-slates. These units differ in thickness, hydrogeological function, and mechanical

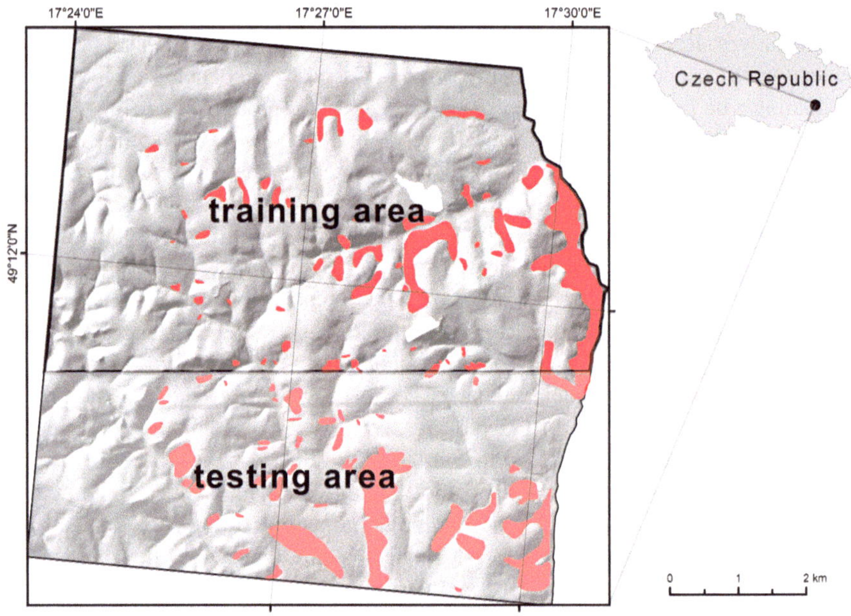

Fig. 2 Location of the study area and example of a manual training/testing splitting (*red* polygons represent the landslide of earth-slide type)

characteristics, which entails occurrence of very different types of slope instabilities. In addition, the eluvial and delluvial soil mantle hosts shallow landslides, especially when it locally thickens to a couple of meters [46, 47].

Shallow earth-slides [48] are the predominant type of slope processes in these terrains, while earth-flows and rock-falls are not as common. They are all triggered by rainfall/snow thaw in combination with the undercutting linear erosion [49, 50]. The area is sparsely populated, thus landslides of such typology and slow displacement rates do not pose a particular threat to the population.

Data

Various terrain attributes have been acquired from different resources. These include thematic terrain attributes, i.e. geological, geomorphological and environmental (Table 1). These attributes are commonly considered significant for landslide assessment. The data have been rasterized and separated in two sets with different resolutions of 10 and 30 m, respectively. Apart from terrain attributes, a landslide inventory has also been compiled at these two resolutions. Only the characteristic earth-slide landslide types are taken into account since they dominate over other

Table 1 List of used terrain attributes ranked by their *Information Gain*

Rank	Terrain attribute	Type, source, resolution/scale[a]	*IG*
1	Channel network base elevations	Morphometric, DTM, 10 + 30 m	0.28775
2	Digital terrain model (DTM)	Morphometric, topographic map, 1:10,000	0.26626
3	Ls-factor	Morphometric, DTM, 10 + 30 m	0.18673
4	Slope	Morphometric, DTM, 10 + 30 m	0.16417
5	Convergence index	Morphometric, DTM, 10 + 30 m	0.08129
6	Land use = arable land	Environmental, orthophoto, 50 cm	0.06773
7	Aspect	Morphometric, DTM, 10 + 30 m	0.05638
8	Elevation above channel network	Morphometric, DTM, 10 + 30 m	0.04708
9	Geology = loess	Geological, geological map, 1:50,000	0.04153
10	Channel network buffer	Morphometric, DTM, 10 + 30 m	0.04139
11	Geology = Solaň subunit	Geological, geological map, 1:50,000	0.02930
12	Land use = sparsely forested areas	Environmental, orthophoto, 50 cm	0.02927
13	Plan curfature	Morphometric, DTM, 10 + 30 m	0.02722
14	Geology = delluvium	Geological, geological map, 1:50,000	0.02660
15	Topographic wetness index	Morphometric, DTM, 10 + 30 m	0.02274
16	Slope length	Morphometric, DTM, 10 + 30 m	0.01917
17	Geology = Zlín subunit	Geological, geological map, 1:50,000	0.01228
18	Land use = forest	Environmental, orthophoto, 50 cm	0.01227
19	Land use = built-up area	Environmental, orthophoto, 50 cm	0.00986
20	Land use = grasslands	Environmental, orthophoto, 50 cm	0.00976
21	Orthophoto PC ratio 32	3rd vs. 2nd Princ. Comp, orthophoto, 50 cm	0.00871
22	Geology = Belovež subunit	Geological, geological map, 1:50,000	0.00695
23	Profile curvature	Morphometric, DTM, 10 + 30 m	0.00678
24	Geology = alluvium	Geological, geological map, 1:50,000	0.00348
25	Land use = orchards and gardens	Environmental, orthophoto, 50 cm	0.00009
26	Land use = water body	Environmental, orthophoto, 50 cm	0.00001

[a]Landslide inventory has been acquired from the 1:10,000 landslide map of Czech Republic

landslide types. Thus, the inventory has been cleaned from the flow-like landslides and rockfall that might compromise the modeling procedure since their behavior is completely different and other terrain attributes might apply.

Data have been preprocessed to suit the SVM training requirements. Numerical data have been normalized and categorical data have been binarized into an according number of dummy variables (binary attributes). For instance, the geological units attribute with six categories was segregated into six binary attributes, whereby each category (each geological unit) represented one new binary terrain attribute in the input dataset. In this way, the categorical data was quantified and fed to the algorithm without any subjective intervention, e.g. scoring/weighting. A side-effect of this processing is that the number of input attributes increased.

There were a few differences in sampling strategy for splitting the training and testing areas since both the susceptibility model and the predictive model types were regarded. Training instances were balanced and randomly sampled for the first model type, while more specific sampling strategies were applied for the second model type (Figs. 2 and 3).

Two processing procedures have been performed over the input data to reduce noise and redundancy in sampling. Firstly, the attribute selection based on the I*nformation Gain* (*IG*) parameter [39] has been performed and the attributes have been ranked accordingly (Table 1). Ranking turned out exactly the same in both 10 m and 30 m datasets. A leave-last-out strategy based on *IG* ranking was then applied. It was proven that the accuracy drops as the last ranked attributes have been removed successively (Fig. 4). Naturally, the accuracy gradually drops from initial 86% as attributes are removed, but at attribute *geology* = *delluvium* (ranked 14) it rises again and reaches nearly 85%. It was therefore justified to remove all of the instances between 15th and 26th rank (Table 1). This procedure demonstrates that some reduction is plausible, but the ranking method is not too reliable, because the threshold estimation with leave-last-out strategy is not very practical (full experiments had to be completed for each leave-last-out step). The strategy to wait for the rise in accuracy was successful in this particular case, although it might not be a good general rule. Second preprocessing procedure included cross-correlation test. Terrain attributes have been tested for autocorrelation against the given landslide inventory, while the Variance Inflation Factor (*VIF*) [51] was examined as an indicator of multicollinearity between predictors (terrain attributes themselves). Two attributes were problematic in this respect because the *VIF* values for *slope* (16.463) and *ls-factor* (14.254) indicate their strong multicolinearity, since for all attributes with *VIF* value greater than 5, multicolinearity is considered to be high. These two attributes have thus been removed from the dataset. However, some experiments have been performed with these attributes included to test the procedure. The results showed no significant change in the model's performance before and after removal of these attributes.

Results

We have chosen the SVM algorithm for the proposed modeling schemes using the two equally preprocessed datasets with 30 and 10 m resolution. Our intention here was somewhat authentic since there were no references on experimenting with different data resolutions before. As indicated before, we structure our findings in respect to two different model types, i.e. landslide susceptibility models and landslide prediction models.

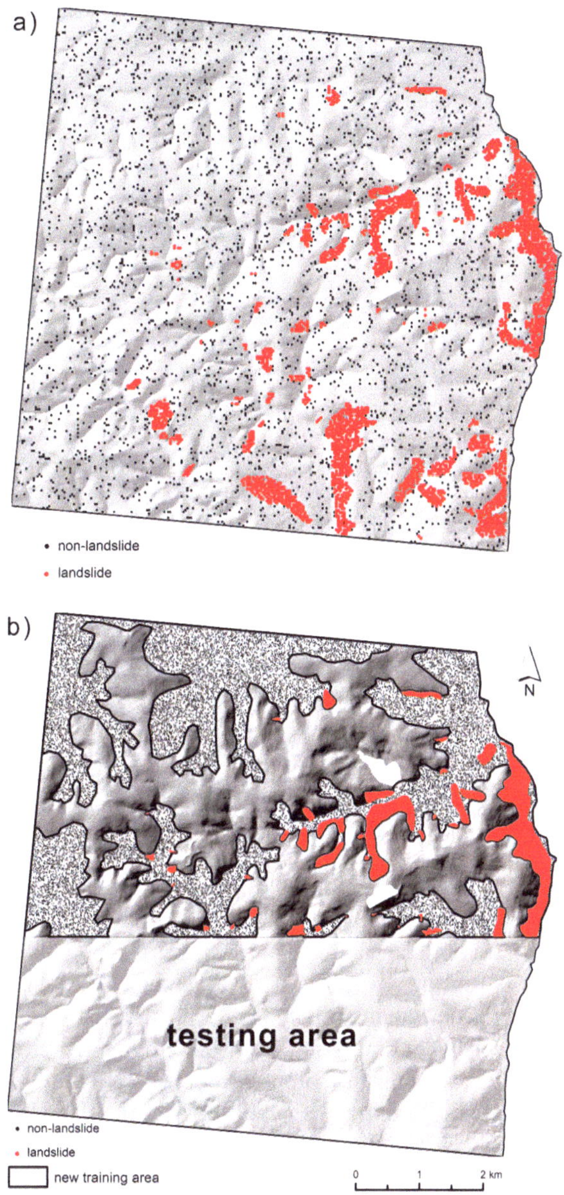

Fig. 3 Sampling strategies for training: (**a**) balanced random sampling for landslide susceptibility mapping; (**b**) balanced sampling manually enhanced by choosing the most instructive non-landslide area (wherein non-landslide instances are selected randomly)

Landslide Susceptibility Models

The landslide susceptibility models are based on averaging of intermediate models, generated after each iteration. They depict distribution of susceptibility zones throughout the spatial extents of the study area. There are usually five High-Low

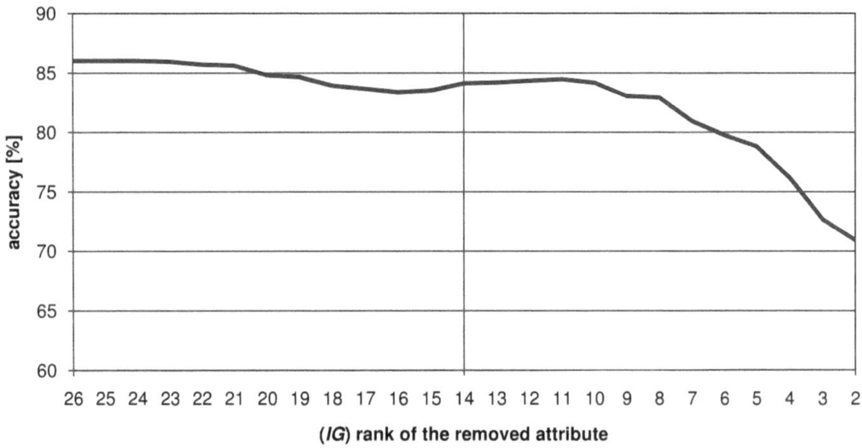

Fig. 4 Leave-last-out diagram (the attributes with the lowest rank have been successively removed from training and the model accuracy has been used to examine the change in model's performance; the attribute's *IG* rank is given on the horizontal axis, see Table 1 for details)

or Very High-Very Low classes of susceptibility, a standard we encountered in authoritative references [3, 5, 37, 48]. Very High and High susceptibility classes are usually used for cross-correlating the model with the existing landslide inventory, which enables evaluation of the model's performance. These two classes can be considered as matching equivalents to e.g. active, dormant or suspended landslides in the inventory. Measuring how much they differ gives the model's performance.

Models from 10 and 30 m sets have been optimized independently. Several C,γ configurations, based on our previous experience with similar case studies [22, 23], have been paired for tenfold CV. First, the general parameter ranges were narrowed down to more plausible combinations (Fig. 5). Afterwards, fine tuning determined the most optimal parameters: for 10 m case $C = 100$ and $\gamma = 30$, while for 30 m case $C = 1$ and $\gamma = 0.5$. The latter case notably lessens the level of penalties (smaller the C, wider the margin and better the generalization), while the kernel width (the need for hyper-dimensioning of the original feature space) is also relatively small, indicating that the algorithm generalizes well with 30 m data.

The first landslide susceptibility model with 10 m data was trained iteratively on the randomly sampled balanced sets. Balanced sets require approximately equal amounts of training classes (landslides and non-landslides) and have proven essential to avoid overfitting. There were 57,792 training instances (i.e. 28,895 of landslide and 28,895 of non-landslide instances) that correspond to 10% of the total number of instances (577931) and make a reasonably small training sample. After a series of 10 iterations, a new randomization of training instances took place so that the final model was achieved by arithmetic averaging. In accordance with the mentioned relation between the susceptibility classes and the landslide inventory, the evaluation regarded only Very High susceptibility class that was compared

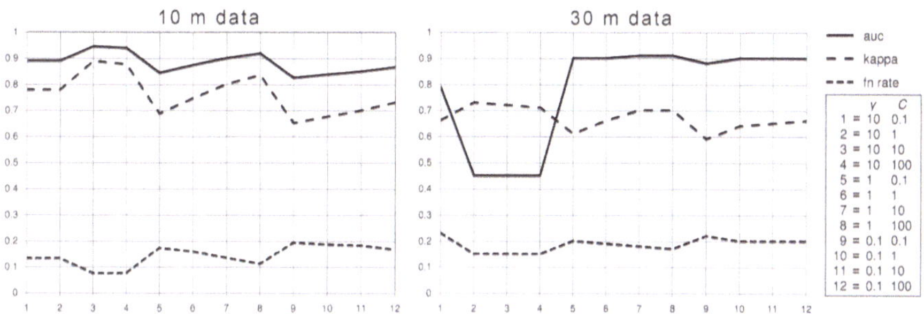

Fig. 5 Optimization of the SVM γ,C parameters on 10 m (*left*) and 30 m data (*right*); note that optimal parameters have been chosen on the basis of three different performance metrics; the most favorable γ, C ranges have been further used for fine tuning and optimal parameter selection

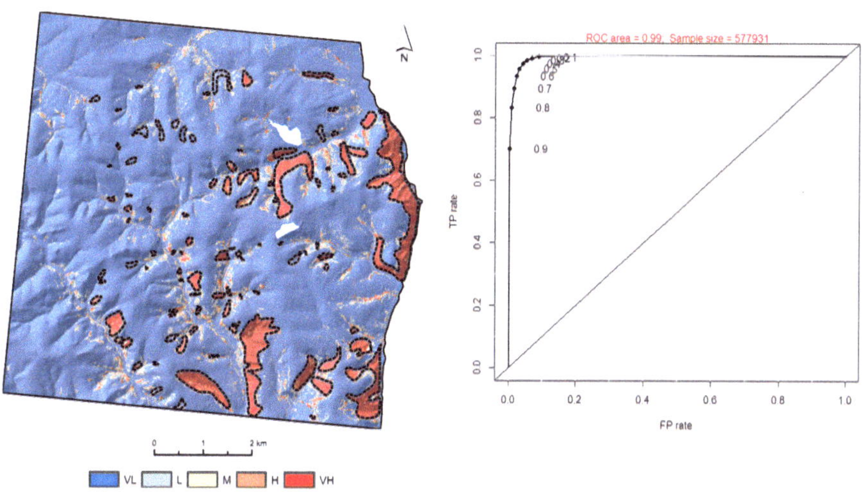

Fig. 6 Landslide susceptibility model built on 10 m data (*left*) and its performance in the ROC space (*right*); *VL* very low, *L* low, *M* medium, *H* high, *VH* very high susceptibility; *dashed contours* represent actual landslides from the inventory

versus unified active and suspended landslides from the inventory. The resulting model (Fig. 6) maps landslide susceptibility exceptionally well.

The second susceptibility model with 30 m data was also trained on randomly sampled balanced sets. Due to the increase of pixel size, the number of instances in the training set has dropped, but the 10% proportion of the training sample size has been preserved. In effect, 6406 of 64,094 instances have entered the training. The performance is slightly poorer than the 10 m model but still exceptional with an *AUC* of 0.96 and more importantly, low false negatives (Fig. 7).

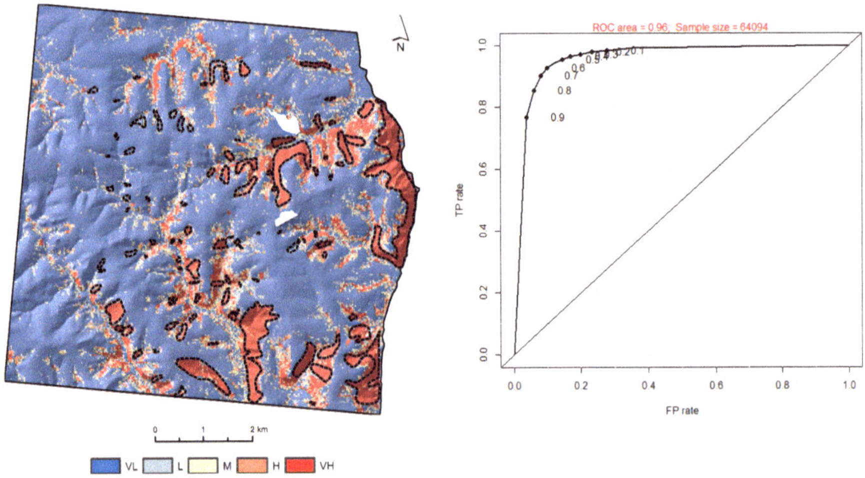

Fig. 7 Landslide susceptibility model built on 30 m data (*left*) and its performance in the ROC space (*right*); *VL* very low, *L* low, *M* medium, *H* high, *VH* very high susceptibility; *dashed contours* represent actual landslides from the inventory

Landslide Prediction Models

These models depict landslides that are predicted outside the spatial extent of the training inventory, and do not involve zoning. Instead, they predict landslides labeled the same way as they were labeled in the training inventory.

Without further optimization, the same C,γ pairs have been applied for predictive modeling (for 10 m data $C = 100$ and $\gamma = 30$, while for 30 m data $C = 1$ and $\gamma = 0.5$). The crucial task was to select the training and the testing area. Led by some earlier experiences [22, 23], we decided to reserve (upper) two-thirds of the entire set for training and the remaining (lower) third for testing (Fig. 2). Separation of these two areas had to be done carefully, because both training and testing parts must contain the same classes of categorical attributes, such as geology, land use. The training area had some further enhancements in sampling strategy (Fig. 3b). A new training area was manually delineated inside the roughly outlined two-third training area. It now included all of the available landslide instances as well as non-landslide instances that are theoretically most appropriate. For instance, all the areas above the level of existing landslide scarps have been disputed due to their potential disturbance in the future (general upslope progression tendency of landslides) and should not be considered as appropriate non-landslide instances. This is in accordance with what is known as Main Scarp Upper Edge method [52]. Non-landslide instances have been randomly sampled across this new training area and their number was balanced to the number of landslide instances. The SVM algorithm was then challenged to make a landslide prediction over the testing area.

The first results for both cases (10 m and 30 m data) were discouraging (Fig. 8), but we noticed slightly better generalization in the 30 m case (Table 2). It is possible

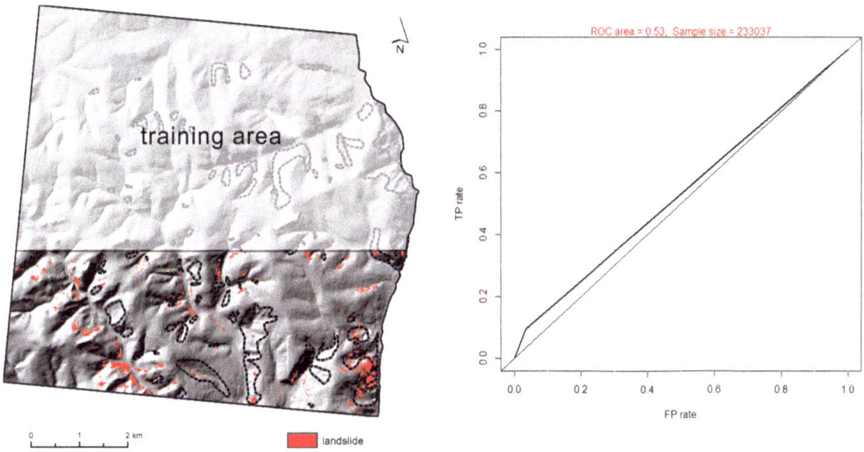

Fig. 8 Landslide prediction model for 10 m data (*dashed contours* represent actual landslides from the inventory) and its performance evaluation

Table 2 Evaluation of predictive models

SVM model variant and its optimal parameters	*AUC*	Kappa index	fn_{rate}
Trained on 10 m tested on 10 m ($\gamma = 30, C = 100$)	0.540	0.097	0.799
Trained on 30 m tested on 30 m ($\gamma = 0.5, C = 1$)	0.671	0.221	0.775
Trained on 10 m tested on 30 m ($\gamma = 30, C = 100$)	0.541	0.091	0.807
Trained on 30 m tested on 10 m ($\gamma = 0.5, C = 1$)	0.696	0.227	0.767
Trained on 30 m tested on 20 m ($\gamma = 0.5, C = 1$)	0.705	0.237	0.761
Trained on 20 m tested on 10 m ($C = 0.1, \gamma = 0.1$)	0.741	0.261	0.752

that the landslide size influences the training in 10 m data and evaluation in 30 m data. Therefore, we decided to cross-scale the models, i.e. to train with 30 m data and test with 10 m data (Fig. 9). Since the initial results were encouraging, we furthered the cross-scaling by creating a 20 m dataset variant (that has been optimized by tenfold CV and optimal parameters were $C = 0.1$ and $\gamma = 0.1$) and applied the same modeling scheme. The model turned out even better (Fig. 10) because its fn_{rate} is much lower than in any other preceding model (Table 2).

Conclusion

High-quality susceptibility maps are now a reality and represent a product that future decision-making and regional planning should rely on. Predictive models are less successful, but still very perspective for further research. They tend to overestimate landslides on behalf of stable areas (lower fn_{rate}). Such conservative maps can be suggested only as a helpful background for the actual landslide mapping campaigns.

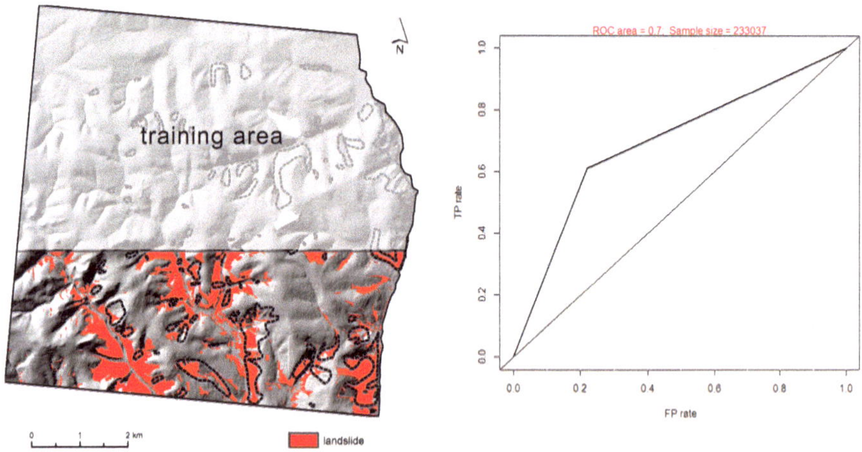

Fig. 9 Landslide prediction for a cross-scaled model trained on 30 m data and tested on 10 m data (*dashed contours* represent actual landslides from the inventory) and its performance evaluation

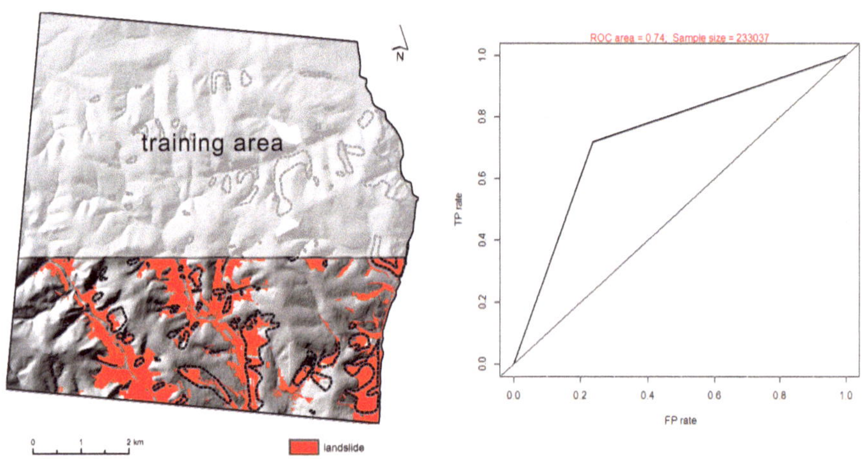

Fig. 10 Landslide prediction for a cross-scaled model trained on 20 m data and tested on 10 m data (*dashed contours* represent actual landslides from the inventory) and its performance evaluation

It is apparent that landslide susceptibility can be modeled very well by an SVM algorithm in the demonstrated example of landslide assessment in the Halenkovice area in Czech Republic (Figs. 6 and 7). However, our findings show that predictive mapping did not achieve as much success as the susceptibility models did. Results indicate that data scale/resolution and landslide size and number plays a very important role and affects the mapping quality. Our crucial findings underline that the prediction can benefit from mixing scales of training and testing datasets, thereby leading to the development of a more meaningful prediction. For instance, meaningful prediction maps the existing landslides well, but proposes locations of future landslides in currently undisturbed areas. In other words, meaningful predictions are those with few false negatives, while other performance parameters are relatively high (Table 2). Such was the case with our best model (Fig. 10) which gives logical estimates of new landslides along the valleys and does not make too many false negatives. In this context, it did not only have fair performance, but visual appeal, too.

Good attribute selection theoretically reduces computation time and improves the model performance, but the practical side of the attribute selection proposed in this work is disputable. Therefore, this issue will remain in our focus in the future. With respect to the implementation of ML, our further research (with Halenkovice and other study areas) could be redirected towards other ML techniques, wherein predictive models seem more challenging and therefore more appealing for ML implementation. We will also seek improvements in predictive models through post-processing techniques that have not been reported in this research. Finally, higher resolution—greater level of detail of input data, new resources and larger number of inputs for this and other study areas will open new challenges and perspectives that will inspire our future work. It is also possible to expect more frequent ML-based modeling examples in this field and in environmental sciences in general as integration with GIS platforms continues. At present, this integration mostly remains loose and drives researchers to use standalone products and communicate with GIS externally.

Acknowledgments This work was supported by the Ministry of Science of the Republic of Serbia (Contracts No. III 47014 and TR 36009).

References

1. Petley D (2012) Global patterns of loss of life from landslides. Geology 40(10):927–930
2. Gokceoglu C, Sezer E (2009) A statistical assessment on international landslide literature (1945–2008). Landslides 6:345–351
3. Chacón J, Irigaray C, Fernández T, El Hamdouni R (2006) Engineering geology maps: landslides and geographical information systems. Bull Eng Geol Environ 65:341–411
4. Nadim F, Kjekstad O, Peduzzi P, Herold C, Jaedicke C (2006) Global landslide and avalanche hotspots. Landslides 3:159–173
5. Guzzetti F, Mondini AC, Cardinali M, Fiorucci F, Santangelo M, Chang KT (2012) Landslide inventory maps: new tools for an old problem. Earth Sci Rev 112:42–66
6. Hwang SG, Guevarra IF, Yu BO (2009) Slope failure prediction using a decision tree: a case study of engineered slopes in South Korea. Eng Geol 104:126–134
7. Saito H, Nakayama D, Matsuyama H (2009) Comparison of landslide susceptibility based on a decision-tree model and actual landslide occurrence: the Akaishi Mountains, Japan. Geomorphology 109:108–121
8. Brus, J, Dobesova Z, Kanok J, Pechanec V (2010) Design of intelligent system in cartography. In: Proceedings of the 9th Roedunet IEEE international conference, Sibiu, Romania, 24–26 June 2010
9. Brenning A (2005) Spatial prediction models for landslide hazards: review, comparison and evaluation. Nat Hazards Earth Syst Sci 5:853–862

10. Lee S, Ryu JH, Kim LS (2007) Landslide susceptibility analysis and its verification using likelihood ratio, logistic regression, and artificial neural network models: case study of Youngin, Korea. Landslides 4:327–338
11. Lee S, Ryu JH, Won JS, Park HJ (2004) Determination and verification of weights for landslide susceptibility mapping using an artificial neural network. Eng Geol 71:289–302
12. Aleotti P, Chowdhury R (1999) Landslide hazard assessment: summary review and new perspectives. Bull Eng Geol Environ 58:21–44
13. Caniani D, Pascale S, Sdao F, Sole A (2008) Neural networks and landslide susceptibility: a case study of the urban area of Potenza. Nat Hazards 45:55–72
14. Ermini L, Catani F, Casagli N (2005) Artificial neural networks applied to landslide susceptibility assessment. Geomorphology 66:327–343
15. Kanungo DP, Arora MK, Sarkar S, Gupta RP (2006) A comparative study of conventional, ANN black box, fuzzy and combined neural and fuzzy weighting procedures for landslide susceptibility zonation in Darjeeling Himalayas. Eng Geol 85:347–366
16. Nefeslioglu HA, Gokceoglu C, Sonmez H (2008) An assessment on the use of logistic regression and artificial neural networks with different sampling strategies for the preparation of landslide susceptibility maps. Eng Geol 97:171–191
17. Yao X, Dai FC (2006) Support vector machine modeling of landslide susceptibility using GIS: a case study. In: Proceedings of the 10th IAEG conference, Nottingham, UK, 6–10 September 2006
18. Yao X, Tham LG, Dai FC (2008) Landslide susceptibility mapping based on support vector machine: a case study on natural slopes of Hong Kong, China. Geomorphology 101:572–582
19. Yuan L, Li W, Zhang Q, Zou L (2006) Debris flow hazard assessment based on support vector machine. In: IEEE International Symposium on Geoscience and Remote Sensing, IGARSS 2006, Denver, 31 July–4 Aug 2006
20. Yilmaz I (2009) Comparison of landslide susceptibility mapping methodologies for Koyulhisar, Turkey: conditional probability, logistic regression, artificial neural networks, and support vector machine. Environ Earth Sci 61(4):821–836
21. Marjanović M, Bajat B, Kovačević M (2009) Landslide susceptibility assessment with machine learning algorithms. In: Proceedings of international conference on intelligent networking and collaborative systems, INCoS 2009, Barcelona, Spain, 4–6 November 2009
22. Marjanović M, Kovačević M, Bajat B, Voženílek V (2011a) Landslide susceptibility assessment using SVM machine learning algorithm. Eng Geol 123:225–234
23. Marjanović M, Kovačević M, Bajat B, Mihalić S, Abolmasov B (2011b) Landslide assessment of the Starča basin (Croatia) using machine learning algorithms. Acta Geotech Slovenica 8(2):45–55
24. Pourghasemi HR, Jirandeh AG, Pradhan B, Xu C, Gokceoglu C (2013) Landslide susceptibility mapping using support vector machine and GIS at the Golestan Province, Iran. J Earth Syst Sci 122(2):349–369
25. Xu C, Dai F, Xu X, Hsi Lee Y (2012a) GIS-based support vector machine modeling of earthquake-triggered landslide susceptibility in the Jianjiang River watershed, China. Geomorphology 145:70–80
26. Xu C, Xu X, Dai F, Saraf AK (2012b) Comparison of different models for susceptibility mapping of earthquake triggered landslides related with the 2008 Wenchuan earthquake in China. Comput Geosci 46:317–329
27. Lai T, Dragićević S (2011) Development of an urban landslide cellular automata model: a case study of North Vancouver, Canada. Earth Sci Inf 4(2):69–80
28. Devkota KC, Regmi AD, Pourghasemi HR, Yoshida K, Pradhan B, Ryu IC, Dhital MR, Althuwaynee OF (2013) Landslide susceptibility mapping using certainty factor, index of entropy and logistic regression models in GIS and their comparison at Mugling–Narayanghat road section in Nepal Himalaya. Nat Hazards 65:135–165
29. Bai SB, Wang J, Lü GN, Zhou PG, Hou SS, Xu SN (2010) GIS-based logistic regression for landslide susceptibility mapping of the Zhoungxian segment in the three gorges area, China. Geomorphology 115:23–31

30. Falaschi F, Giacomelli F, Fedrici PR, Pucinelli A, D'Amato Avanzi G, Pochini A, Ribolini A (2009) Logistic regression versus artificial neural networks: landslide susceptibility evaluation in a sample area of the Serchio River valley, Italy. Nat Hazards 50:551–569
31. Erener A, Düzgün H (2010) Improvement of statistical landslide susceptibility mapping by using spatial and global regression methods in the case of more and Romsdal (Norway). Landslides 7(1):55–68
32. Fotheringham AS, Brunsdon C, Charlton M (2002) Geographically weighted regression: the analysis of spatially varying relationships. Wiley, Chichester
33. Voženílek V (2009) Artificial intelligence and GIS: mutual meeting and passing. In: Proceedings of international conference on intelligent networking and collaborative systems, INCoS 2009, Barcelona, Spain, 4–6 November 2009
34. Development Core Team R (2008) R: a language and environment for statistical computing. R Foundation for Statistical Computing, Vienna
35. Kanevski M, Pozdnoukhov A, Timonin V (2009) Machine learning for spatial environmental data: theory, applications and software. EPFL Press, Lausanne
36. Witten IH, Frank E, Hall MA (2011) Data mining practical machine learning tools and techniques. Elsevier, Burlington
37. Gerath R, Jakob M, Mitchell P, Van Dine D (2010) Guidelines for legislated landslide assessment for proposed residential developments in BC. Association of Professional Engineers and Geoscientists of British Columbia (APEGBC), British Columbia
38. Lee EM, Jones DKC (2004) Landslide risk assessment. Thomas Thelford, London
39. Mitchell TM (1997) Machine learning. McGraw Hill, New York
40. Quinlan JR (1993) C4.5: programs for machine learning. Morgan-Caufman, San Mateo
41. Woods KS, Bowyer KW (1997) Generating ROC curves for artificial neural networks. IEEE Trans Med Imaging 16(3):329–337
42. Fawcett T (2006) An introduction to ROC analysis. Pattern Recogn Lett 27:861–874
43. Marjanović M (2013) Comparing the performance of different landslide susceptibility models in ROC space. In: Margottini C, Canuti P, Sassa K (eds) Landslide science and practice. Landslide inventory and susceptibility and hazard zoning, vol 1. Springer, Berlin, pp 579–584
44. Dvorský J, Snášel V, Voženílek V (2010) On maps comparison methods. In: Proceedings of the international conference on computer information systems and industrial management applications, CISIM 2010, Krakow, Poland, 8–10 October 2010
45. Hagen A (2003) Fuzzy set approach to assessing similarity of categorical maps. Int J Geogr Inf Sci 17(3):235–249
46. Klimeš J, Baroň I, Pánek T, Kosačík T, Burda J, Kresta F, Hradecký J (2009) Investigation of recent catastrophic landslides in the flysch belt of Outer Western Carpathians (Czech Republic): progress towards better hazard assessment. Nat Hazards Earth Syst Sci 9:119–128
47. Marjanović M (2014) Conventional and machine learning methods for landslide assessment in GIS. Palacky University in Olomouc, Olomouc
48. Varnes DJ (1984) Landslide hazard zonation: a review of principles and practice. International Association for Engineering Geology, Paris
49. Bíl M, Müller I (2008) The origin of shallow landslides in Moravia (Czech Republic) in the spring of 2006. Geomorphology 99:246–253
50. Kircher K, Krejčí O, Máčka Z, Bíl M (2000) Slope deformations in eastern Moravia, Vsetín District (Outer Western Carpathians). Acta Univ Carol Geogr 35:133–143
51. Fox J (2008) Applied regression analysis and generalized linear models, 2nd edn. Sage, Thousand Oaks
52. Clerici A, Perego S, Tellini C, Vescovi P (2006) A GIS-based automated procedure for landslide susceptibility mapping by the conditional analysis method: the Baganza valley case study (Italian northern Apennines). Environ Geol 50:941–961

Influence of DEM Uncertainty on the Individual-Based Modeling of Dispersal Behavior: A Simple Experiment

Vincent B. Robinson

Introduction

Movement behavior has become an important topic in dispersal ecology with dispersal being central to the development of spatially explicit population models [1]. Dispersal is an important component of many vertebrate behavioral systems in that it contributes to the maintenance of a metapopulation in fragmented landscapes as well contributing to the spread of a species. In most dispersing individuals of a species, dispersal takes place before first reproduction and is termed natal dispersal (Howard [46]). It plays an essential role in the spatial dynamics of patchy populations as well as metapopulation dynamics, including population spread, recolonization [2], and gene flow [3, 4]. Landscape heterogeneity affects how animals are spatially distributed [5] as well as influencing the change a habitat patch is colonized [6]. Stevenson et al. [7] have used global positioning system (GPS) telemetry to validate modeling of gray squirrel (*Sciurus carolinensis*) movement within a fragmented landscape. It is therefore widely recognized that to understand animal dispersal it is important to consider the complex interaction between the animal's behavior and the surrounding landscape [8, 9].

The modeling of animal dispersal provides a useful paradigm for investigating the complex interactions between animal behavior and landscapes [10]. Modeling of animal dispersal behavior in relation to landscape is also viewed as a useful conceptual tool for landscape conservation planning [11–13]. Due to the difficulty in gathering and analyzing results on animal dispersal processes, simulation models have become a common, cost-effective approach to studying various aspects of dispersal dynamics [14, 15]. Simulation models with spatially explicit landscapes

V.B. Robinson (✉)
University of Toronto Mississauga, 3359 Mississauga Rd, Mississauga, ON, Canada L5L 1C6
e-mail: doc.robinson@utoronto.ca

J.-C. Thill, S. Dragicevic (eds.), *GeoComputational Analysis and Modeling of Regional Systems*, Advances in Geographic Information Science,
DOI 10.1007/978-3-319-59511-5_12

enable the integration of the relationships between species and the landscape thus providing explicit representation of the spatial elements that promote or constrain dispersal. Also, such simulations can be used to suggest habitat management strategies for focal species (e.g., [16]).

In dispersal modeling the spatial representation of the landscape is usually based on grid models where the landscape is represented by a finite number of equally sized cells [17–21]. Each cell contains one or more values, which represent attributes of the landscape such as vegetation types, land cover, or topography. Representing the landscape in this way enables flexibility in spatial analysis and mathematical modelling (Burrough and McDonnell [45]). Most importantly, it is used to formalize the concept of the perceptual range of an individual. It is through this perceptual range that an individual is able to gather information about its surroundings. Using that information the individual makes movement decisions [9, 22].

The perceptual range refers to the maximum distance from which an individual animal can perceive the presence of remote landscape elements such as habitat [22]. In a grid-based model this is usually implemented as a window of fixed size that represents the portion of the landscape that falls within the perceptual range. It is generally accepted that the perceptual range of an animal towards different landscape elements can influence its movement through heterogeneous landscapes. It has been commonly assumed in such models that animals exhibit fixed isotropic perceptual ranges that are independent of any environmental stimuli. Due to variations in environmental stimuli, such as topography, perceptual ranges should take the context into account so that the anisotropic nature of a perceptual range can be represented in an individual-based simulation model [23]. Topographic heterogeneity can be a significant source of landscape heterogeneity that influences the nature of the information about the landscape that falls within the perceptual range of an individual [24].

There are only a few studies using individual-based simulation models (IBM) integrating topography to specify anistropic perceptual ranges. Pe'er and Kramer-Schadt [24] used context-dependent and varying perceptual ranges to study hill-topping behavior in butterflies. Graf et al. [18] used an IBM that incorporated the context of topography in a mountainous landscape to study the potential dispersal behavior of capercaillie (*Tetrao urogallus*) in central Europe. Robinson [25] presented an approach based on fuzzy logic that used line-of-sight combined with landscape heterogeneity to specify a context-dependent perceptual range that integrated the effect of topography on the dispersal. His model specifications were used to simulate the dispersal of gray squirrels (*Sciurus carolinensis*) in a fragmented forest landscape [20].

The issue of uncertainty is sometimes addressed when simulating dispersal movements. Ruckelshaus et al. [26] showed through simulations that errors in dispersal parameters may have significant effects on predicted dispersal success. The uncertainty related to model parameters such as the perceptual range may be addressed by sensitivity analysis [18, 27]. Movement rules such as random walks versus correlated random walks may also be used to study variations in dispersal possibilities [28]. Given the importance of dispersal modeling and the underlying uncertainty regarding the parameters of dispersal models, Robinson [25] theorized how the logic of fuzzy spatial relations may be used to control the movement of animal objects in simulations of movement about a landscape.

Although many studies make use of geographic information systems and/or GIS-based data to provide landscape information (e.g., [18, 19, 28]), it is relatively uncommon for the uncertainty of that data to be assessed. Ruckelshaus et al. [26] assumed the landscape classification errors would be relatively minor. In their simulation study the errors in classification of habitat quality rarely produced prediction errors that exceeded 15%. However, like many other studies, their work did not use an anisotropic perceptual range that was a function of topographic variability which has until recently received little attention in modeling dispersal (Pe'er et al. [48]; [18]).

Although Pe'er et al. [48] studied the potential effects of topography on butterfly dispersal, Robinson and Graniero [20] present the only known case of modeling individual animal dispersal movements using both a fuzzy decision model and an anisotropic perceptual range in the context of a GIS-based simulation. They accomplish this by incorporating digital elevation model (DEM) data in the representation of the landscape and combining the concept of the perceptual range with line-of-sight analysis. Elevation data in a DEM are not without some level of uncertainty. This simple experiment uses the Monte Carlo simulation of DEM uncertainty methodology of Weschler and Kroll [29] to study the effect DEM uncertainty may have on the simulated movements of grey squirrels (*Sciurus carolinensis*). This study explores whether the uncertainties would lead to variations in the simulated movement behavior. Of special interest to population modeling would be whether the individuals ended in the same habitat patch or not.

Methodology

The approach taken in this study is to use Monte Carlo simulation of DEM uncertainty to generate 100 DEMs each of which represent one possible realization of the true elevation surface [29]. Then the DEMs plus land cover data is used to generate 100 landscapes. Upon each of those landscapes the dispersal movements of a small population of squirrels is simulated (Fig. 1). The results are analyzed with regards to how the movement patterns and ending habitat patch correspond to the results of the landscape using the original DEM. The ending habitat patch is the forest patch where the individual is located at the end of each simulation.

This study simulates movement of the eastern gray squirrel (*Sciurus carolinensis*) because it has desirable qualities as a modeling subject; such as an extensive knowledge base about its ecological behavior especially its perceptual abilities [22] and it may be of some conservation management interest. In some regions of North America dispersal of gray squirrels is an important issue because fragmentation of

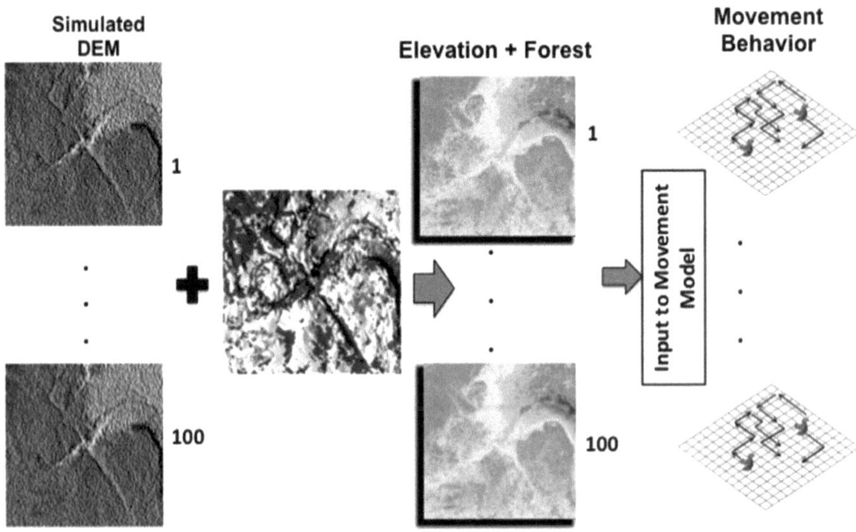

Fig. 1 An overview of the experiment construction where each simulated DEM is created using Monte Carlo simulation. One hundred realizations of a DEM are created. There is only one layer of forest height. It is added to each DEM to obtain the combined elevation plus forest heights that is an input to the dispersal model

habitat has led to a noticeable decrease in their population level [30]. In addition, the dispersal of this species can have effects on other sciurids such as the red squirrel [19, 31].

Study Area

The study area encompasses a section of the Niagara Escarpment that lies to the west of Milton, Ontario Canada with an extent of approximately 11.1 km. east/west and north/south. It is in the Mount Nemo area where Bronte Creek cuts through the escarpment (Fig. 2). The elevation varies from a maximum of 309 m above sea level to a minimum elevation of 151 m. The lower elevations to the north-northeast of the escarpment are characterized by a landscape dominated by cropland with highly fragmented relatively small forest patches. In contrast the higher elevations on the escarpment are characterized by a cropland-forest landscape where there is more area covered by larger forest patches, yet also quite fragmented. All the simulated individuals start in the lowland so that the escarpment and valleys provide potentially significant influence on their movement behavior.

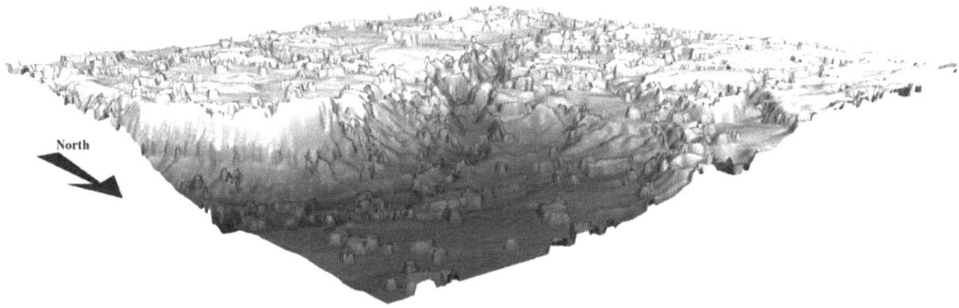

Fig. 2 A three-dimensional depiction of the terrain in the study where the elevation of the forest land cover has been added to elevation from the DEM. This provides a view of the terrain used by the model when performing line-of-sight analysis in the construction of the visible perceptual range. This study area is 11.1 km on each side. Note that the surface varies from a maximum of 324 m to a minimum of 151 m

Individual-Based Simulation Model

The Extensible Component Objects for Constructing Observable Simulation Models (ECO-COSM) [32] is used to conduct the spatially explicit individual-based simulations. An overview of the dispersal model is presented here with an emphasis only on those elements relevant to this experiment. Robinson [25] first presented the detailed conceptual discussion of the fuzzy decision model that forms the basis of this model. A subsequent implementation used to simulate dispersal of squirrels in a fragmented landscape [20] contains more details on the ECO-COSM framework and how it relates to simulating squirrel dispersal movements based on Robinson [25].

The movement behavior of an individual is modeled as a function of two decisions—a movement decision and a residence decision. When an individual is to move from its current location it assesses its surroundings to decide on a target location. Once at the target location it again evaluates its surroundings by gathering information to be used in the residence decision. If it finds the location suitable for taking up residence, it has found a home. Otherwise, it must engage in movement decision making once again. Both the movement and residence decisions are determined by an aggregation of fuzzy sets that represent relevant goals and constraints (Bellman and Zadeh [44]). Each step in the simulation is composed of both a movement and residence decision.

For the purposes of this experiment the number of steps has been limited to 15. This provides a definite end so that the individual does not wander an unrealistic number of steps. Since the normal dispersal range of this species is within just a few kilometers [33, 34] the 15 steps allows for such a dispersal range. In addition, Wolff [35] notes that the mean dispersal distance of this species is 0.5 km. Similar to Lurz et al. [19] it is assumed that an individual that has not found a home range within the set maximum has died.

The movement decision model (Table 1) constraints consist of locations within the visible perceptual range and spatially separated from conspecifics. The goal

Table 1 Movement decision sets

Equation	Description
$C^M = \Psi \cap F$	**Constraint Set** (C^M) constraining the search to those locations that are in the visible perceptual range (Ψ) and far from competing conspecifics (F)
$\Psi = P \cap L$	**Visible Perceptual Range** (Ψ) The degree to which a cell is both visible and falls within the perceptual range
$P = \mu_p(x) = \begin{cases} 1 & \text{if} \quad d_x^c \le \beta \\ \theta\left(\beta - d_x^c\right) + 1 & \text{if} \quad \beta < d_x^c < \beta + 1/\theta \\ 0 & \text{if} \quad \beta + 1/\theta \le d_x^c \end{cases}$	The fuzzy set defining the 'ideal' perceptual range for a single individual. $X = \{x\}$ is a finite set of locations bounded by the limits of the study area. d_x^c is the Euclidean distance from the location of the dispersing animal object, c, to location x. The point at which $\mu_p = 1$ is represented by β and the parameter θ controls the rate at which $\mu_p \to 0$
$L = \mu_L(x) = \max\left(\min\left(\frac{los_x^c - \alpha}{\beta - \alpha}, \frac{\gamma - los_x^c}{\gamma - \beta}\right), 0\right)$	The fuzzy set describing the degree to which location x is visible from a particular squirrel. The membership function for L is defined by a closed-form triangular function where los_x^c is the angle at which location x is visible from location c. If the local terrain creates a physical obstruction to visibility between c and x, then $L = 0$
$F(x) = \mu_F(x) = 1.0 - \left(\bigcup_{k=1}^{c} \mu_{NC}^k(x)\right)$	The fuzzy membership of each location in the set of **far_from_conspecific** where if a conspecific is within the visible perceptual range (*i.e.*, $k \in^{0+} \Psi$.) then d_i^k is the distance from conspecific k to location i
$NC^k(x; \alpha, \beta) = \mu_{NC}^k(x) = \begin{cases} \frac{\beta - d_x^k}{\beta - \alpha} & \alpha \le d_x^k \le \beta \\ 0 & \mathit{otherwise} \end{cases}$	The fuzzy set **near_conspecific** k where $\mu_{NC}^k(x)$ is the degree to which x is near conspecific k and d_x^k is the distance from conspecific k to x
$G^M = A \cap I$	**Goal Set** (G^M) degree to which a location is as near the edge of the perceptual range as possible and is forested

$A(x) = \mu_A(x) = \begin{cases} 1 & if\ \ forest \\ 0 & if\ \ nonforest \end{cases}$	**Habitat**. In the case of this species that habitat would be forest. We use the crisp classification because it is unlikely, especially towards the edge of the perceptual range, that squirrels can evaluate vegetation in any detailed manner. Once an individual has moved to a location then, through exploratory movement, an evaluation of the habitat becomes more detailed
$\mathrm{I}(x) = \mu_I(x) = \max\left(\min\left(1, \frac{d_x^c-\alpha}{\beta-\alpha}\right), 0\right)$	**Dispersal Imperative** membership function where $\alpha = 0$ and β is the distance of the farthest location in Ψ that has a non-zero membership value. Reflects the imperative of finding a home as far from the current location as possible, given constraint of perceptual range
$D^M = C^M \cap G^M$	**Decision set** on first move, movement is to location with highest value. In case of ties, the first one in the list is chosen
$B = \mu_B(x) = \left[\left(\left(\frac{\cos(q_p)+\cos(q(x))}{2}\right)^2 + \left(\frac{\sin(q_p)+\sin\left(q(q(x)\right)}{2}\right)^2\right)^{0.5}\right]^{\rho}$	The fuzzy set representing the degree to which a location falls within the set of **direction_to_move**. where q_p is the direction, in radians, of the move to the current location and $q(x)$ the direction, in radians, from the current location (κ) to location x and exponent ρ functions like a *hedge*, we assume $\rho = 2$
$D^M = (C^M \cap G^M) \cap B$	**Decision set** on subsequent moves. Movement is to location with highest value. In case of ties, the first one in the list is chosen

This table is based on Robinson and Graniero [20] where the rationale for specific parameters and function forms is discussed in detail. It is reproduced with permission from Graniero and Robinson [32]

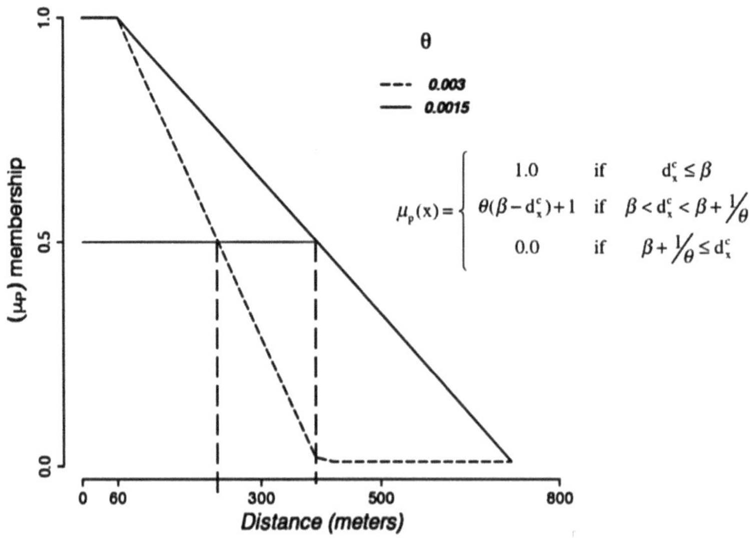

Fig. 3 This is the function that assigns a membership value to a location to define the extent of the perceptual range as a fuzzy set. The distance from the individual is dx. Out to a distance of β the membership of a location in the perceptual range is 1.0. Membership in the perceptual range declines as a function of the value of θ. The crossover point (membership = 0.5) as well the membership curve for the two values of θ used in the experiment

is to find a location as near the edge of the visible perceptual range as possible that is considered acceptable as habitat and fits within the set of constraints. An important social constraint is distance from conspecifics. The goal set is therefore a function of the spatial arrangement of habitat and the dispersal imperative. The visible perceptual range is an important influence on the movement decision since all information used as the basis of this decision is a function of it.

The visible perceptual range is modeled as an aggregation of two fuzzy sets where one determines the extent of a fuzzy perceptual range while the other represents the degree to which a location within the extent is visible to the individual. The perceptual range function assigns a membership value to a location as a function of distance from the individual (Fig. 3). It is controlled by two parameters β and θ. In this study β = 60 which means that out to 60 m from the individual membership in the perceptual range is 1.0. It is the rate at which the membership declines from β is controlled by θ. It determines how large the fuzzy extent of the perceptual range will be. In this experiment θ = {0.003, 0.0015}. At about 390 m the membership in the perceptual range has fallen to near 0.0 for θ = 0.003 while it is near 0.5 for θ = 0.0015. Both membership curves fall close to lower and upper bounds of perceptual ranges noted for this species (Mech and Zollner [47]). This is combined with a second fuzzy set that is a function of a line-of-sight analysis to provide a combined fuzzy membership for locations within a visible perceptual range (Fig. 4).

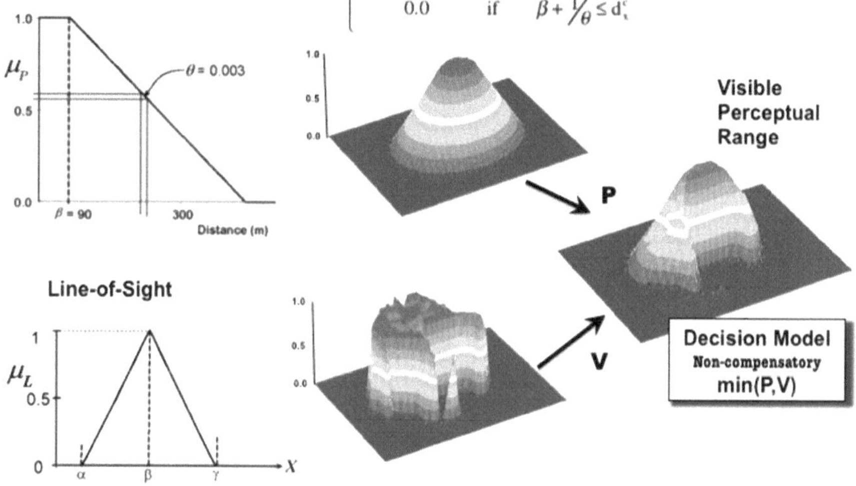

$$\mu_p(x) = \begin{cases} 1.0 & \text{if} \quad d_x^c \le \beta \\ \theta(\beta - d_x^c) + 1 & \text{if} \quad \beta < d_x^c < \beta + \frac{1}{\theta} \\ 0.0 & \text{if} \quad \beta + \frac{1}{\theta} \le d_x^c \end{cases}$$

Fig. 4 The fuzzy set of perceptual range (P) is combined with the fuzzy set of visibility (V) to determine the visible perceptual range. The visibility (V) set is a function of the line-of-sight. The fuzzy sets P and V are combined using the non-compensatory operator (i.e., min(P,V) which is the most commonly used fuzzy AND connective

Once a move is made an individual assesses whether the location is suitable for establishing a home range (Table 2). At this point the density of conspecifics is a very important parameter as territoriality affects the ease with which an individual may find a suitable location. A social fence is created when the density of conspecifics is at such a level that all suitable home range sites are already occupied. Since this experiment is concerned solely with movement behavior, the density of conspecifics is represented as a social fence. This forces movement until the maximum number of steps is reached or the individual moves out of bounds. The study area is modeled with an absorbing boundary so that when an individual reaches a zone near the edge of the study area it is considered out of bounds and the simulation for that individual ends.

Landscape Data

The landscape data used in the model is derived from both elevation and a land cover classification. The elevation data is the Canadian Digital Elevation Data (030M12) from Natural Resources Canada at a spatial resolution of 30 m [36]. The land cover data is from the Ontario Land Cover Database [37] that was produced between

Table 2 Residence decision sets

Equation	Description
$C^R = \bigcap_c \mu^c_{Far}(\kappa)$	The **Constraint Set** (C^R) is a function of the spatial separation from surrounding conspecifics
$\mu^c_{Far}(\kappa) = \begin{cases} 1.0 - \left\{ \frac{1.0}{\left[1+\frac{d_c(\kappa)-\beta^c_{Far}}{\theta^c_{Far}-\beta^c_{Far}}\right]} \right\} & if\ d_c(\kappa) \geq \beta^c_{Far} \\ 0.0 & if\ d_c(\kappa) < \beta^c_{Far} \end{cases}$	The membership of location κ in the fuzzy set **Far_from_conspecific** c where $d_c(\kappa)$ is the distance from conspecific c ($c = 1 \ldots$ k) and the current location (κ) of the Agent, β^c_{Far} represents the limit of a hypothetical core and θ^c_{Far} is the distance at which membership $= 0.5$
$G^R = LC \cap HA$	The degree to which location κ falls in the goal set G^R. In effect a measure of the degree to which the current animal location is habitat and contained within a large enough patch of habitat
$LC(\kappa) = \mu_{LC}(\kappa) = \begin{cases} 1.0\ if & oak \\ 0.9\ if & oak/deciduous\,bottomland \\ 0.75\ if & deciduous \\ 0.0\ if & conifer \\ 0.0\ if & early\,successional\,deciduous \\ 0.0\ if & wetland, pasture, grassland, ag. \\ 0.0\ if & water \end{cases}$	The degree to which a land cover type found in our GIS database can be considered quality habitat for a gray squirrel. The Agent uses the land cover at the location κ, where the squirrel has moved
$HA(\kappa) = \mu_{HA}(\kappa) = \max\left(0, \min\left(1, \left[\frac{farea(\kappa)-\alpha_{HA}}{\beta_{HA}-\alpha_{HA}}\right]\right)\right)$	The degree to which location κ falls within the class of **minimum habitat area.** By setting the parameters $\alpha_{HA} = 0.3$ and $\beta_{HA} = 2.0$ any patch less than 0.3 ha is clearly too small while any patch greater than 2 ha is clearly large enough
$D^R = G^R \cap C^R$	The membership of location κ in the **residence_location** set
IF $D^R \geq 0.5$ THEN **reside** ELSE **move**	The **decision rule** for residence versus move

This table is based on Robinson and Graniero [20] where the rationale for specific parameters and function forms is discussed in more detail. It is reproduced with permission from Graniero and Robinson [32]

1991 and 1998 with an original spatial resolution of 25 m that was converted to 30 m to match the resolution of the DEM. It is derived from digital, multispectral Landsat Thematic Mapper data. The land cover classification was performed using a supervised classification method, informed by extensive field knowledge of land cover conditions throughout Ontario. Interactive editing was used extensively to map certain classes that could not be positively identified without taking pattern and/or context into account, in addition to spectral values.

Simulating DEM Uncertainty

A methodology based on Weschler and Kroll [29] is used to simulate DEM uncertainty. They apply a stochastic approach to representing DEM error through random fields and Monte Carlo simulation by considering four random field methods—unfiltered, neighborhood autocorrelation, mean spatial dependence, and weighted spatial dependence. In this study the neighborhood autocorrelation filter method is used. The steps to generate a single realization of a DEM surface incorporating the simulated error are:

First: Generate a random field with a mean of zero and a standard deviation equal to the root mean square error (RMSE) for the DEM. In this case a RMSE of 5 m was used.

Second: Pass a mean 3×3 low-pass filter over the surface of the random field. In this manner, each cell in the random field is replaced by the mean of the value of the center cell in the filter's nine-cell window.

Third: The filtered random field is rescaled to a mean of zero and a standard deviation equal to the RMSE.

Fourth: The rescaled random field is applied to the original DEM.

One hundred simulated DEMs were generated in this manner. The heights of trees and other landscape features were then added to each DEM. The resulting landscape terrain is then used in the individual-based simulations of dispersal movement (Fig. 1).

Dispersal Simulations

All simulations were done using the same starting positions for each of 15 individuals all of whom are located in the fragmented landscape of the lowlands. The conspecific landscape consisted of a "social fence" which effectively constrains an individual to search until the maximum number of steps is reached or goes out of bounds. There were two versions of the perceptual range used. One is very conservative ($\theta = 0.003$) and the other more liberal in its definition of perceptual range ($\theta = 0.0015$). Not all individuals in both scenarios completed their movement up to the maximum number of steps (Table 3). One individual where $\theta = 0.0015$

Table 3 For each individual the total number of simulations out of the 100 realizations where the individual completed 15 steps entirely within bounds

	Total number of simulations inbounds	
Individual	θ=0.0030	θ=0.0015
1	81	0
2	100	94
3	100	92
4	100	100
5	100	100
6	100	100
7	100	23
8	100	97
9	100	100
10	100	86
11	100	100
12	100	46
13	91	30
14	98	100
15	100	100

was always out of bounds no matter which realization was used. It is clear that the case of $\theta = 0.0015$ led to more cases where individuals did not complete their movements within the bounds of the study area.

Movement Behavior

In the analysis of the resulting effects of the uncertain DEMs on the simulation of individuals' movements two aspects of their movement behavior were measured. One aspect concerned the movement behavior in terms of walk taken. The other aspect concerned where the individuals ended, especially in relation to forest patches. Of the indices of movement behavior that Almeida et al. [38] discuss, the mean displacement distance (MDD) [39] and straightness (ST) index [40] were used to provide an indication of how movement behavior might vary as a function of DEM uncertainty. In addition to comparing the simulated DEM results to baseline simulations, the degree of variation in results using the two perceptual ranges (i.e., $\theta = \{0.003, 0.0015\}$) can be assessed using ST and MDD well as the similarity of ending locations.

ST is calculated for each individual for each of the 100 realizations. It is defined as D/L where D is the Euclidean distance between the beginning and end of the walk. L is the total length of the path. Each step of an individual's simulation produces a leg in the walk. The sum of the lengths of each leg provides the value of L. The values of ST should vary little, or none at all, within an individual's set of simulations if the uncertainty in the DEM has little effect on movement behavior.

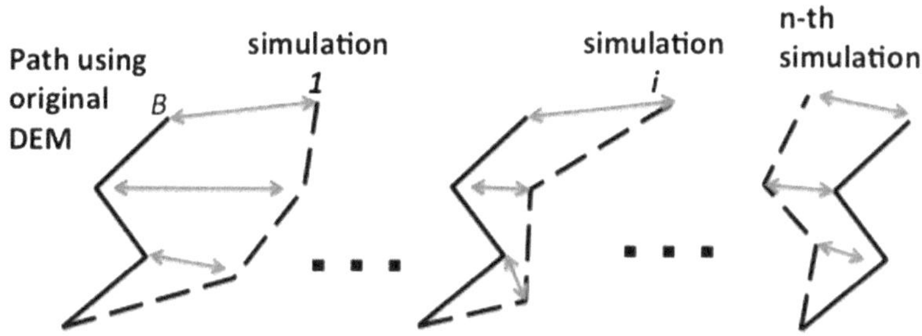

Fig. 5 Concept of how the mean displacement distance (MDD) for each individual is calculated. Each vertex on the simulated walks of an individual is compared to the corresponding vertex in the walk of the individual using the baseline simulation

The MDD of Swihart and Slade [39] is used to compare the walk taken using the original DEM versus the DEMs generated through the Monte Carlo simulation. In this case MDD is defined for each of k simulations for an individual as:

$$MDD_k = \frac{1}{n}\sum_{i}^{n}\sqrt{(x_{iB} - x_{ik})^2 + (y_{iB} - y_{ik})^2} \tag{1}$$

Where n is the number of vertices in the walk. The coordinates x_{iB}/y_{iB} are those of the point in the walk taken in the simulation based on the original DEM while x_{ik}/y_{ik} are the coordinates for the corresponding point in simulation k (Fig. 5). If there are no effects of DEM uncertainty then the difference between the baseline and uncertainty-based simulations should be zero. Since a social fence is used all individuals that do not go out of bounds will have the same number of steps.

Forest patches were defined using the eight neighbor-tracing rule rather than the four-neighbor rule [41]. This seems most appropriate given the scale of the raster layer (i.e., 30 × 30 m) in relation to the typical body size of the individuals being simulated. Each patch was assigned a unique identifier. Thus, if an individual ended at any location with the patch then it was noted as having ended its movement in that forest patch. It is therefore quite possible for individuals to end at different locations depending on the realization of a DEM being used yet still be in the same patch. If the DEM uncertainty has no effect then it would be expected that an individual would always end in the same patch. In this manner, the uncertainty would be demonstrated as having little effect on a spatially explicit population patch-based model.

Software Note

Open source software was used whenever possible to prepare the raster layers for input to the simulation model of movement and subsequent analysis. The Geographic Resources Analysis Support System (GRASS) geographic information system (GIS) [42] was used to generate the error fields and subsequent simulated DEM layers. It was also used to prepare all raster layers needed by the ECOCOSM model for simulating squirrel movement. The R package [43] was used for analyzing the results of this experiment as well as creating barplots and histograms.

Results and Discussion

There are two major aspects of movement behavior to consider. One is the pattern of movement behavior as measured by ST and MDD. ST is a basic measure of sinuosity whereas MDD in this case is a measure of deviation of the walk from the baseline. The other aspect is patch-based to determine if individuals always end in the same patch regardless of the realization of the DEM in the simulation.

Movement Behavior

The ST results show that walks in the $\theta = 0.0015$ case tend to vary more as well as deviate more often from the baseline than in the $\theta = 0.003$ case (Figs. 6 and 7). This is not entirely unexpected since the case of $\theta = 0.0015$ provides more options for movement due to the larger potential area of visible perceptual difference. For individual five with ST = 0.73 that is as close as any baseline results came to a straight-line walk (i.e., when ST = 1.0). All the results from the simulations resulted in walks with ST values less than 0.73 so that the sinuosity of the walks were always greater for the simulations. Although ST results in the $\theta = 0.003$ appear to tend towards agreement with baseline, it is still the case that search paths did deviate from the baseline results in all cases. However, individual six is notable by its lack of variation as well the tendency for the simulation results to cluster tightly within 0.005 of the baseline simulation.

It appears that the uncertainty-based simulations generally differ from the same walk as the baseline simulation (Figs. 8 and 9), some more than others. Like the ST results there appears to be more variation in values of MDD for the case of $\theta = 0.0015$. In the case of individuals like five and nine there appears a distinct concentration of results but at some deviation from the baseline walk. Although there is clearly some variation in the results for individuals 6 and 15, there is a distinct clustering of values near zero for both individuals. This indicates that their respective walks tended to be similar to the baseline walk for many of the

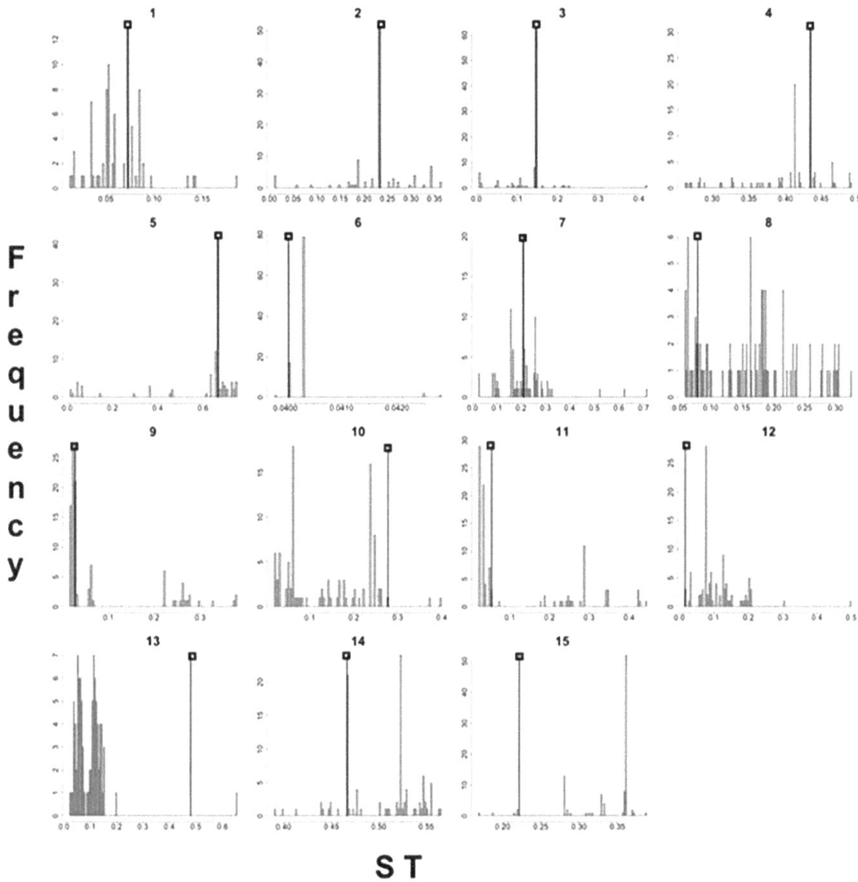

Fig. 6 Histograms showing the results of ST for each individual by perceptual range where $\theta = 0.0030$. Lines with the box at the top represent the ST for the baseline simulation

simulations. On the other hand, individuals such as two and ten illustrate cases with substantial variety in values of MDD. Individuals five and nine show a tendency for walks to cluster at values that differ quite a bit from the baseline. The case of $\theta = 0.0030$ is a function of a more limited perceptual range. This is a possible reason some individuals such as two, three, four, and five, frently have walks that are very close or the same as the baseline. In the case of individual two this contrasts sharply with the movement behavior when $\theta = 0.0015$ even though the landscape configuration is the same for each realization in both cases. This illustrates well how differences in perceptual range can affect movement behavior in spatially explicit simulations of individuals.

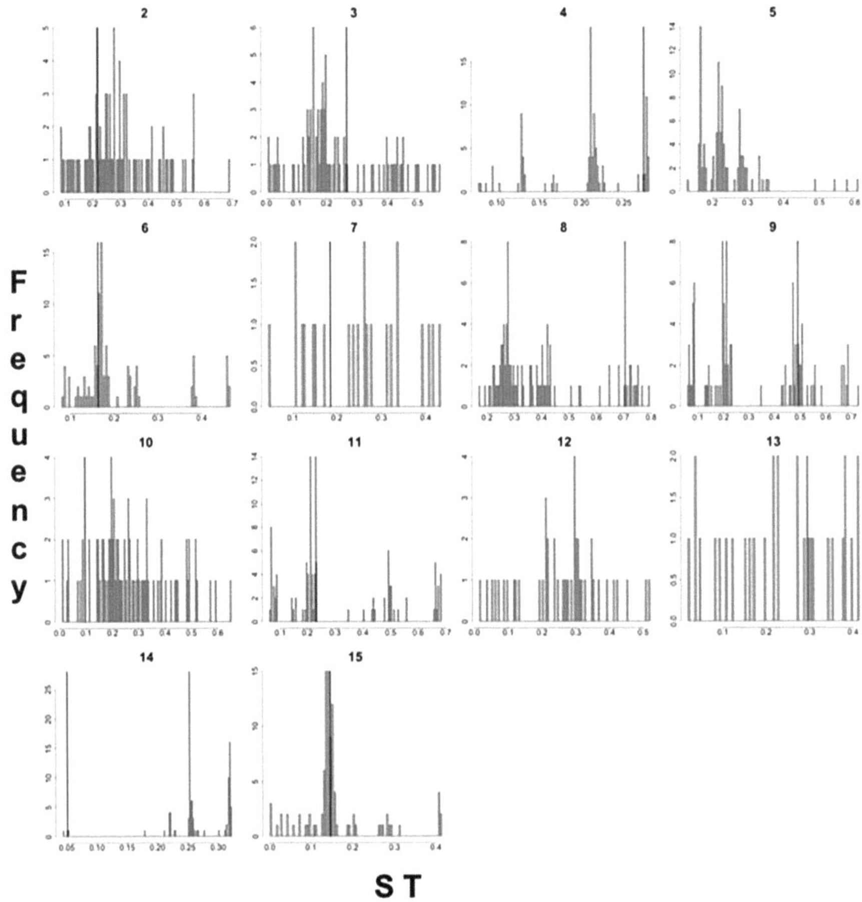

Fig. 7 Histograms showing the results of ST for each individual by perceptual range where $\theta = 0.0015$. Lines with the box at the top represent the ST for the baseline simulation. Note that only 14 individuals are shown because one individual went out of bounds in all 100 simulations

Patch Terminus

Although ST and MDD may provide an indication of how the different realizations of the DEM may affect movement behavior, what may matter more in spatially explicit population models is whether or not the individuals end in the same habitat patch as the baseline simulation. It is the patch terminus that will ultimately affect the results of a spatially explicit population model. The effects of DEM uncertainty varied greatly between the two versions of perceptual range (Fig. 10). When $\theta = 0.0030$ many individuals tended to end their walk in the same habitat patch as they did in the baseline simulation, whereas there was a noticeable tendency for them to end their walk in different a different patch when $\theta = 0.0015$. That is

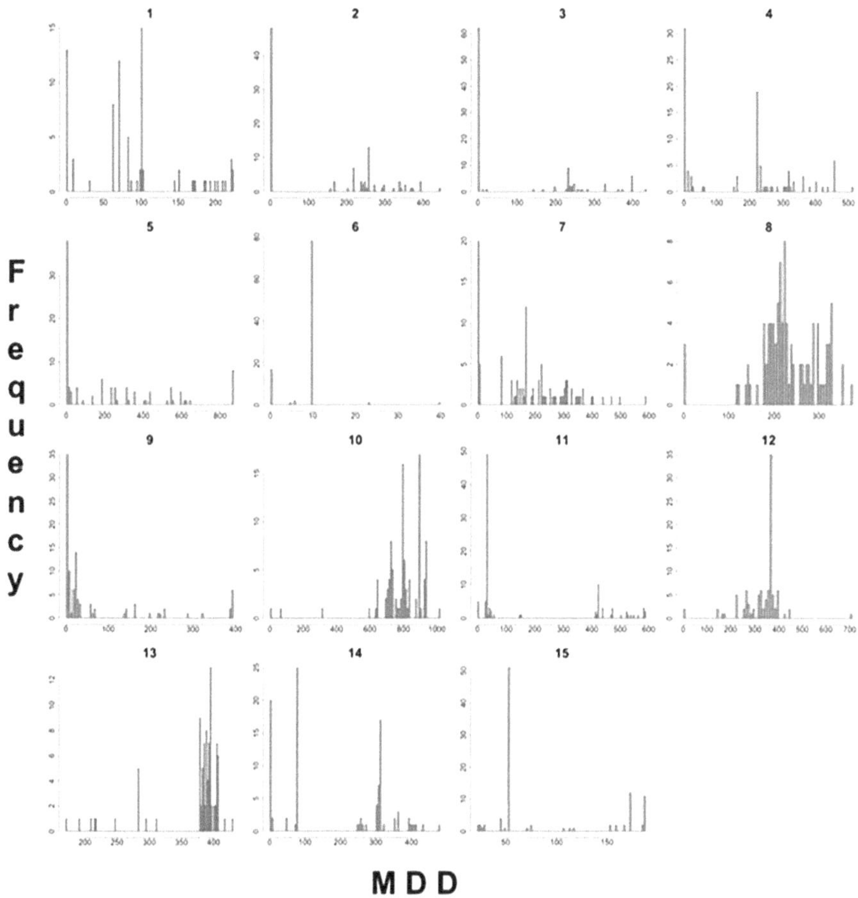

Fig. 8 Histograms showing the results of MDD for each individual by perceptual range where $\theta = 0.0030$

not to say there was no variance from the baseline results in the case of $\theta = 0.0030$. There are some individuals that did tend to have a patch terminus that differed from that in the baseline simulation. In a similar manner, there were a few individuals that tended to have the same patch terminus as in the baseline simulation.

Perceptual range defines the spatial extent of the landscape for which information is available to make movement decisions. As the extent increases so does the amount of information about the landscape that becomes part of the movement decision process. Also, as the extent increases so does the chance that an important environmental cue (i.e., forest patch) may become evident to the individual. Given the increased area and chance of new environmental cue's, it is not entirely surprising that the patch terminus of individuals showed a greater tendency to differ from the baseline terminus with $\theta = 0.0015$ in comparison to $\theta = 0.0030$. Although

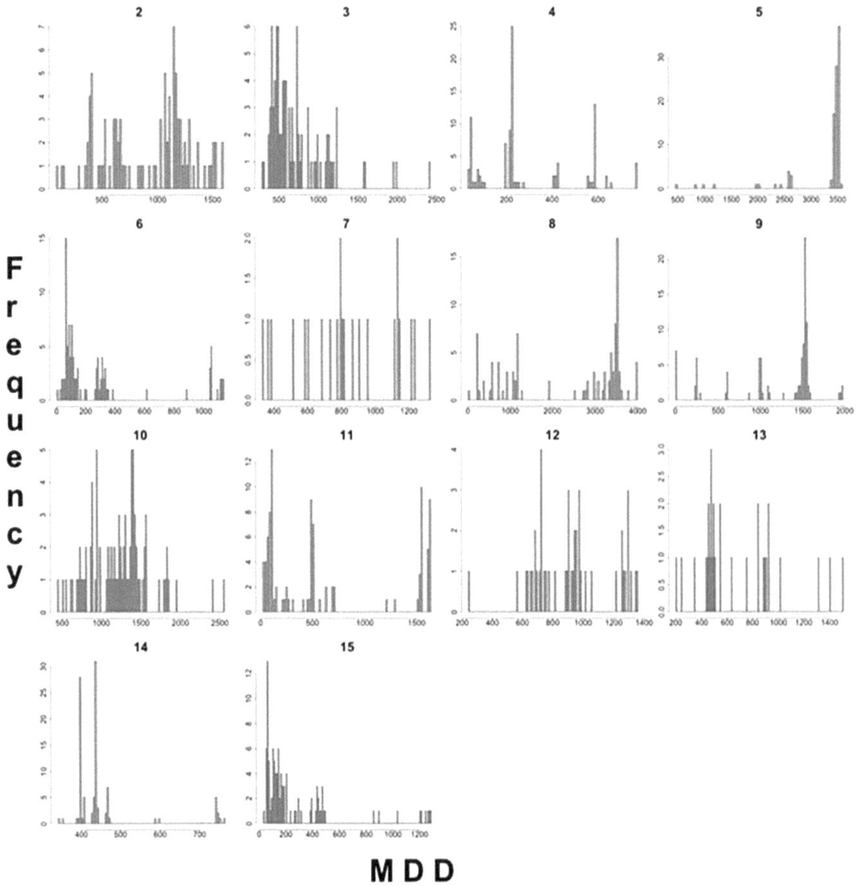

Fig. 9 Histograms showing the results of MDD for each individual by perceptual range where $\theta = 0.0015$. Note that only 14 individuals are shown for $\theta = 0.0015$ because individual one went out of bounds in all 100 simulations

this is a simple experiment with a small number of individuals, the results suggest that as the perceptual range increases, that the overall effect of DEM uncertainty on a spatially explicit population model would be somewhat greater. In one of the few studies to simulate animal movement incorporating terrain, they noted how the variation in the perceptual range extent could affect movement behavior of simulated individuals especially in relation to terrain features [18]. Although they varied the extent of the perceptual range to arrive at their observation, there was no attempt to assess how sensitive the results were to DEM uncertainty.

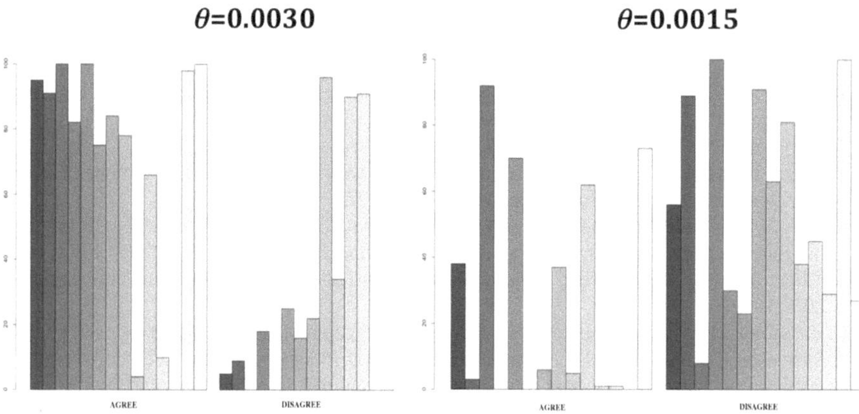

Fig. 10 Barplots showing the results of the patch terminus analysis for each perceptual range. The plots are arranged so individual results are arranged from left to right. Since individual one was out of bounds for all simulations under $\theta = 0.0015$ only individuals two through 15 for both cases are shown

Concluding Comments

It was shown that the influence of DEM uncertainty on simulations of movement behavior was affected by the spatial extent of the perceptual range. For each realization it was determined which forest patch each individual ended in and compared with the result for the baseline DEM. Significant in relation to spatially explicit population simulations is the observation that as perceptual range increased so did the variation in patch-level outcomes that were likely a result of the variations in movement behavior as measured by ST and MDD. Whether or not individuals may reach the same forest patch in each simulation holds implications for modeling metapopulation dynamics. Overall the results are consistent with others that found that the extent of the perceptual range could have an effect on simulations of movement behavior since it has an effect on landscape connectivity [16, 18, 23].

The results of this simple experiment indicate that DEM uncertainty can have an influence on the results of GIS-based simulations using an individual-based model to study the interplay between landscape and small mammal dispersal. Even with a small population of individuals it is evident that the use of a DEM is but one realization of movement behavior. Other realizations were evident based on Monte Carlo simulations of how elevation values may vary depending on DEM error. This implies that results of spatially explicit population models incorporating terrain to construct anisotropic perceptual ranges that drive individual movements may have results that may vary depending on the realization of DEM that is used.

Acknowledgments The partial support of a Discovery Grant RGPIN-386183 from the Natural Sciences and Engineering Research Council (NSERC) is gratefully acknowledged.

References

1. Hawkes C (2009) Linking movement behavior, dispersal and population processes: is individual variation a key? J Anim Ecol 78:894–906
2. Hengeveld R (1994) Small step invasion research. Trends Ecol Evol 9:339–342
3. Neigel JE, Avise JC (1993) Application of a random walk to geographic distributions of animal mitochondrial DNA variation. Genetics 135:1209–1220
4. Peakall R, Ruibal M, Lindenmayer DB (2003) Spatial autocorrelation analysis offers new insights into gene flow in the Australian bush rat, *Rattus fuscipes*. Evolution 57(5):1182–1195
5. Turner MG (1989) Landscape ecology: the effect of pattern on process. Annu Rev Ecol Syst 20:171–197
6. Hansson L (1991) Dispersal and connectivity in metapopulations. Biol J Linn Soc 42:89–103
7. Stevenson CD, Ferryman M, Nevin OT, Ramsey AD, Bailey S, Watts K (2013) Using GPS telemetry to validate least-cost modeling of gray squirrel (*Sciurus carolinensis*) movement within a fragmented landscape. Ecol Evol 3(7):2350–2361
8. Anthony LL, Blumstein DT (2000) Integrating behaviour into wildlife conservation: the multiple ways that behaviour can reduce N-e. Biol Conserv 95:303–315
9. Lima SL, Zollner PA (1996) Towards a behavioral ecology of ecological landscapes. Trends Ecol Evol 11:131–135
10. Vuilleumier S, Metzger R (2006) Animal dispersal modeling: Handling landscape features and related animal choices. Ecol Model 190:159–170
11. Kareiva P, Wennergren U (1995) Connecting landscape patterns to ecosystem and population processes. Nature 373:299–302
12. King AW, With KA (2002) Dispersal success on spatially structured landscapes: when do dispersal pattern and dispersal behavior really matter? Ecol Model 147:23–39
13. Kramer-Schadt S, Revilla E, Wiegand T, Breitenmoser U (2004) Fragmented landscapes, road mortality, and patch connectivity: modelling dispersal for the Eurasian lynx in Germany. J Appl Ecol 41:711–723
14. Tischendorf L, Fahrig L (2000) How should we measure landscape connectivity? Landsc Ecol 15:633–641
15. Wiegand T, Moloney K, Naves J, Knauer F (1999) Finding the missing link between landscape structure and population dynamics: a spatially explicit perspective. Am Nat 154:605–627
16. Alderman J, McCollin D, Hinsley SA, Bellamy PE, Picton P, Crockett R (2005) Modelling the effects of dispersal and landscape configuration on population distribution and viability in fragmented habitat. Landsc Ecol 20:857–870
17. Gardner RH, Gustafson EJ (2004) Simulating dispersal of reintroduced species within heterogenous landscapes. Ecol Model 171:339–358
18. Graf RF, Kramer-Schadt S, Fernandez N, Grimm V (2007) What you see is where you go? Modeling dispersal in mountainous landscapes. Landsc Ecol 22:853–866
19. Lurz PWW, Rushton SP, Wauters LA, Bertolino S, Currado I, Massoglio P, Shirley MDF (2001) Predicting grey squirrel expansion in North Italy: a spatially explicit modelling approach. Landsc Ecol 16:407–420
20. Robinson VB, Graniero PA (2005) Spatially explicit individual-based ecological modeling with mobile fuzzy agents. In: Petry FE, Robinson VB, Cobb MA (eds) Fuzzy modeling with spatial information for geographic problems. Springer, Heidelberg, pp 299–334
21. South A (1999) Extrapolating from individual movement behavior to population spacing patterns in a ranging mammal. Ecol Model 117:343–360
22. Zollner PA (2000) Comparing the landscape level perceptual abilities of forest sciurids in fragmented agricultural landscapes. Ecology 80:1019–1030
23. Olden JD, Schooley RL, Monroe JB, Poff NP (2004) Context-dependent perceptual ranges and their relevance to animal movements in landscapes. J Anim Ecol 73(6):1190–1194
24. Pe'er G, Kramer-Schadt S (2008) Incorporating the perceptual range of animals into connectivity models. Ecol Model 213(1):73–85

25. Robinson VB (2002) Using fuzzy spatial relations to control movement behavior of mobile objects in spatially explicit ecological models. In: Matsakis P, Sztandera LM (eds) Applying soft computing in defining spatial relations. Physica-Verlag, Heidelberg, pp 158–178
26. Ruckelshaus M, Hartway C, Kareiva P (1997) Assessing the data requirements of spatially explicit dispersal models. Conserv Biol 11:1298–1306
27. Robinson VB (2010) Exploring the sensitivity of fuzzy decision models to landscape information inputs in a spatially explicit individual-based ecological model. In: Kacprzyk J, Petry FE, Yazici A (eds) Uncertainty approaches for spatial data modeling and processing: a decision support perspective. Springer, Heidelberg, pp 29–42
28. La Morgia V, Malenotti E, Badino G, Bona F (2011) Where do we go from here? Dispersal simulations shed light on the role of landscape structure in determining animal redistribution after reintroduction. Landsc Ecol 26:969–981
29. Weschler SP, Kroll CN (2006) Quantifying DEM uncertainty and its effect on topographic parameters. Photogramm Eng Remote Sens 72(9):1081–1090
30. Nixon CM, Havera SP, Greenberg RE (1978) Distribution and abundance of the gray squirrel in Illinois. Illinois Natural History Survey Biological Notes 105, Illinois, Champaign, p 55
31. Wauters A, Lurz PW, Gurrell J (2000) Interspecific effects of grey squirrels (*Sciurus carolinensis*) on the space use and population demography of red squirrels (*Sciurus vulgaris*). Ecol Res 15:271–284
32. Graniero PA, Robinson VB (2006) A probe mechanism to couple spatially explicit agents and landscape models in an integrated modeling framework. Int J Geogr Inf Sci 20(9):965–990
33. Koprowski JL (1994) *Sciurus carolinensis*. Mamm species 480:1–9
34. Thompson DC (1978) The social system of the grey squirrel. Behaviour 64:305–328
35. Wolff JO (1999) Behavioral model systems. In: Barrett GW, Peles JD (eds) Landscape ecology of small mammals. Springer-Verlag, Heidelberg, pp 11–40
36. Natural Resources Canada (2006) Canadian Digital Elevation Data – 030M12 [computer file]. Sherbrooke, Quebec
37. Ontario Ministry of Natural Resources (2002) Ontario Land Cover Data [computer file]. Toronto, Ontario
38. Almeida PJAL, Vieira MV, Kajin M, Forero-Medina G, Cerqueira R (2010) Indices of movement behavior: conceptual background, effects of scale, and location errors. Zoologica 27(5):674–680
39. Swihart RK, Slade NA (1985) Testing for independence of observations in animal movements. Ecology 66(4):1176–1184
40. Batschelet E (1981) Circular statistics in biology. Academic, London
41. Baker WL, Cai Y (1992) The r.le programs for multiscale analysis of landscape structure using the GRASS geographical information system. Landsc Ecol 7(4):291–302
42. GRASS Development Team (2010) Geographic Resources Analysis Support System (GRASS) Software version 6.4.0RC6. Open Source Geospatial Foundation Project. http://grass.osgeo.org
43. R Development Core Team (2009) R: A language and environment for statistical computing (version 2.10.1). R Foundation for Statistical Computing, Vienna, Austria. http://www.R-project.org
44. Bellman RE, Zadeh LA (1970) Decision-making in a fuzzy environment. Manag Sci 17:141–164
45. Burrough PA, McDonnell RA (1998) Principles of geographical information systems. Oxford University Press, New York
46. Howard WE (1960) Innate and environmental dispersal of individual vertebrates. Am Midl Nat 63:152–161
47. Mech SG, Zollner PA (2002) Using body size to predict perceptual range. Oikos 98:47–52
48. Pe'er G, Heinz SK, Frank K (2006) Connectivity in heterogeneous landscapes: analyzing the effect of topography. Landsc Ecol 21:47–61

A Semi-Automated Software Framework Using GEOBIA and GIS for Delineating Oil and Well Pad Footprints in Alberta, Canada

Verda Kocabas

Introduction

An anthropogenic footprint can be defined as any disturbance on the natural landscape that is caused by human activity, such as well sites, pipelines, seismic lines and cut blocks, among others. Delineation of anthropogenic footprints such as oil and gas well pads, pipelines, and access roads play an essential role in many forms of spatial analysis on the human impact of energy industry activities. Accurate, detailed and timely mapping of these features over vast geographical areas is a major challenge in resource management and regulation. The information is crucial to government administrations, industry operations, and environmental monitoring. However, the scale and complexity of geospatial data to be acquired and analyzed can be formidable due to the required manpower, costs, and technology involved.

Among the anthropogenic footprints, this study's main focus is to delineate the disturbance of oil and gas well pads. Recent advances in remote sensing (RS) and geographic information science (GIScience) provide a potentially low-cost alternative but require the development of methods to easily and accurately extract the required information. The current method of mapping oil and gas well pad footprint requires extensive manual interpretation and it requires assumptions based on typical or average areas. This draws into question the accuracy of current products as additional clearings around the well pads cannot be captured with average size buffers around the wells. For example, Leu et al. [1] have calculated

V. Kocabas (✉)
Planet Labs Geomatics, 3528 30th St. N., Lethbridge, AB, Canada, T1H 6Z4

Southern Alberta Institute of Technology, 1301 16th Ave NW, Calgary, AB, Canada, T2M 0L4
e-mail: verdakcb@gmail.com

J.-C. Thill, S. Dragicevic (eds.), *GeoComputational Analysis and Modeling of Regional Systems*, Advances in Geographic Information Science,
DOI 10.1007/978-3-319-59511-5_13

the physical effect area and spatial extent of anthropogenic features including oil and gas pads by multiplying each feature by predetermined coefficients and by calculating well point densities within a certain radius. Weller et al. [2] used manually digitized physical footprint from oil and gas development in their spatial analysis of the ecological footprints.

Pasher et al. [3] developed a manual approach for mapping disturbances and was selected over any automatic classification, image segmentation, or object-based methodology across Canada's boreal ecosystem. They also concluded that the Landsat 30 m interpretation detected up only 38% of the seismic lines that were visible in the 2.5 m resolution imagery for their small test area.

While many generalizations can be made about the well pad footprints, there is significant variation across the half million oil and gas wells in Alberta. These features can range in size from square meters to several hectares and can have various shapes such as rectangular, circular, tear drop, and non-linear. Adding to the complexity, there are many land cover types around the well sites, different well types, various stages of reclamation and vegetation encroachment, adjacency to other anthropogenic footprint features, and data reporting accuracy.

In general, the task of feature delineation could be solved by techniques based on individual pixel values image classification [4]. However, the pixel based methods neglects the spatial elements in the image such as shape, context, and texture. In addition, the high resolution images have more information than medium resolution ones which result in less accurate classification results when traditional classifiers are used.

Chen et al. [5] has developed a methodology to delineate linear disturbances such as roads, seismic lines, and pipelines using time series Landsat 7 images. They have encountered the limitation of low resolution and low contrast of Landsat imagery.

The proposed methodology utilizes SPOT 5 panchromatic satellite imagery mosaic of the Western Canadian Sedimentary Basin (WCSB) as it is the largest, most recent and cloud free high resolution mosaic in Canada for the region of interest. The individual images are normalized to each other to create a seamless mosaic. As a result, the well pad detection and delineation algorithm is designed specifically for SPOT 5 2.5 m panchromatic imagery of the WCSB region. This creates challenges in the pixel based image classification methods as the panchromatic images have limited information and they rely on thresholds derived from spectral information during the detection process [6]. One of the important properties commonly observed in panchromatic images is the increase in brightness of the area affected by mining and other anthropogenic causes due to loss of vegetation and exposure of rock and soil on the well pad area. Therefore, most of the anthropogenic features show similar spectral characteristics in the panchromatic images which makes the delineation process challenging as it is difficult to differentiate each anthropogenic footprint from each other.

Apart from using panchromatic imagery, there are also footprint detection challenges which can be divided into two categories: image-related and location-related. Satellite imagery within the WCSB satellite image mosaic varies in data quality due to a range of acquisition dates which forms the *image-related challenges*.

This results in poor accuracy in low image quality areas where it can be very difficult to detect well pad boundaries, and in areas where the well pads are very small in size. *Location-based challenges* can consist of two or more well pads being very close in proximity, roads or pipeline corridors near the well pad that are too wide to be detected. As a result, image segmentation or classification on their own is not enough to overcome all the challenges for high accuracy shape detection.

Therefore, there is a need for a semi-automated software framework for use in an operational mapping context to overcome the research problem outlined above. The proposed methodology employs geographic object-based image analysis (GEOBIA) to achieve the desired results in an intelligent mapping system. Hay and Castilla [7] describes GEOBIA as sub-discipline of GIScience which develops theory, methods, and tools to replicate the human interpretation of RS images in automated/semi-automated ways by partitioning the imagery into meaningful image-objects.

GEOBIA is receiving more popularity in the remote sensing literature due to the fact that "spatial location" is the key component of the analysis [8]. GEOBIA does not solely rely on the single pixel spectral values but also on its texture and pixel spatial continuity. Powers et al. [9] introduced a multi-scale geographic object-based image analysis (GEOBIA) approach that incorporates new object-based texture measures and a decision-tree classifier to assess wetlands. Martha et al. [6] have applied object-oriented image analysis to detect and classify landslides using Cartosat-1 (2.5 m) and IRS-1D (5.8 m) panchromatic images. Consequently, adding spatial characteristics (e.g. image texture, contextual information, pixel proximity and geometric attributes of the features, etc.) to the image analysis give advantages over pixel –based methods [10, 11].

The overall objective of this study is to create a method to efficiently and accurately map the well pad and gas plant footprint in Alberta. This paper utilizes geographic object-based image analysis methodology to extract and delineate well pads from SPOT 5 2.5 m panchromatic satellite imagery. The proposed methodology employs unique combination of multiple RS methodologies together with topological, geometric and geographic properties of the delineated objects as a part of geographic object-based image analysis. Spatial information of the objects is utilized by linking the pixels to objects to delineate meaningful objects. Some of the remote sensing methodologies include image segmentation using watershed transformation, standard Hough transform, region growing, edge delineation, polygon simplification, texture and contrast differential techniques.

Methodology

Study Area

The region of interest is the province of Alberta which is located in western Canada with a total area of 661,848 km^2. It is the largest producer of conventional crude oil,

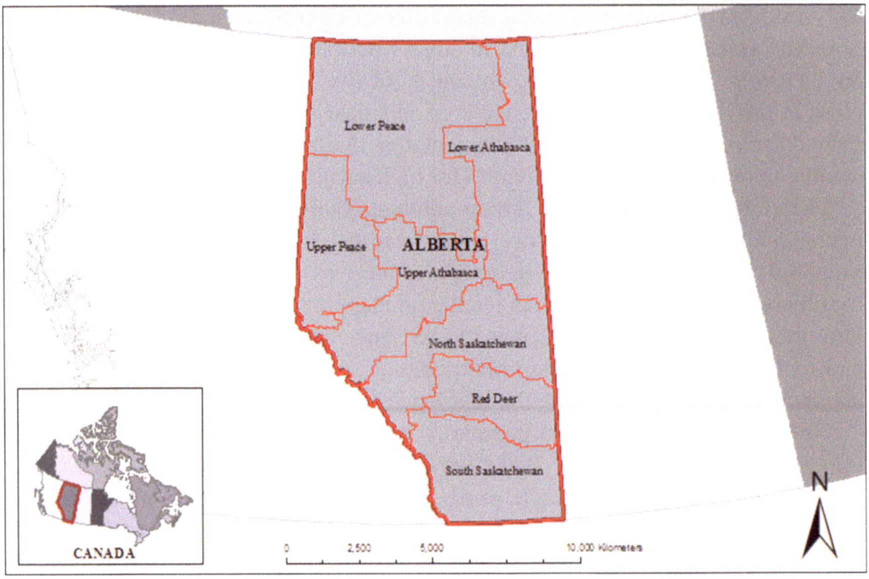

Fig. 1 Province of Alberta with seven Land Use Framework regions

synthetic crude, natural gas, and gas products in Canada. Approximately 70% of Canada's natural gas production is from Alberta [12]. Between 10,000 and 15,000 new wells are drilled in Alberta each year [13], and nearly 120 ha of land per day is being industrialized for oil and gas production. There are over 500,000 oil and gas well pads and thousands of gas plants in Alberta as of 2010. Each of these developments has resulted in a footprint on Alberta's land surface. The Government of Alberta has started "The Land-Use Framework (LUF)" which is a strategic planning initiative to manage Alberta's land and natural resources to achieve long-term economic, environmental and social goals [14]. The LUF establishes seven land-use regions (Fig. 1) and calls for the development of a regional plan for each. As a result, oil and gas footprint layers that this study aims can be as input layers into modelling and mapping initiatives within Alberta Energy's operational activities related to the LUF.

Satellite Image Data

In this study, a well pad detection and delineation algorithm designed specifically for SPOT 5 2.5 m panchromatic image mosaic of Alberta for the year of 2010 is presented and rigorously evaluated. This mosaic is the largest high resolution mosaic in Canada which is cloud free, seamless and up to date, and mostly uses spring and summer images.

The mosaic dataset is derived from SPOT 5 2.5 m panchromatic Level 1A (raw imagery) products. It has been processed using the most accurate control available, such as Canadian National Road Network, Alberta Government Access Vectors, and LANDSAT 7 orthorectified imagery supplied by the Government of Canada. The Canadian Digital Elevation Data (CDED) is used as the Digital Elevation Model (DEM) source for the orthorectification. The PCI Geomatics High Resolution Satellite Ortho Package was used to orthorectify image data. The orthorectification process computes a Rational Functions Math Model. Each individual scene contains coefficients, called Rational Polynomial Coefficients (RPC), which are used to define the math model together with the collected high accuracy ground control points (GCPs). The methodology ensures adequate distribution of the control points within the image.

Ancillary Datasets

When a well site is drilled, there are two locations associated with the site: Surface Hole Location (SHL) and Bottom Hole Location (BHL). The surface hole is what is seen at ground level, the bottom is at its deepest point. If the hole is drilled vertically, both locations are the same; however, many wells are drilled directionally and some even horizontally. As this study aims to capture the disturbance of the well on the ground level, a vector point file containing the surface hole locations for oil and gas well sites were used as one of the inputs to the system. This file contains: (a) sub-meter accuracy point locations for the surface hole locations; (b) the latitude longitude coordinate of the well, activity type, spud date, status of the well and unique well identifiers (UWI). Therefore, the extraction of oil and gas well pads is only performed where well site points exist.

The second input to the system was agricultural land cover data for 2006. This land cover data has nine classes: annual cropland, native posture, improved posture, hay, forest, wetland, water, barren, and built-up. As the northern and southern parts of Alberta show different characteristics of land cover, the well pad footprint characteristics also change. In the north where there is more boreal forest, the footprints tend to be larger and have more orthogonal shapes. On the other hand, within the southern region well pad footprints tend to become tear drop shaped from agricultural activities and/or from natural grassland growth, both of which encroach on the well pad relatively quickly. Therefore, this study utilizes different delineation rules based on the land cover data.

Footprint Delineation Rules

A significant challenge in defining the footprint of well pads and gas plants will be the shape variability caused by both land cover type and the encroachment of

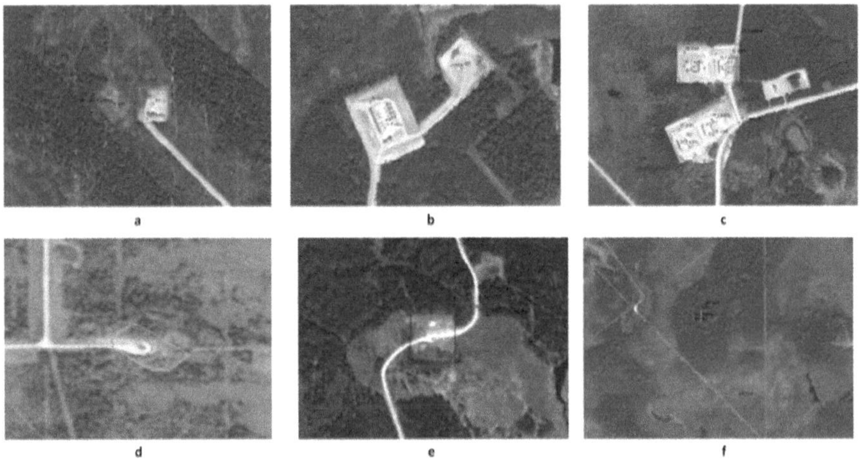

Fig. 2 Footprint delineation rule examples: (**a**) Rule 1 (**b**) Rule 2 (**c**) Rule 3 (**d**) Rule 4 (**e**) Rule 5 (**f**) Rule (6)

vegetation which dictates how long a footprint will remain visible. For example, a site in a forested area could have visible footprint for decades. However, in a grassland area the original footprint might be reduced after only few years. Also challenging are the varied appearances of the well pads in the mosaic. Many of them have a clear two-tone appearance, with very bright but irregular central part (most likely soil surface), and a more subdued original shape (encroaching vegetation). In some areas, the satellite imagery has a low sun angle which results in less identifiable well pad boundaries. As a consequence, there is a need to create general "rules" to dictate how the automated algorithm will decide what to capture as a well pad footprint. These rules were formed in partnership with regulators, and industry partners to define the size, shape, and special situations of various well pad types. They govern the decision flow in the well pad detection and delineation system.

Figure 2 shows six examples from the predefined rules.

(a) ***Rule 1*** (Fig. 2a): Area that is visible around well centers, non-treed (some sort of a clearing), and a human disturbance should be captured. If there is no well centre (point) in the clearing, do not capture.
(b) ***Rule 2*** (Fig. 2b): Access roads that are becoming part of the well pad should be included. If the access road is only passing by and continuing, do not include as a part of the well pad.
(c) ***Rule 3*** (Fig. 2c): Continuous well pads, i.e. well pads that are next to each other should be combined as one polygon.
(d) ***Rule 4*** (Fig. 2d): Over time the as built pad area and shape will change because of ingress from surrounding vegetation. Ingress depends on the age of the well (spud date on attribute table). The older the well, the greater the ingress of surrounding vegetation. The older the well the more irregular the pad shape.

The system will consider the relationship spud year and irregular shapes. In the figure the well site was drilled in 1968 and the original "as built" pads is barely visible (green line) but the current foot print is much smaller (red line). The older the spud date on any given point the higher the irregularity of the captured well pad. In these cases, the system should capture the red line.

(e) ***Rule 5*** *(*Fig. 2e*):* If the well pad location is not distinguishable from its surrounding, i.e. no definitive tree line, then a square in average (this is predefined) size of a well pad should be captured around the well site. However, if the well site metadata indicated "Reclaimed/Pre63/Exempt", then do not capture a polygon at all as these are reclaimed and abandoned wells.

(f) ***Rule 6*** (Fig. 2f*):* If there is a well point and the well is active (i.e. that is not Abandoned or Reclaimed) but no visible pad on the imagery consider the following:

- Is the well spud date after the acquisition date of the imagery?
- Is the well site spud date 1912 and ingress from surrounding vegetation has completely render the original pad indistinguishable from the surrounding vegetation?
- Is the original pad now within a larger facility location or within a pipeline corridor?
- Is the pad under permanent water and not visible?

The rule in all of the above cases, where the point data indicates no presence of a pad and no pad is visible in the imagery, is do not capture any footprints. However, if the well point is active, but no visible well pad and it does not fall into the above cases, then a standard polygon needs to be placed.

Automated Footprint Mapping System

The proposed system partitions the SPOT 5 image mosaic into well pad boundary objects similar to the way humans conceptually organize the landscape to understand it.

Figure 3 shows the designed system workflow for footprint delineation. The system has two main components: Feature Extraction System and Automated Quality Control System.

Feature Extraction System

Feature extraction system finds the candidates for each well pad point and sends the results to the automated quality control system. The system goes through each well point one by one and defines three neighborhoods around the well point. The

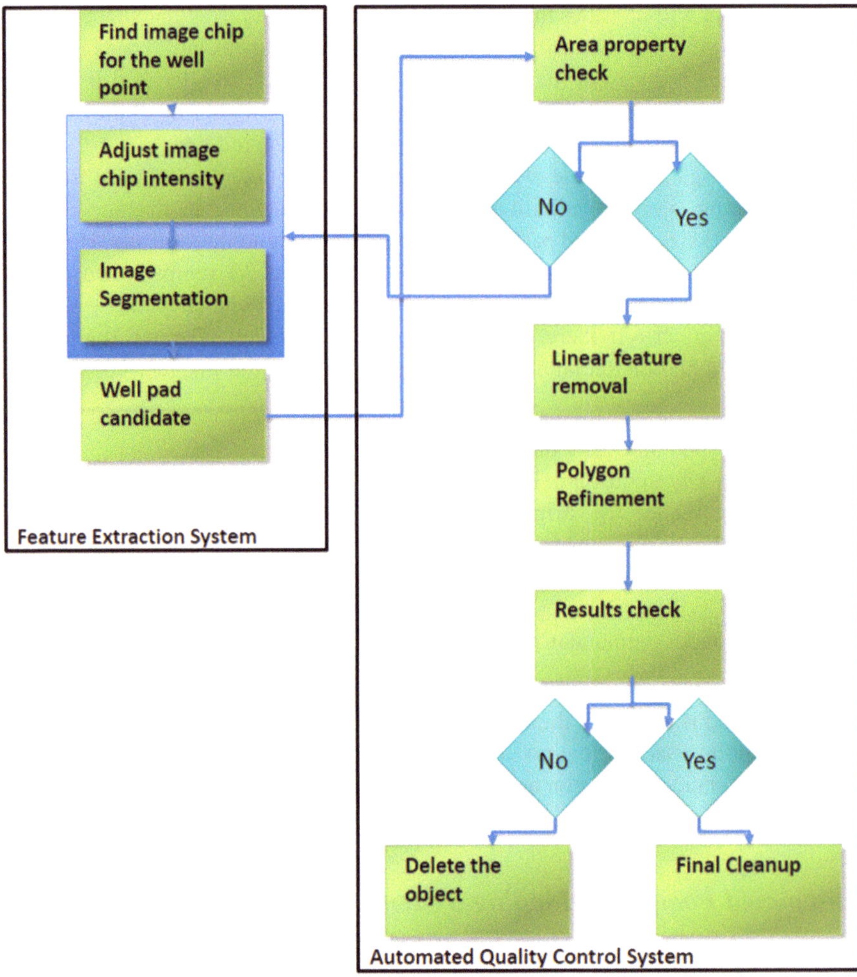

Fig. 3 System flowchart

defined neighborhood sizes are 150 m (n1), 200 m (n2) and 250 m (n3) (Fig. 4). The steps described below uses these three neighborhoods for the calculations.

Find panchromatic imagery: The system starts with a well point and then clips the panchromatic imagery for the 250 m neighborhood around that point.

After finding the imagery, the next step is to **adjust the image intensity**. This step involves changing the original pixel values of the image so that more of the available range is used, which increases the contrast between features and their backgrounds. This step makes the next step, image segmentation work better to differentiate well pads from forest, and other land use/cover types. The parameters of the intensity adjustment changes based on the image characteristics and also the

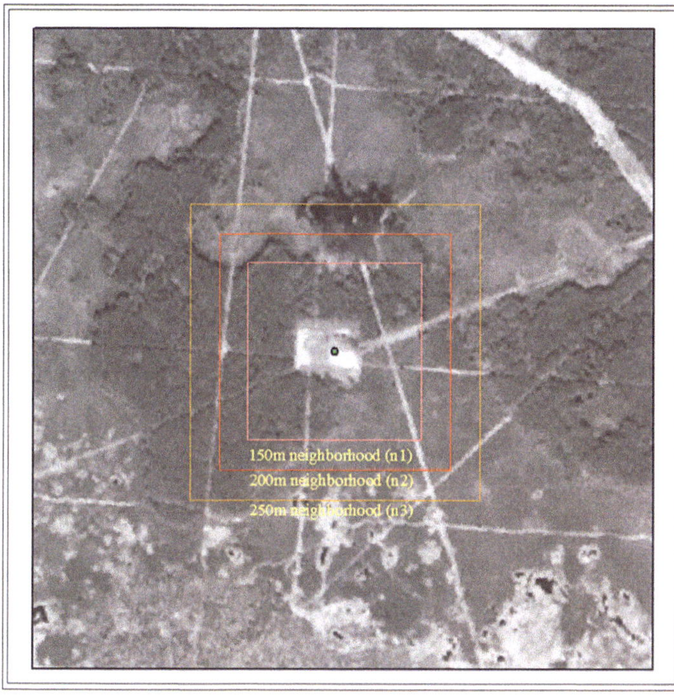

Fig. 4 Three neighborhoods defined around the oil and gas well points

input from the checking steps of the automatic quality checking system. The system chooses the parameters for the intensity adjustment based on image characteristics, image date or if the previously selected parameters were failed and new ones needed.

Image segmentation: The contrast enhanced image chips then enter into the segmentation function. The objective of the segmentation is to divide the image into relatively homogeneous and semantically significant groups of pixels. In this study, watershed algorithm based on mathematical morphology is employed. Watershed algorithm in image processing acquires the basics from topography as it treats the image pixels as the morphological landforms, the peak correspond to the maximum in the grey scale images, and valley corresponds to the minimum [15, 16].

First, the original image is transformed to a gradient image which represents the edge strength of each pixel. Thus, the gradient will be high at the borders of the objects. Second, background and foreground markers are calculated to separate the main objects from the background objects. When calculating the foreground markers, the system creates flat maxima inside of each main object. To find the background objects, the system utilizes both gradient magnitude image and the foreground markers. The aim is to find background markers not to be too close to the edges of the objects that are being segmented as "candidate objects". Third, the system modifies the gradient magnitude so that the regional minima only occur

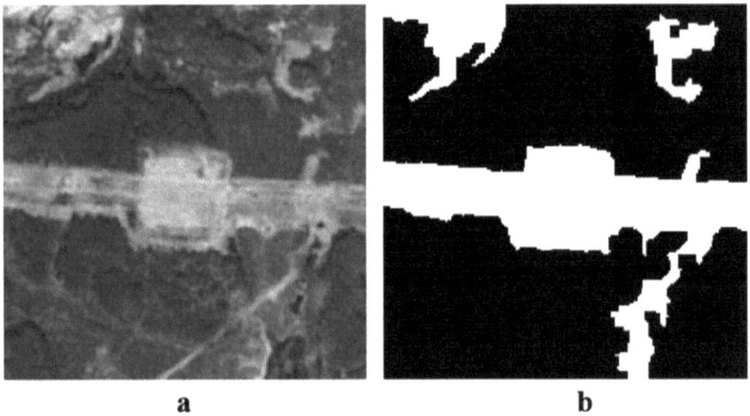

Fig. 5 Example from image segmentation (a) original SPOT 5 panchromatic image (b) segmented well pad candidate objects

at foreground and background pixels. Last, the system creates a binary image of candidate objects after cleaning the isolated pixels and the edges of the objects. Figure 5 shows an example of a segmented candidate objects for a well site.

In this step, well pad boundaries for each well center are identified roughly due to the fact that some of the features (such as roads) around the well pads have a similar radiometry and classified with the actual well pad area by the algorithm. After the classification step, the developed algorithm works on the individual classified pixels labeled as well pads to remove unwanted features in the candidate objects which will be explored in the next steps.

It was noted that the 2010 Alberta SPOT 5 mosaic contained late season imagery in some parts of the province which resulted in poor quality data. Figure 6 shows an example of a well site with 2009 and 2010 imagery. 2009 imagery clearly shows the well pad boundary as it has a better acquisition date and higher sun angle than the 2010 imagery. Lower sun angle, reduced dynamic range and reduced contrast were a few of the key challenges which required a unique set of steps for proper feature extraction. As a result, the feature extraction system incorporates a few basic imagery parameters to determine the quality of the imagery in use and then select the correct sub-algorithm to perform the feature extraction. These parameters include image spectral statistics, acquisition date of the imagery, and sun angle at the time of acquisition. Based on these parameters, different segmentation sub-algorithms were designed which would be called upon when the specific imagery conditions present. This leads to a more efficient and consistent feature extraction and the feature extraction system will have greater adoptability from year to year or region to region.

Fig. 6 Comparison of the same well site with 2010 and 2009 imagery (**a**) 2010 imagery with low image quality (**b**) 2009 imagery with high image quality

Automated Quality Control System

Feature extraction system passes the resultant objects to the automated quality control (QC) system. The resultant objects are checked against rules and certain criteria. If they pass the QC, they are converted to polygons. If they don't pass the QC, the system either sends them back to feature extraction system for different segmentation or the objects are rejected and the system works on the next well point.

Area properties check: At this step, the system checks the area characteristics of the candidate object to decide if the first segmentation was successful or not. The candidate objects go through two check rules, and depending on the results, the system might run the segmentation steps again with different parameters to achieve the desired candidate object.

Check rule 1: If the total area of the candidate object is larger than 30,000 pixels (18.75 ha) in the n3 neighborhood and the well pad status is reclaimed, then the segmentation result is too large to be a well pad candidate object. Thus, the system applies the segmentation step with a different image intensity adjustment. If the new resultant candidate object's area is still larger than 30,000 pixels (18.75 ha) in the n3 neighborhood and the well pad status is reclaimed, then the system skips this point with no polygon generated.

Check rule 2: The second rule checks if the area of the candidate object is larger than 18,000 pixels (11.25 ha) in the n1 neighborhood or if the area of the object is 0 in the n1 neighborhood. This states that the system found another object that is far away from the well pad point; hence the system applies the segmentation step with enhancing the panchromatic image by contrast limited adaptive histogram equalization [17].

Linear feature removal, edge and skinny pixel group cleanup: After the first checks, the algorithm works on the candidate well pad objects and the individual pixels in them which form those objects to remove unwanted linear features. These linear features are roads that are adjacent to the well pads which are labelled together

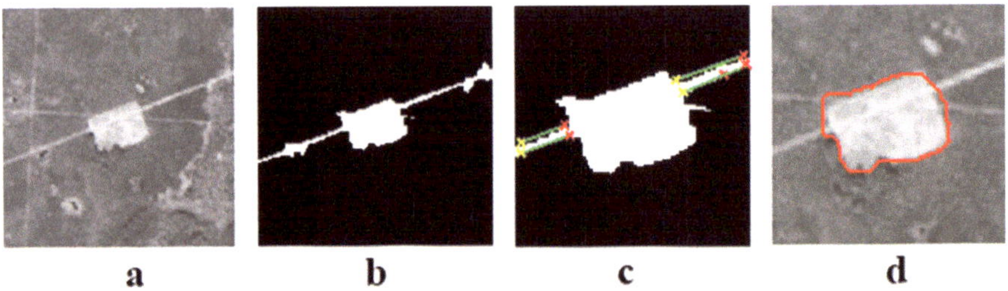

Fig. 7 Linear feature removal (**a**) original image in n3 (250 m) neighborhood (**b**) segmented image in n3 (250 m) neighborhood (**c**) linear features identified in n1 (150 m) neighborhood (**d**) final object boundary after linear feature removal in n1 (150 m) neighborhood

with the well pads as their spectral characteristics are close together. Therefore, this step finds the linear features in the classified well pad boundary and removes them from the well pad boundary. Standard Hough Transform (SHT) has been used to detect the lines in the classified well pad polygons. The SHT is considered as a very powerful tool in edge linking for line extraction due to its capability to extract lines even in areas with pixel absence (pixel gaps). SHT proposed by Duda and Hart [18] is widely applied for line extraction in image processing.

As some well pads have straight orientation and others have diagonal orientation, the algorithm first finds the orientation of the well pad candidate. In order to do that, it compares every line that was calculated by the SHT. By calculating the slopes of each line, it decides on the orientation of the well pad. Then, based on the orientation and direction, the algorithm rotates the object until it is straight as it eases the linear feature removal process.

After the rotation, the algorithm focuses on the linear objects that were identified by the SHT and removes them one by one from the well pad object. For each pixel belonging to the line, the system calculates its location relative to the rest of the pixels in the object and its relation to it, i.e. is it still part of the line or does it belong to a larger group that form the main well pad area. Thus, the pixels belonging to the linear object are removed until they belong to the main well pad area.

After the removal of the linear features, the resultant well pad object is checked for any unsmooth edges caused by the removal. The algorithm checks also the area between n2 and n3 neighborhood to find any skinny pixel groups that are attached to the well pad. These skinny features are usually the pixels that are incorrectly segmented or remains of the linear feature removal. Figure 7 shows an example of before and after the removal process. Figure 7c illustrates the lines that are identified by the algorithm. Then, these lines were removed and the final well pad object is obtained in Fig. 7d.

Polygon refinement: Linear feature removal is followed by refining the polygons so that their shape is close to the actual well pad boundary, i.e. eliminating unwanted pixels that do not belong to the well pads. This step employs several different methods to find the unwanted pixels by analyzing three neighbourhoods of the well center. The algorithm analyses individual cells for each neighbourhood (n1, n2 and n3) and runs several algorithms to extract and analyze the regional characteristics of the cells and decides if the cell belong to the well pad in question or not. These characteristics include location of the cell compared to the well pad, orientation of the well pad, extent, area, perimeter, solidity of the well pad, and distance of the cell to the well pad center.

Results check: The algorithm now checks if there should be polygon based on the well pad spud year, reclamation status, and the spectral characteristics of the image chip for the n3 neighborhood. If the algorithm decides that there shouldn't be a polygon, then the created object is deleted. For example, if "the well type is abandoned" and "the reclamation status is reclaimed" and "the spud year is from 1990s" and "the average spectral characteristics around the well point show less bright areas", then there shouldn't be a polygon captured for that well.

At this step, the algorithm also does a final check on the regional characteristics of the candidate object. It calculates area, perimeter, solidity and extent. By calculating the regional characteristics of the candidate object, the algorithm decides if the candidate should or should not have been classified. Extent is the ratio of pixels in the candidate object to the pixels in the total bounding box of the object. Solidity is the proportion of the pixels in the convex hull that are also in the candidate object. If the area and the perimeter of an object is large but the extent and the solidity are smaller than 0.80, this shows a largely classified object and most likely a wrong one. Thus, this object should not be classified and deleted from the results.

Final Cleanup: At this step, the algorithm checks if there are more than one object in the neighborhood n1. If there are, then the algorithm eliminates the objects that are away from the center. It also evaluates the area proportion of the object in the neighborhoods. The area proportion check decides if the polygon is small or large. Small polygons also go through another check and the cleanup procedure that is designed for their size. Candidate objects that pass are then converted to vector polygons. The final resultant polygon goes through the shape cleanup. After the first polygon refinement, the system checks if the resultant polygon has a proper shape. For example, it checks the number of vertices, and if the shape is orthogonal. The final polygon refinement algorithms apply generalization operation on the generated polygons. The algorithm reduces details in the boundaries of well pads, while maintaining the essential shape and size of the well pads. The simplification process preserves and enhances the orthogonality for the well pads that are in the non-agricultural areas. Figure 8 shows an example of a polygon refinement and polygon simplification. In this example, algorithm refined the polygon so that it is orthogonal.

Well pads falling in either crop or grass land appear to be consistent as they are usually a tear drop shape. When the well is drilled, the well pad is a typical square shape but over time the surrounding vegetation (crop or grass) ingresses and eventually forms a mature pad in a tear drop shape regardless of cover type. A mature pad (tear drop) is consistent with the turning radius of well inspection vehicles. Road maintenance prevents ingress of surrounding vegetation and as a

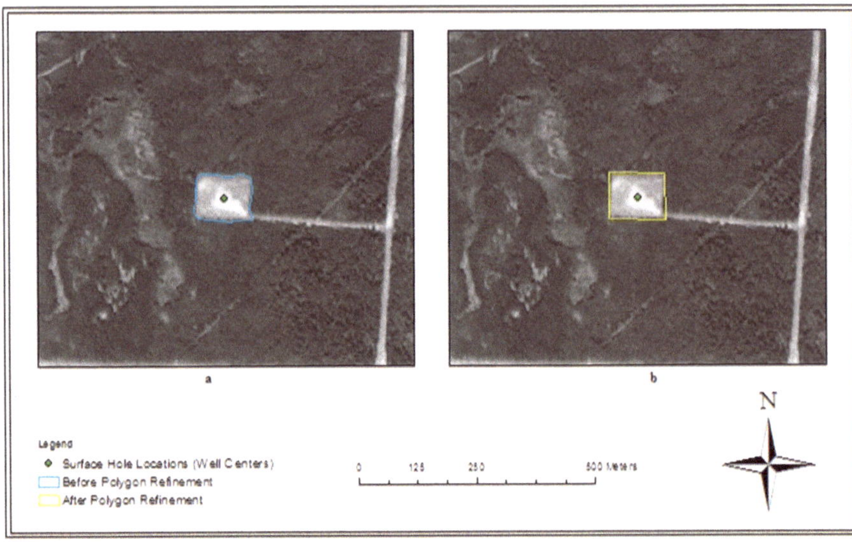

Fig. 8 Polygon refinement and simplification (**a**) before polygon refinement (**b**) after polygon refinement

result the tear drop shape is maintained. Thus, for the well pads that are in the agricultural areas, the algorithm simplifies them so that they are not very complex in shape.

Results and Discussion

All the algorithm codes were written in Python 2.7 with object-oriented programming using geospatial data abstraction library (GDAL) and Numpy libraries. The proposed methodology was developed and tested using networked workstations, multi-core servers and GPU units for high-speed processing. The feature extraction algorithm processes a minimum of 30 well pads per minute. The proposed system processes 694 individual well points a day for extracting features and performs QC whereas using a manual method in standard GIS software, ArcGIS, can process only 321 points a day.

The feature extraction was employed for the whole Alberta region. A total of 409,417 points were processed using the automated system. Table 1 shows the number of well points per each LUF region. The well point dataset included all well points that were drilled before the end of 2010. Land cover data for the year 2006 was also employed in the system to help the automation process in the decision rules as different rules and checks were defined for different land cover types around the well points.

Table 1 Total number of well points per each LUF region

LUF region	Total number of well points
Lower peace	28,538
Lower Athabasca	58,328
Upper peace	30,818
Upper Athabasca	31,632
North Saskatchewan	87,437
Red Deer	73,520
South Saskatchewan	99,144
Total	**409,417**

From the well point dataset, 20% of the wells were selected to access the accuracy of the results. Reference inventories were created manually using visual interpretation technique for those preselected well sites. The accuracy of the results were calculated through the error Eq. (1).

$$\%Error = \left[{}^{|ShapeArea - ActualArea|}\Big/_{ActualArea}\right] \times 100 \tag{1}$$

Figure 9 shows some example of the comparison between reference inventory (manually digitized) polygons with automated process results. Based on the comparisons of the selected wells to that of manually digitized validation polygons (reference inventories), the automated feature extraction system captured the 82.37% of the well pad footprint polygons.

The performance of the methodology to detect the oil and gas well pad footprint is moderately good.

Figure 10 shows some final results. Among all the well site areas, the northern part of Alberta shows better results than the rest due to the forest land cover type. The results show that the well sites are more visible when the surrounding land cover is forested.

As there are cases that multiple well points are within a single well pad area, well pad attributing rules has to be defined in order to label each well pad polygon with correct well point attribute. As set of rules were created to attribute the well pad based on criteria from the various well point attributes. Factors included the type of well—gas, bitumen, abandoned, reclaimed; date of the well point; and status of the well point. From the criteria hierarchy, the well pad is attributed automatically with the correct information. If a single well point intersects a single pad then the pad should reflect the attribute/commodity tagged to the well point. If a single pad has two or more well point intersections then the following rules/priority should apply. The well types priority order is (1) Bitumen (2) Oil (3) Coal based methane (CBM) (4) Gas (5) Drilled & Cased (6) Other (7) Abandoned (ABD). Highest priority goes to bitumen and the lowest priority goes to ABD regardless of the number of well point intersects. If there are more than one well points for the well pad and all points have the same well type (for example, two oil wells or three Bitumen), then the well point that has the oldest year takes priority.

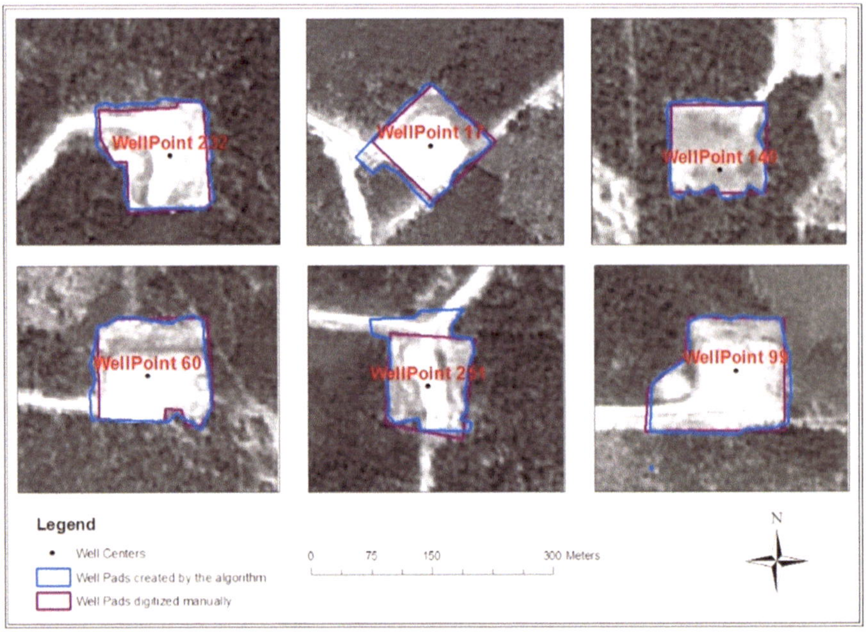

Fig. 9 Comparison of manually digitized polygons and automated process resultant polygons

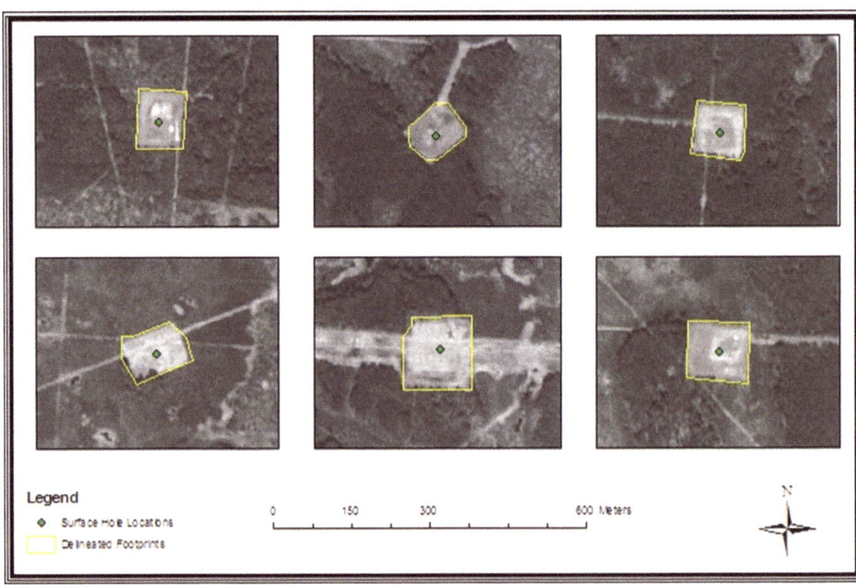

Fig. 10 Some of the delineated footprints by the automated footprint system

Table 2 summarizes the average well pad sizes per each well type for each LUF region which was generated after the well pad attributing rules were applied. It is obvious from the table that southern regions of the province has smaller well pads than the northern region as a consequence of the agricultural and grass lands. Lower Athabasca region also demonstrates similar characteristics as the southern regions by having smaller size well pads except bitumen well types. The well pads in the Upper Peace and Upper Athabasca regions incorporate a greater size on average than in the Lower Peace and Lower Athabasca regions. Although there is oil development within the Upper Peace and Upper Athabasca, the main development is natural gas. On the other hand, the development within the Lower Peace and Lower Athabasca is gas, bitumen and oil. In more recent years horizontal well development with several horizontal legs being drilled form a single pad has become more common in the upper regions which result in larger well pads.

In low image quality areas, areas that are very difficult to differentiate the well pad boundary, and in areas where the well pads are very small in size, feature extraction system has an accuracy of 77.57%. In the southern region of Alberta where agricultural and grass land areas exist, the system has a rate of 56.89%. The challenges in the southern region come from problems in the agricultural areas, issues caused by saturated imagery (high or low gain), and difficulties due to the fact that well pads that are too small. Using the land cover dataset in these problematic areas and defining different parameters in the system was helpful in delineating some of the problematic well pads.

Conclusion

This study has developed a method to efficiently and accurately map the well pad and gas plant footprint in Alberta from satellite imagery; and to integrate this method into a complete semi-automated software solution for the production of anthropogenic footprint map layers. Multiple remote sensing methodologies in a GEOBIA framework were used to obtain the footprints in an automated manner, such as image classification, standard Hough transform, region growing, edge delineation, polygon simplification, texture and contrast differential techniques. The results show that the combination of several spatial characteristics such as image texture, contextual information, pixel proximity and geometric attributes of the features gives advantages over typical pixel based methods. The algorithm currently processes a minimum of 30 well pads per minute with an accuracy greater than 80%.

There are area specific situations in which the developed automated feature extraction process has lower accuracy in delineating the well pad footprints. For example, scenarios with two or more well pads being very close to each other, or roads or pipeline corridors near the well pad being too wide for the algorithm to differentiate from actual well pads are some remaining technical challenges. Continued development of the algorithm will resolve these challenges and improve the accuracy results. The innovative approach proposed in this study provides a

Table 2 Average well pad sizes in hectares per well type category for each LUF region

	Land-use framework (LUF) region							
	Average well pad sizes in hectares							
		Lower Athabasca	Lower Peace	Upper Athabasca	Upper Peace	North Saskatchewan	Red Deer	South Saskatchewan
Well type	Abandoned	0.4269	1.0894	1.0282	1.2643	0.6947	0.4078	0.3375
	Bitumen	1.8424	1.7966	0.9933	1.3955	0.7885	–	–
	Coal based methane	–	–	1.6164	1.2100	0.1681	0.1765	0.2442
	Drilled & Cased	0.7240	1.0432	1.1801	1.3009	0.7075	0.3420	0.3171
	Gas	0.7692	1.0159	1.2338	1.3130	0.5443	0.2355	0.2050
	Oil		1.3518	1.0763	1.2177	0.5371	0.3680	0.3562
	Other	0.6512	1.1778	0.9038	0.9825	0.3679	0.2765	0.2049

standardized methodology for footprint mapping with quantifiable accuracy, eliminating the need for estimation, which can be useful to private sector, NGO markets and government ministries. The developed technology can easily be adjusted to map additional anthropogenic footprint features, such as pipelines, forestry cut blocks, gravel pits, and resource access roads.

Acknowledgments The author is thankful to The National Research Council-Industrial Research Assistance Program (NRC-IRAP) of Canada and Alberta Innovates Technology Futures for the financial support of this study. Alberta Department of Energy has provided some of the spatial data used in this study. The author gratefully acknowledges the assistance of Tom Churchill for his expertise on energy activities and well pads in Alberta.

References

1. Leu M, Hanser SE, Knick ST (2008) The human footprint in the west: a large-scale analysis of anthropogenic impacts. Ecol Appl 18(5):1119–1139
2. Weller C, Thomson J, Morton P, Aplet G (2009) Fragmenting our lands: the ecological footprint from oil and gas development. http://wilderness.org/resource/fragmenting-our-lands-ecological-footprint-oil-and-gas-development. Accessed 15 Aug 2013
3. Pasher J, Seed E, Duffe J (2013) Development of boreal ecosystem anthropogenic disturbance layers for Canada based on 2008 to 2010 Landsat imagery. Can J Remote Sens 39(1):42–58. doi:10.5589/m13-007
4. Castillejo-González IL, López-Granados F, García-Ferrer A, Peña-Barragán JM, Jurado-Expósito M, de la Orden MS, González-Audicana M (2009) Object- and pixel-based analysis for mapping crops and their agro-environmental associated measures using QuickBird imagery. Comput Electron Agric 68(2):207–215. doi:10.1016/j.compag.2009.06.004
5. Chen Z, Jefferies B, Adlakha P, Salehi B, Power D (2014) Monitoring linear disturbance footprint based on dense time series Landsat imagery. Can J Remote Sens 40(5):348–361. doi:10.1080/07038992.2014.987375
6. Martha TR, Kerle N, van Westen CJ, Jetten V, Vinod Kumar K (2012) Object-oriented analysis of multi-temporal panchromatic images for creation of historical landslide inventories. ISPRS J Photogramm Remote Sens 67:105–119. doi:10.1016/j.isprsjprs.2011.11.004
7. Hay G, Castilla G (2008) Geographic object-based image analysis (GEOBIA): a new name for a new discipline. In: Blaschke T, Lang S, Hay G (eds) Object-based image analysis spatial concepts for knowledge-driven remote sensing applications. Springer, Berlin.
8. Blaschke T, Lang S, Hay G, SpringerLink (Online service) (2008) Object-based image analysis spatial concepts for knowledge-driven remote sensing applications Lecture notes in geoinformation and cartography
9. Powers RP, Hay GJ, Chen G (2012) How wetland type and area differ through scale: a GEOBIA case study in Alberta's Boreal plains. Remote Sens Environ 117:135–145. doi:10.1016/j.rse.2011.07.009
10. Duro DC, Franklin SE, Dubé MG (2012) A comparison of pixel-based and object-based image analysis with selected machine learning algorithms for the classification of agricultural landscapes using SPOT-5 HRG imagery. Remote Sens Environ 118:259–272. doi:10.1016/j.rse.2011.11.020
11. Jobin B, Labrecque S, Grenier M, Falardeau G (2008) Object-based classification as an alternative approach to the traditional pixel-based classification to identify potential habitat of the grasshopper sparrow. Environ Manag 41(1):20–31. doi:10.1007/s00267-007-9031-0

12. AlbertaEnergy (2013a) Natural gas facts. http://www.energy.alberta.ca/NaturalGas/726.asp. Accessed 15 Aug 2013
13. AlbertaEnvironment (2013) Oil and gas. http://environment.alberta.ca/02242.html. Accessed 15 Aug 2013
14. AlbertaEnergy (2013b) Regional plans. http://www.energy.alberta.ca/Initiatives/3433.asp. Accessed 15 Aug 2013
15. Gonzalez RC, Woods RE (2008) Digital image processing, 3rd edn. Pearson/Prentice Hall, Harlow
16. Vincent L, Soille P (1991) Watershed in digital spaces: an efficient algorithm based on immersion simulations. IEEE Trans Pattern Anal Mach Intell 13(6):583–598
17. Zuiderveld K (1994) Contrast limited adaptive histograph equalization. Graphic Gems 4:474–485
18. Duda RO, Hart PE (1972) Use of the Hough transformation to detect lines and curves in pictures. Commun ACM 15(1):11–15. doi:10.1145/361237.361242

Modeling Urban Land-Use Suitability with Soft Computing: The GIS-LSP Method

Suzana Dragićević, Jozo Dujmović, and Richard Minardi

Introduction

Spatial Decision Support Systems (SDSSs) facilitate spatial decision-making using a hybrid computational and expert knowledge approach for semi-structured decision tasks [1, 2]. SDSS frameworks combine spatial data management, analytical modeling, visualization, and require the interaction of a decision-maker, analyst or group of stakeholders. SDSSs may function as stand-alone software tools customized for a narrow application domain or integrated within a Geographic Information Systems (GIS) framework [3]. A GIS-SDSS integration methodology builds on the visualization, data processing, and database management capabilities of fully developed GIS applications. Integration is recognized as an appropriate approach for implementing SDSS methods and more particular multicriteria evaluation (MCE) procedures [4].

Spatial multicriteria evaluation (MCE) is a general term given to decision modeling approaches that can be used within SDSSs operating on geospatial data [5]. A spatial MCE approach consists of a set of mapped choice alternatives (locations), a set of preference criteria, and a means of evaluating each choice alternative based on the criteria set [6]. Alternatives are given cumulative suitability scores presented cartographically as a mapped suitability index or suitability map.

S. Dragićević (✉) • R. Minardi
Spatial Analysis and Modeling Laboratory, Department of Geography, Simon Fraser University, 8888 University Drive, Burnaby, BC, Canada V5A 1S6
e-mail: suzanad@sfu.ca; rdm4@sfu.ca

J. Dujmović
Department of Computer Science, San Francisco State University, 1600 Holloway Avenue, San Francisco, CA 94132, USA
e-mail: jozo@sfsu.edu

J.-C. Thill, S. Dragicevic (eds.), *GeoComputational Analysis and Modeling of Regional Systems*, Advances in Geographic Information Science,
DOI 10.1007/978-3-319-59511-5_14

Suitability maps are the product of a suitability analysis used for the visualization of preference, likelihood, or consequences surrounding a phenomenon, or activity [7]. For this study, the term MCE refers specifically to a soft computing evaluation logic approach that is implemented in a raster GIS environment. Soft computing MCE approaches use inputs encoded as continuous variables ranging from a minimum of 0 (not suitable) to a maximum of 1 (most suitable) [8].

The scientific literature on spatial MCE approaches can be broadly classified as addressing either applications or theory development. Land use suitability analysis is one common application area [9–12]. Spatial MCE in land use planning also forms a significant application area [13, 14]. A prototype urban planning support tool based on MCE is developed for the Queensland region of Australia [15]. Joerin et al. [16] examined residential suitability under conditions of environmental noise pollution. Hill et al. [17] presented a decision support system (ASSESS) used for agricultural land use policy analysis in Australia. Moreover, other application approaches represent a wide range of spatial decision-making situations including habitat suitability [18], agricultural suitability analysis [19], risk and hazard assessment [20], urban landslide susceptibility [21], infrastructure planning [22, 23], environmental sustainability [24, 25], and socio-economic analysis [26].

In addition to the various application domains, researchers have focused on enhancing three theoretical aspects of spatial MCE—decision operators and preferences, hybrid systems, and uncertainty analysis. Rinner and Taranu [27] developed an interactive tool for MCE-based decision making dealing with preference evaluations. Proctor and Drechsler [28] as well as Feick and Hall [29] developed and tested an MCE approach for collaborative planning. A GIS-MCE approach is used to study competing goals in forest conservation planning in Malaysia [30]. Wood and Dragicevic [31] used a multi-objective GIS decision support framework to identify optimal marine protection locations based on criteria representing the conflicting objectives of conservation and resource extraction. Recent work has also dealt with developing hybrid spatial MCE systems [32, 33] and addressing uncertainty [34, 35].

Spatial MCE approaches based on linear models are conceptually limited since they produce an oversimplified representation of human reasoning and decision making [36–38]. These linearized MCE have two key shortcomings: (1) the number of inputs that may be combined is limited, and (2) the decision logic does not reflect logic conditions needed for decision problems. These two issues are related to the linear aggregation process used in the MCE process. A linear additive combination is used, known as the *Weighted Linear Combination* (*WLC*) rule. In the WLC, criteria are first assigned a weight and subsequently summed returning suitability scores used to make a suitability map [8]:

$$S = \sum_{i=1}^{n} w_i x_i, \quad 0 < w_i < 1, \quad i = 1, \ldots, n, \quad \sum_{i=1}^{n} w_i = 1, \tag{1}$$

where S is the aggregated overall suitability, w is an array of positive normalized weights representing the relative importance of elementary decision criteria used to generate an array of n scores $(x_1, \ldots, x_n)$. The WLC rule is compensatory; a low input score may *always* be compensated by other higher criteria scores in the same location. In other words, it is not possible to model mandatory requirements where the absence of a mandatory input $(x_i = 0)$ must not be compensatory and must yield $S = 0$.

A second key shortcoming of the WLC systems is their limitation in the number of data inputs n. As the number of input attributes increases, the significance of each input decreases. Because the total sum of attribute weights must sum to unity, the mean value of weights is $1/n$ and it can become insignificant for large number of input attributes. For WLC systems the total impact of input x_i is:

$$\delta_i = S(x_1, \ldots, x_{i-1},\ 1, x_{i+1}, \ldots, x_n) - S(x_1, \ldots, x_{i-1},\ 0, x_{i+1}, \ldots, x_n) = w_i \quad (2)$$

The average impact is:

$$\overline{\delta} = (\delta_1 + \cdots + \delta_n)/n = 1/n \quad (3)$$

and as n increases the average impact becomes negligible. In a general case, that is not a desirable property.

A third key shortcoming of WLC systems is their limitations with respect to decision logic. In many cases decisions require the application of logic conditions and requirements to compare and select a set of locations over the alternatives. The most frequent logic operators are models of simultaneity (AND) and replaceability (OR). The AND operator denotes partial or full conjunction and is similar to a minimum function, whereas the OR operator denotes partial or full disjunction and is similar to a maximum function. The traditional WLC approach yields a fully *neutral* decision logic that is neither AND nor OR. Neutral logic is only one of several necessary logic aggregators. According to Dujmović et al. [39], logical requirements such as *mandatory, nonmandatory, sufficient, mandatory-optional, sufficient-optional*, and others are necessary for real-world decision-making. In order to practically implement these logic aggregators, the Logic Scoring of Preference (LSP) method is proposed.

The LSP method is based on the multicriteria decision-making approach but has origins in soft computing where variables are treated as a matter of degree. A key feature of the LSP method is the nonlinear attribute criteria and aggregation structures that model decision requirements. These features make LSP a more well-suited method to handle complex spatial problems that require numerous attributes and a high level of detail. The LSP has previously been used as a method for evaluating software and web interfaces [40–42]. Spatial applications and LSP suitability map have also been proposed by Dujmović et al. [39] at theoretical level and hypothetical spatial datasets. The integration of LSP method within GIS and with use of real geospatial data is at initial stages [43, 44] and yet is to be fully implemented.

Therefore the main objective of this study is to develop an integrated GIS and LSP method for the purpose of defining land use suitability. Suitability is expressed as raster suitability maps representing a real geographic study area. The model is built to test the LSP approach in a spatial context using geospatial data in a raster GIS framework. The LSP approach is also compared to a MCE/WLC structured approach. The comparison served to: (1) identify the limitations of MCE/WLC suitability maps, and (2) highlight the relevant qualities of LSP such as nonlinear aggregation and flexible logic aggregators to address the WLC limitations. The LSP method is applied to a residential land use suitability analysis procedure for the Bowen Island Municipality, Canada.

Properties of the Logic Scoring of Preference (LSP) Method

Theoretical Background

The LSP method is outlined below as a novel approach to investigate semi-structured spatial decision problems in a GIS framework. The LSP is originally conceived as a general multicriteria approach and used for evaluation of software systems, web browsers and user interfaces [40]. The approach has also been extended to the evaluation of complex spatial systems [39]. De Tré et al. [45] described a framework for building LSP suitability maps, or *s-maps*. Approaches for generating LSP-derived suitability maps have been described using empirically derived data for a hypothetical optimal home location sitting [46]. Dujmović and De Tre [47] investigated this problem and described an interactive, dynamic web-based LSP system integrated with Google Maps (publicly available at seas.com suitability maps) enabling non-expert users to parameterize and customize the system. A spatial LSP system may be configured to provide analysis of the financial components and costs related to a decision strategy [45]. De Tré et al. [48] have proposed suitability maps that express bipolar satisfaction (degrees of satisfaction and dissatisfaction) of decision criteria incorporating LSP aggregators.

Elements of the LSP

The LSP criterion structure is presented in Fig. 1. LSP criteria are built in three steps: (1) an attribute tree used to derive n input attributes ($a_1, \ldots, a_n$), (2) a set of n elementary (attribute) criteria that are functions for evaluation of individual attributes ($x_1 = g_1(a_1), \ldots, x_n = g_n(a_n)$), and (3) an aggregation structure to combine the attribute suitability scores $x_1, \ldots, x_n$ and generate the overall suitability of the system $x = L(x_1, \ldots, x_n)$. This criterion structure is consistent with observable properties of human evaluation reasoning and can be used in each point of a two-dimensional map to create a suitability score.

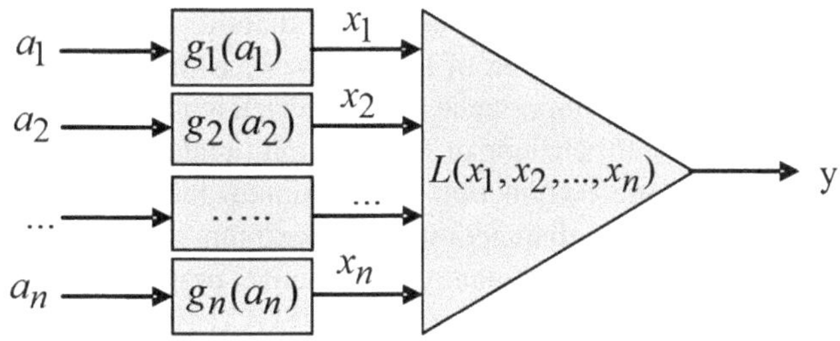

Fig. 1 Structure of the LSP criterion function

A key feature in LSP suitability modeling is the expression of logic requirements. The basic logic requirements are *simultaneity* (partial conjunction function) and *replaceability* (partial disjunction function). In many cases criteria may require *mandatory* satisfaction, or reflect a requirement that is merely *optional*. For example, a slope criterion must be satisfied for many land use decision problems. However, satisfying various "view criteria" may be entirely optional. Ultimately, if the mandatory slope criterion is not satisfied (assigned a 0 score on a scale of 0 to 1) in any location, the result will be a null score for that location. A view criterion with an *optional* satisfaction requirement is more replaceable; a null score of an optional criterion will not disqualify a choice alternative.

Partial conjunction and partial disjunction are fundamental LSP aggregation operators. The partial conjunction is similar to the traditional (full) conjunction (*and* function, or minimum) and the partial disjunction is similar to the traditional (full) disjunction (*or* function, or maximum). The degree of similarity between the partial conjunction and the full conjunction is called *andness* (α) and it satisfies $0 \leq \alpha \leq 1$. The degree of similarity between the partial disjunction and the full disjunction is called *orness* (ω) and it satisfies $0 \leq \omega \leq 1$. Furthermore, $\alpha + \omega = 1$ and the pure conjunction is denoted using $\alpha = 1$, $\omega = 0$, and the pure disjunction is denoted using $\alpha = 0$, $\omega = 1$. By selecting andness and orness between 0 and 1 we get a continuous transition from conjunction to disjunction and can select the desired degree of simultaneity and replaceability. For example, if $\alpha > 1/2$, then we have a model of simultaneity and by increasing the value of α we can increase the level of penalizing systems that do not simultaneously satisfy a set of criteria. The arithmetic mean and all MCE/WLC models are a special case characterized as neutrality, $\alpha = \omega = 1/2$.

The partial conjunction, disjunction, and neutrality can be interpreted as special cases of the Generalized Conjunction/disjunction (GCD) function which is usually implemented as a weighted power mean [39, 49] as follows:

$$S = \left(\sum_{i=1}^{n} w_i x_i^r \right)^{1/r}, \quad 0 < w_i < 1, \quad 0 \leq x_i \leq 1, \quad i = 1, \ldots, n, \quad \sum_{i=1}^{n} w_i = 1,$$
$$-\infty \leq r \leq +\infty, \quad 0 \leq S \leq 1. \tag{4}$$

Here S represents the aggregated degree of suitability (or a suitability score), x_i denotes an input attribute degree of suitability, w_i is the user defined attribute weight reflecting the relative importance of the selected input, and r is the parameter that determines the logical behavior of the function (andness/orness). If $r = -\infty$ then GCD becomes a pure conjunction (the minimum function) and if $r = +\infty$ then GCD becomes a pure disjunction (the maximum function). In the range $-\infty \leq r < 1$ GCD has predominantly conjunctive properties and is used for modeling simultaneity. In the range $-\infty < r \leq 0$ GCD is called the *hard partial conjunction* (HPC) and used for modeling mandatory requirements. For example, if $r = 0$ then GCD becomes a geometric mean as follows:

$$S = \prod_{i=1}^{n} x_i^{w_i} \tag{5}$$

Obviously all inputs are mandatory; if any input is not satisfied ($x_i = 0$), then $S = 0$, proving that the satisfaction of all inputs is indeed mandatory. If $0 < r < 1$ then GCD becomes a *soft partial conjunction* (SPC) which has conjunctive properties, but a single positive input is sufficient to produce the positive output. In the range $1 < r < +\infty$ GCD based on weighted power mean has predominantly disjunctive properties and is used for modeling soft partial disjunction. The hard partial disjunction can be modeled as a De Morgan dual of hard partial conjunction [42, 49]. For $r = 1$ the resulting GCD becomes the neutral arithmetic mean which has a perfect balance of conjunctive and disjunctive properties. In this research we use the GCD symbols and parameters presented in Table 1. The soft partial conjunction with increasing strength is implemented using aggregators C−−, C−, and the hard partial conjunction using aggregators C−+, CA, C+−, C+, and C++. Similarly, the partial disjunction is implemented using aggregators D−−, D−, D−+, DA, D+−, D+, and D++.

The GCD has a spectrum of properties and it is up to decision maker to select those properties that reflect the desired behavior (the intensity of simultaneity/replaceability) of the suitability aggregation function. By combining various forms of GCD it is possible to create advanced compound functions. The most frequently used compound function is the *conjunctive partial absorption* (CPA) that combines two asymmetric inputs: a mandatory input x and an optional input y.

Table 1 Symbols and parameters of GCD (n = 2)

Orness (ω)	1	$\frac{15}{16}$	$\frac{7}{8}$	$\frac{13}{16}$	$\frac{3}{4}$	$\frac{11}{16}$	$\frac{5}{8}$	$\frac{9}{16}$	$\frac{1}{2}$
Symbol	D	D++	D+	D+−	DA	D−+	D−	D−−	A
r	+∞	20.6	9.52	5.8	3.93	2.79	2.02	1.45	1
Andness (α)	1	$\frac{15}{16}$	$\frac{7}{8}$	$\frac{13}{16}$	$\frac{3}{4}$	$\frac{11}{16}$	$\frac{5}{8}$	$\frac{9}{16}$	$\frac{1}{2}$
Symbol	C	C++	C+	C+−	CA	C−+	C−	C−−	A
r	−∞	−9.06	−3.51	−1.66	−0.72	−0.15	0.26	0.62	1

The output value $z = f(x,y)$ has the following properties: $f(0, y) = 0$, $y \geq 0$, $f(x, 0) = x\text{-}penalty$, $x > 0$, and $f(x, 1) = x + reward$, $0 < x < 1$. Both the penalty and the reward are functions of input x and the decision makers determine parameters of the CPA function by selecting the average desired penalty P and the average desired reward R. Most frequently the average penalty is selected in the range [10–40%] and the average reward is selected from $R < P$. More details are available in Dujmović [42].

A less frequently used asymmetric aggregation function is the *disjunctive partial absorption* (DPA) that combines two asymmetric inputs: a sufficient input x and an optional input y. The output value $z = f(x,y)$ has the following properties: $f(1, y) = 1$, $y \geq 0$, $f(x, 0) = x—penalty$, and $f(x, 1) = x + reward$, $0 < x < 1$.

The degree of suitability can be interpreted simply as a score, but there are two other more precise interpretations. Each degree of suitability can be interpreted as the degree of fuzzy membership in the fuzzy set of perfectly suitable systems, or we can interpret the degree of suitability as the degree of truth of the statement claiming perfect satisfaction of requirements. All these interpretations are equivalent and equally convenient in the area of suitability maps. For simplicity, we assume that "suitability" denotes the degree of suitability (or the suitability score) with either fuzzy or logic interpretation. The LSP suitability maps rank each geographic location with a score ranging from 0 (not suitable) to 1 or 100% (most suitable).

The LSP approach and logic aggregation structures process input data more effectively than an MCE/WLC approach and also provide more data-rich and expressive suitability models. The weights in LSP models can be determined using neural network training methods [50], AHP [51], and various auxiliary software tools. The LSP aggregation process is based on systematic use of hard partial conjunction, soft partial conjunction, hard partial disjunction, soft partial disjunction, neutrality, conjunctive partial absorption and disjunctive partial absorption in a way illustrated in the next section. These fundamental aggregators, as well as hierarchical aggregation structures built using the superposition of seven fundamental LSP aggregator types are unique features of the LSP approach and provide more flexibility than previously used techniques for making criteria based on WLC, AHP, and OWA [52].

The hierarchical method of aggregation permits a large number of relevant inputs to be included in an evaluation with minimized data loss. The next section outlines the approach for creating GIS-LSP suitability maps.

Approach for Designing GIS-LSP Urban Land Suitability Maps

This section presents the framework for an integrated GIS-LSP prototype model used to evaluate the urban land use suitability in the Bowen Island, British Columbia, Canada using raster GIS data sets. The extent of the study is a 14 × 14 km area comprised of rugged bedrock-dominated terrain situated at the entrance of Howe Sound, Canada [53, 54]. Under the pressures of intensified residential development, Bowen Island has experienced significant changes in its natural environments. Raster GIS data sets and digital elevation model (DEM) sets acquired from the Province of British Columbia are used in this study. The main goal is to illustrate all steps in the process for developing GIS-LSP suitability maps.

Integrating Raster GIS and the LSP Method

The LSP approach assumes a tight integration with raster GIS. A GIS-LSP framework uses geographically referenced database map layers as input, and relies on GIS operations to standardize and define elementary attributes, evaluate their suitability, implement LSP operators, and calculate suitability scores for each choice alternative. The GIS may be used to formulate a suitability index, or alternatively select the top ranked locations in the study site. It is assumed that an overall justifiable LSP suitability score can be computed in each (x,y) point of the analyzed area. In such a case, inputs and output results can be visualized in map form. With the support of map visualization, users may perform model validation, a sensitivity analysis, or create a series of alternative decision scenarios. By changing different features (the attribute tree, factor weights, LSP aggregators, and the aggregation structure) of the LSP system a new or series of new output maps can be generated. The integrated GIS and LSP model involves systematic development the following stages: (1) development of an attribute tree, (2) elementary criteria definition, (3) aggregation structure selection, and (4) computation of a global suitability score.

The Attribute Tree

The first step in creating an LSP suitability map involves constructing an attribute tree that organizes the decision problem and contains all relevant attributes. There are two considerations when selecting the attributes: (1) the attributes are restricted to only the data in the GIS database, and (2) the selected attributes must be sufficient to completely and correctly describe the suitability map criterion based on the needs and interests of stakeholders.

In the Bowen Island study, the available GIS data included the following parameters: (1) slope, (2) aspect, (3) road access, (4) water access, (5) ferry terminal access, (6) natural park access, (7) elevation, (8) stream location, (9) wetland areas, (10) lakes, (11) public transit locations, (12) forestry data (tree age, average tree volume), and (13) watershed unit locations. For simplicity, the first six components from the above list to evaluate the suitability for residential development have been chosen. Then the structure of the attribute tree is shown in Fig. 2.

1. Suitability for residential development
    - 1.1. Suitability of site (+)
        - 1.1.1 Basic site properties (+)
            - 1.1.1.1 Slope (+)
            - 1.1.1.2 Aspect (-)
        - 1.1.2 Accessibility and transportation (+)
            - 1.1.2.1 Proximity to roads (+)
            - 1.1.2.2 Proximity to ferry terminal (+)
    - 1.2. Amenities (-)
        - 1.2.1 Proximity to coast (-)
        - 1.2.2 Proximity to parks (-)

Fig. 2 Basic attribute tree for residential development suitability analysis

In Fig. 2, the symbol (+) denotes mandatory inputs and the symbol (−) denotes optional inputs. Whenever we aggregate two components using weighted power means there are the following four possibilities:

- Both components are mandatory (if any input suitability is zero, the output suitability is zero). In such cases the aggregator must be a HPC;
- Both components are optional (the output suitability is zero only if all inputs are zero). In such cases the aggregator can be a SPC, neutrality, or a partial disjunction;
- One input is mandatory and the other input is optional (if the mandatory input is zero the output is zero regardless the value of the optional input; if the mandatory input x is positive and the optional input y satisfies $y > x$ the output z satisfies $z > x$; otherwise, if $y < x$ then $z < x$). In such cases the aggregator is a CPA;
- One input is sufficient and the other input is optional (if the sufficient input is 1, the output is 1 regardless the value of the optional input; if the sufficient input x is positive and the optional input y satisfies $y > x$ the output z satisfies $z > x$; otherwise, if $y < x$ then $z < x$). In such cases the aggregator is a DPA.

A similar reasoning can also be used in cases with more than two inputs. As shown in Fig. 2, it is possible and useful to identify mandatory and optional (desired, but nonmandatory) attributes in the earliest stage of designing an LSP criterion. In our case, the slope is mandatory and the aspect is optional. Similarly, the analyzed site must be suitable, but the amenities are optional. These decisions must be justifiable and correctly reflect the stakeholder's standpoint. In Fig. 2 we defined six input attributes (1.1.1.1, 1.1.1.2, 1.1.2.1, 1.1.2.2, 1.2.1, and 1.2.2). All other attributes are compound. Each input attribute should be relevant to the problem at hand and non-redundant.

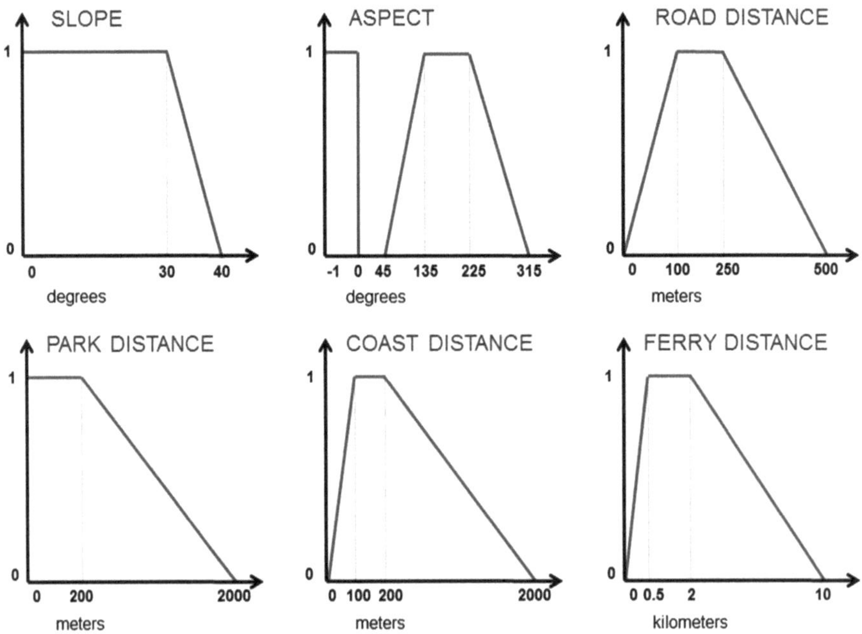

Fig. 3 Elementary criteria for case study evaluation of residential development suitability. (The variable −1 refers to locations without measurable degree of slope)

Elementary Criteria Definition

The second step of the LSP method defines the elementary criteria with one criterion function for each input attribute. Each elementary criterion is a function that shows the degree of satisfaction that corresponds to specific values of the input attribute. The six elementary criteria for the evaluation of the suitability for residential development in the Bowen Island study are shown in Fig. 3. The selection of criteria has been based on the household point of view. The range of suitability is from a minimum of 0 to a maximum of 1. The suitability score can also be interpreted as a degree of fuzzy membership or the degree of truth of a value statement. The descriptions of the criterion logic are provided in Table 2.

The Suitability Aggregation Structure

The third step of the LSP method involves selecting an appropriate aggregation structure to combine attribute suitability scores. Elementary criteria structured within the attribute tree described above generate attribute suitability scores that are used as input to a user-designed aggregation structure. The structure employs user-

Table 2 Detailed description of attribute criterion logic

Logical requirement		Criteria
Mandatory	Site (+)	**Slope** (+): 0–30° = 100%, 40° = 0%; Reflects the relative costs of residential development on steep graded slopes as opposed to level surfaces. Slopes from level (0°) up to 30° are considered suitable, with monotonically decreasing suitability to 40°. Grades above 40° are considered too costly and unsuitable
		Aspect (−): 0–45° = 0%, 135–225° = 100%, 315–360° = 0%; Reflects the desirability of south-facing sites for development for the objective of maximizing sunlight exposure. Aspect refers to the direction in which a slope faces measured in decimal degrees (180° = south). The −1 value refers to level ground
	Transportation (+)	**Road access** (+): 0 m = 0%,100–250 m = 100%, 500 m = 0%; Locations adjacent to roads are unsuitable with increased suitability to 100 m. Car ownership is high and many residents use them for daily commute. Areas with the highest suitability are within 100 and 250 m of roads with monotonically decreasing suitability from 250 to 500 m. Areas at a distance of 500 m are the least suitable and locations beyond that distance are considered unsuitable
		Ferry terminal access (+): 0 km = 0%, 0.5–2 km = 100%, 10 km = 0% Locations with a greater access to the ferry terminal under 0.5 km are less suitable and coincide with higher noise and traffic levels. Locations between 0.5 and 2 km are highly suitable. Distances beyond 2 km decrease in suitability to a maximum of 10 km. Distances farther than 10 km are considered unsuitable. The ferry terminal is the Municipality's link to Metro Vancouver for residential commuters
Optional	Amenities (−)	**Park access** (−): 0–200 m = 100%, 2 km = 0% .Reflects a desirability of sites with access to parks for recreational purposes. Locations within walking distance (0–100 m) are valued higher than areas at a greater distance that would require cycling or motor vehicles
		Coast access (−): 100–200 m = 100%, 2 km = 0% Relates to the desirability and value placed on residential locations with access to coastlines and related amenities. Locations adjacent to the coast are considered high risk in terms of home construction, with increasing suitability up to 100 m. Locations ranging from 100 to 200 m represent optimal distances with a gradual decrease in suitability up to 2 km

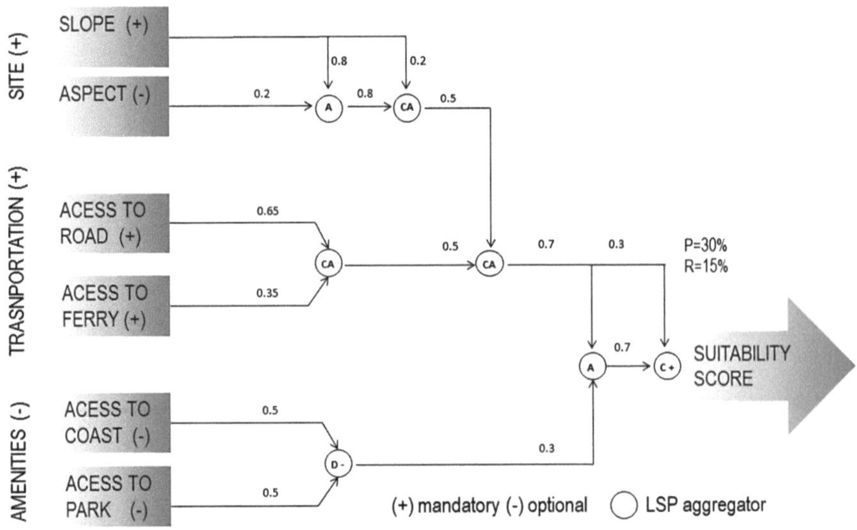

Fig. 4 Suitability aggregation structure for the residential development criterion

selected LSP aggregators based on the GCD function and implemented using the WPM. The aggregation structure combines criteria according to the categorization of the attribute tree. Categorically grouped inputs are then aggregated. Aggregation structure design must follow patterns defined in the attribute tree and must reflect other stakeholder requirements (relative importance, simultaneity, replaceability, etc.).

The proposed suitability aggregation structure for the residential development criterion is presented in Fig. 4. Inputs of all aggregators show the percent weights that reflect the degrees of relative importance. We use two conjunctive partial absorption (CPA) aggregators; their parameters are derived from the desired average penalty (*P*) and reward (*R*) that are also shown in Fig. 4.

Generally, it is theorized that system evaluations follow identifiable patterns. Dujmović and De Tre [47] refer to these evaluation patterns as *canonical aggregation structures* (*CAS*). In an LSP system, criteria are aggregated and combined in a stepwise, non-linear fashion. As more input attributes are combined into subgroups, their collective importance and logical strength increases. A *conjunctive CAS,* for example, uses less conjunctive logical operators at lower levels in the aggregation structure. As additional criteria become absorbed into larger aggregate layers, the level of *andness* increases in the system. Stronger conjunctive aggregators are needed to reflect stronger requirements, requiring a hard partial conjunction operator to derive the final solution.

A series of connected LSP aggregators are implemented with the WPM to combine each elementary criteria into a comprehensive suitability score. Parameter values *r* correspond to the different levels of modeled logical requirements and are: −3.510 (C+ or hard partial conjunction); 1 (A or mean average); and 9.521

(D+ or hard partial disjunction). The land-use suitability assessment is done with an *aggregated mandatory/optional CAS* and used as a template for modeling the decision problem.

First, the categories defined by problem domain are aggregated. The *park access* and *coast access* categories are aggregated using an LSP aggregator representing *SPD* (D−) to reflect the optional/replaceable nature of these inputs. The road and the ferry access criteria are aggregated with a *HPC* (CA) aggregator reflecting a mandatory requirement. Aggregation of the *slope* and the *aspect* criterion is accomplished with the application of a *CPA* structure built from a neutral aggregator (A) and a *HPC* aggregator (CA). In this case, the full satisfaction of the optional criterion (*aspect*) augments the non-zero score of the mandatory criterion with a *reward*, and a null *aspect* score assigns a *penalty* to the mandatory criterion. Ultimately, if the *mandatory* input (slope) is not satisfied, the *optional* input has no compensatory power, and the aggregator returns a zero value. The final aggregator applied is another *CPA* structure to combine the mandatory and optional categories using the A and C+ aggregators, as shown in Fig. 4.

GIS-Based LSP Suitability Maps

The input criteria used for the LSP evaluation are derived from GIS data at 25 m spatial resolution and standardized. Slope and aspect are obtained from a DEM data set. The road access, ferry terminal access, and park access are obtained by applying a raster-based GIS Euclidian distance function. Piecewise linear trapezoidal functions are applied for designing elementary criteria and computing attribute suitability. As this study investigates novel methodological approaches, criteria variables are chosen primarily to illustrate the methodology. Each attribute suitability map is presented in Fig. 5. The final suitability map of Bowen Island Municipality with obtained scores for urban land-use development is presented in Fig. 6. The obtained values of suitability scores are based on the aggregation structure presented in Fig. 4.Values closer to 1 (dark grey and black) indicate the locations with highest level of suitability based on GIS-LSP model while the lighter colors (up to 0) indicate unsuitable locations.

It is important to emphasize that the suitability map in Fig. 6 reflects logic conditions specified in the logic aggregation structure (Fig. 4) where the slope, the road distance and the ferry terminal distance are the mandatory requirements. Therefore, it is not acceptable to propose development in areas that have unacceptable slope, or are too far from roads, or too far from the ferry terminal. That is expressed as white areas in the LSP suitability map presented in Fig. 6. On the contrary, if the suitability map is based on an equivalent linear WLC-MCE criterion, the result is shown in Fig. 7, where all areas are grey, indicating that there is no location that is unsuitable for urban development. The assigned weights with the WLC are selected for each criteria layer to be as follows: *slope* (.30), *aspect* (.10), *road access* (.30), *coast access* (.15), *park access* (.05), *ferry terminal access* (.10). These results are

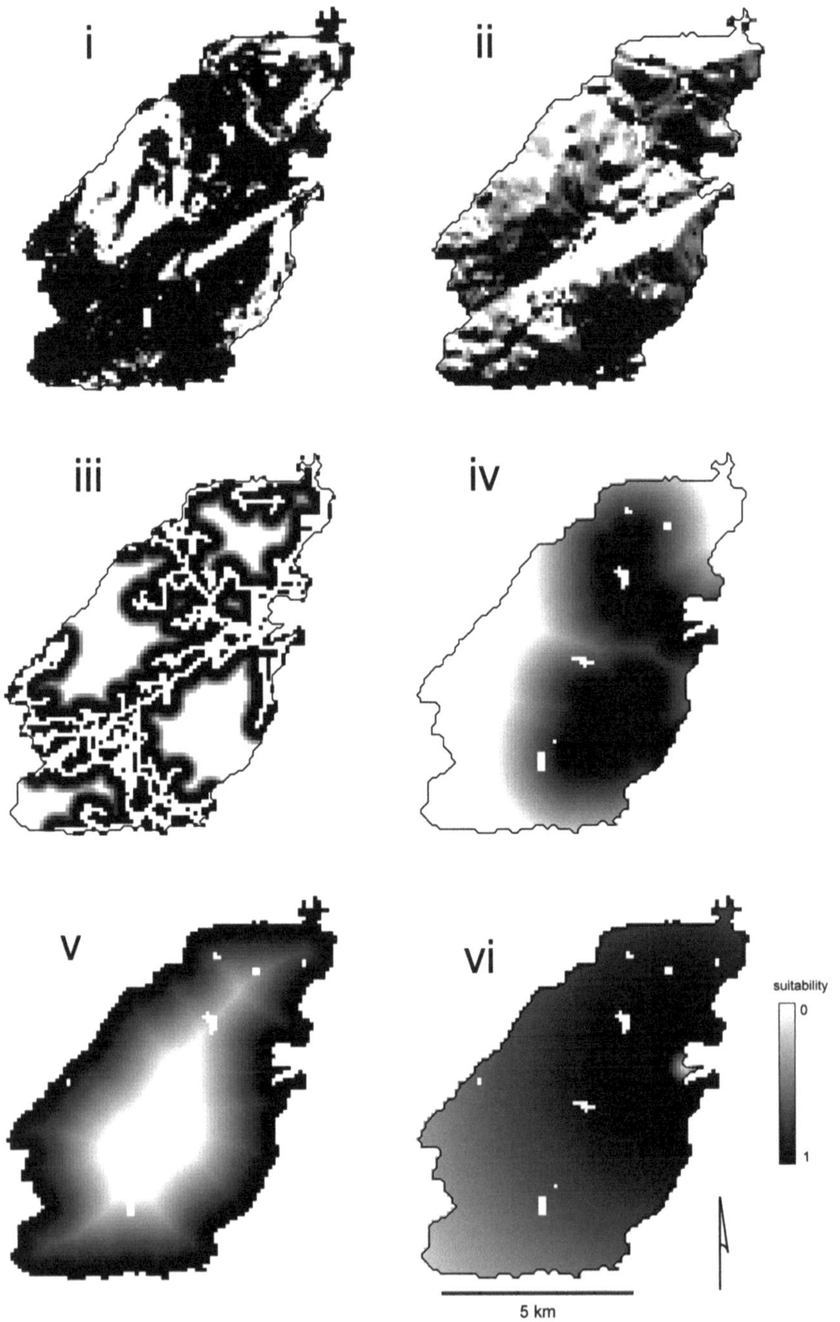

Fig. 5 Maps for suitability of input attributes based on the selected criteria: (*i*) slope, (*ii*) aspect, (*iii*) road access, (*iv*) park access, (*v*) coast access, and (*vi*) ferry terminal access

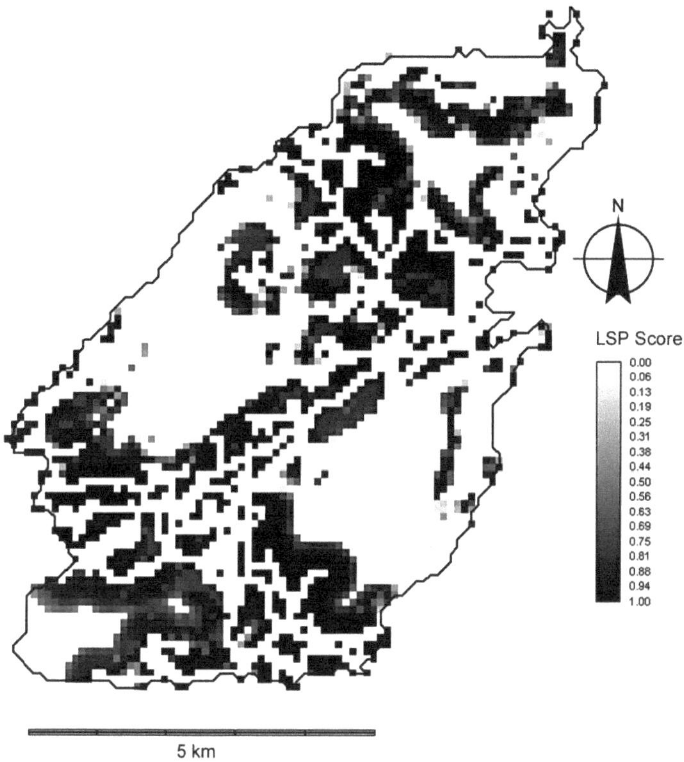

Fig. 6 GIS-based LSP suitability map with suitability scores

less selective then the ones obtained by the LSP method. For example, suitability scores sometimes can be meaningless when the linear model would claim that the location with a very high (or vertical) slope (for example close to 90°) is suitable for urban development only because some other properties (e.g. a distance from the ferry terminal, or the distance from a road) are partially satisfied. Such errors are the consequence of additive compensatory features of the linear model and do not occur in the nonlinear LSP suitability maps.

Conclusions

The selection of logic qualifiers and aggregators is by definition subjective and unique to the decision problem and expertise of the decision-maker. While the model output is sensitive to the decision maker's choice of aggregation structure, weights, and aggregators, the use of the GIS-LSP method provides a tool that decision makers can efficiently use to precisely express a spectrum of justifiable requirements. Basically, our study has demonstrated that: (1) the soft computing LSP criteria

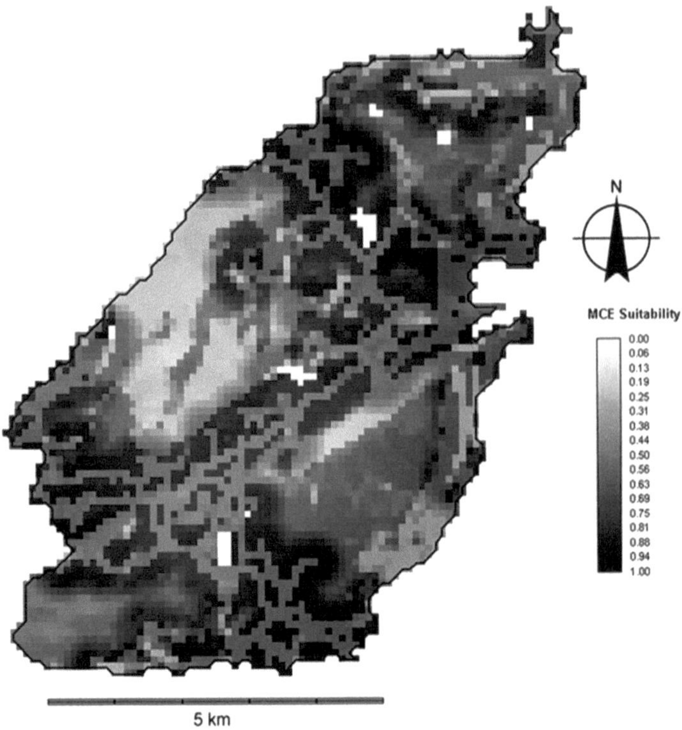

Fig. 7 A linear WLC-MCE criterion and the corresponding suitability map

can be integrated in a GIS for processing data representing real study sites, (2) the GIS-LSP approach overcomes the unreliable results of WLC-MCE through the application of modeled logic requirements resulting in more selective and justifiable suitability maps, and (3) the LSP maps are produced through nonlinear aggregation and result in more data-rich results than WLC-MCE can provide. Dujmović and De Tre [47] have started a theoretical study of comparison of different MCE methods. However, in the GIS area, more detailed work is necessary to perform the comparisons of LSP method with other ones based on analytical hierarchy process (AHP), weighted linear combination (WLC), Ordered Weighted Averaging (OWA), and with the use of geospatial data in real life problems, particularly those with large number of attributes. This work indicates that the GIS-based LSP approach offers many opportunities to implement highly complex relationships among factors and multiple objectives in the analysis of complex spatial suitability problems.

Acknowledgements The Natural Sciences and Engineering Research Council (NSERC) of Canada provided full support for this study under a Research Discovery Grant awarded to Dr. Suzana Dragićević. The data used in this study are provided thanks to Biodiversity BC and the Nature Conservancy of Canada and through the collaborative Hectares BC pilot project. Authors are also thankful for valuable comments of the anonymous reviewer.

References

1. Densham PJ (1991) Spatial decision support systems. In: Maguire DJ, Goodchild MS, Rhind DW (eds) Geographical information systems: principles and applications. Longman, London, pp 403–412
2. Malczewski J (1999) GIS and multicriteria decision analysis. John Wiley, New York
3. Church RL, Murray AT, Figueroa MA, Barber KH (2000) Support system development for forest ecosystem management. Eur J Oper Res 121(2):247–258
4. Carver SJ (1991) Integrating multicriteria evaluation with geographical information systems. Int J Geogr Inf Syst 5(3):321–339
5. Malczewski J (2010) Multiple criteria decision analysis and geographic information systems. In: Ehrgott M, Figueira JR, Greco S (eds) Trends in multiple criteria decision analysis. Springer, Boston
6. Jankowski P (1995) Integrating geographical information systems and multiple criteria decision-making methods. Int J Geogr Inf Syst 9(3):251–273
7. Hopkins LD (1977) Method for generating land suitability maps—comparative evaluation. J Am Inst Plann 43(4):386–400
8. Eastman JR, Jin WG, Kyem PAK, Toledano J (1995) Raster procedures for multi-criteria multi-objective decisions. Photogramm Eng Remote Sens 61(5):539–547
9. Abbaspour M, Mahiny AS, Arjmandy R, Naimi B (2011) Integrated approach for land use suitability analysis. Int Agrophys 25(4):311–318
10. Bagdanaviciute I, Valiunas J (2013) GIS-based land suitability analysis integrating multi-criteria evaluation for the allocation of potential pollution sources. Environ Earth Sci 68(6):1797–1812
11. Feizizadeh B, Blaschke T (2013) Land suitability analysis for Tabriz County, Iran: a multi-criteria evaluation approach using GIS. J Environ Plan Manag 56(1):1–23
12. Malczewski J (2004) GIS-based land-use suitability analysis: a critical overview. Prog Plan 62:3–65
13. Fitzsimons J, Pearson CJ, Lawson J, Hill MJ (2012) Evaluation of land-use planning in greenbelts based on intrinsic characteristics and stakeholder values. Landsc Urban Plan 106(1):23–34
14. Zucca A, Sharifi AM, Fabbri AG (2008) Application of spatial multi-criteria analysis to site selection for a local park: a case study in the Bergamo Province, Italy. J Environ Manage 88(4):752–769
15. Pettit C, Pullar D (1999) An integrated planning tool based upon multiple criteria evaluation of spatial information. Comput Environ Urban Syst 23(5):339–357
16. Joerin F, Theriault M, Musy A (2001) Using GIS and outranking multicriteria analysis for land-use suitability assessment. Int J Geogr Inf Sci 15(2):153–174
17. Hill MJ, Braaten R, Veitch SM, Lees BG, Sharma S (2005) Multi-criteria decision analysis in spatial decision support: the ASSESS analytic hierarchy process and the role of quantitative methods and spatially explicit analysis. Environ Model Softw 20(7):955–976
18. Store R, Kangas J (2001) Integrating spatial multi-criteria evaluation and expert knowledge for GIS-based habitat suitability modelling. Landsc Urban Plan 55(2):79–93
19. Ceballos-Silva A, Lopez-Blanco J (2003) Evaluating biophysical variables to identify suitable areas for oat in Central Mexico: a multi-criteria and GIS approach. Agric Ecosyst Environ 95(1):371–377
20. Aceves-Quesada JF, Diaz-Salgado J, Lopez-Blanco J (2007) Vulnerability assessment in a volcanic risk evaluation in Central Mexico through a multi-criteria-GIS approach. Nat Hazards 40(2):339–356
21. Dragicevic S, Lai T, Balram S (2015) GIS-based multicriteria evaluation and multiscale analysis to characterize urban landslide susceptibility in data scarce environments. Habitat Int 45(2):114–125

22. Jankowski P, Richard L (1994) Integration of GIS-based suitability analysis and multicriteria evaluation in a spatial decision support system for route selection. Environ Plann B 21(3): 323–340
23. Rybarczyk G, Wu CS (2010) Bicycle facility planning using GIS and multi-criteria decision analysis. Appl Geogr 30(2):282–293
24. Gorsevski PV, Donevska KR, Mitrovski CD, Frizado JP (2012) Integrating multi-criteria evaluation techniques with geographic information systems for landfill site selection: a case study using ordered weighted average. Waste Manag 32(2):287–296
25. Liu G, Rasul MG, Amanullah MTO, Khan MMK (2012) Sustainability indicator of renewable energy system based on fuzzy multi-criteria decision making methods. In: Xu QJ, Ge HH, Zhang JX (eds) Natural resources and sustainable development, Pts 1-3, vol 361–363, pp 1263–1273
26. Gamboa G (2006) Social multi-criteria evaluation of different development scenarios of the Aysen region, Chile. Ecol Econ 59(1):157–170
27. Rinner C, Taranu JP (2006) Map-based exploratory evalution of non-medical determinants of population health. Trans GIS 10(4):633–649
28. Proctor W, Drechsler M (2006) Deliberative multicriteria evaluation. Environ Plann C-Gov Policy 24(2):169–190
29. Feick R, Hall BG (2002) Balancing consensus and conflict with a GIS-based multi-participant, multi-criteria decision support tool. Geo J 53:391–406
30. Phua MH, Minowa M (2005) A GIS-based multi-criteria decision making approach to forest conservation planning at a landscape scale: a case study in the Kinabalu Area, Sabah, Malaysia. Landsc Urban Plan 71(2–4):207–222
31. Wood LJ, Dragicevic S (2007) GIS-Based multicriteria evaluation and fuzzy sets to identify priority sites for marine protection. Biodivers Conserv 16(9):2539–2558
32. Mahiny AS, Clarke KC (2012) Guiding SLEUTH land-use/land-cover change modeling using multicriteria evaluation: towards dynamic sustainable land-use planning. Environ Plann B Plann Design 39(5):925–944
33. Yu J, Chen Y, Wu JP, Khan S (2011) Cellular automata-based spatial multi-criteria land suitability simulation for irrigated agriculture. Int J Geogr Inf Sci 25(1):131–148
34. Ligmann-Zielinska A, Jankowski P (2012) Impact of proximity-adjusted preferences on rank-order stability in geographical multicriteria decision analysis. J Geogr Syst 14(2):167–187
35. Soltani SR, Mahiny AS, Monavari SM, Alesheikh AA (2013) Sustainability through uncertainty management in Urban land suitability assessment. Environ Eng Sci 30(4):170–178
36. Dujmović JJ, De Tré G, Dragićević S (2009) Comparison of multicriteria methods for land-use suitability assessment. In: IFSA World Congress 8 EVSFLAT Conference, Lisbon, Portugal
37. Malczewski J (2006) Ordered weighted averaging with fuzzy quantifiers: GIS-based multicriteria evaluation for land-use suitability analysis. Int J Appl Earth Obs Geoinf 8(4):270–277
38. Malczewski J (2011) Local weighted linear combination. Trans GIS 15(4):439–455
39. Dujmović JJ, De Tre G, Van de Weghe N (2010) LSP suitability maps. Soft Comput 14(5): 421–434
40. Dujmović JJ, Nagashima H (2006) LSP method and its use for evaluation of Java IDEs. Int J Approx Reason 41(1):3–22
41. Dujmović J, Bai H (2006) Evaluation and comparison of search engines using the LSP method. Comput Sci Inf Syst 3(2):31–56
42. Dujmović JJ (2007) Preference logic for system evaluation. IEEE Trans Fuzzy Syst 15(6):1082–1099
43. Hatch K, Dragicevic S, Dujmovic J (2014) Logic scoring of preference and spatial multicriteria evaluation for urban residential land use analysis. In: Porceedings of GIScience 2014 conference. Lect Notes Comput Sci, vol 8728, pp 64–80
44. Montgomery B, Dragicevic S, Dujmovic J, Schmidt M (2016) A GIS-based Logic Scoring of Preference method for evaluation of land capability and suitability for agriculture. Comput Electron Agric 124:340–353

45. De Tré G, Dujmović J, Weghe N (2010) Supporting spatial decision making by means of suitability maps. In: Kacprzyk J, Petry FE, Yazici A (eds) Uncertainty approaches for spatial data modeling and processing. Springer, Berlin, pp 9–27
46. Dujmović JJ, Scheer D (2010) Logic aggregation of suitability maps. In: Proceedings of the 2010 IEEE world congress on computational intelligence, Barcelona, Spain, July 18–23, pp 2222–2229
47. Dujmović J, De Tre G (2011) Multicriteria methods and logic aggregation in suitability maps. Int J Intell Syst 26(10):971–1001
48. De Tré G, Dujmović JJ, Van de Weghe N, Matthé T, Charlier N (2009) Heterogeneous bipolar criteria satisfaction handling in geographical decision support systems: an LSP based approach. ACM, Honolulu, pp 1704–1708
49. Dujmović JJ, Larsen HL (2007) Generalized conjunction/disjunction. Int J Approx Reason 46(3):423–446
50. Dujmović J (1991) Preferential neural networks. In: Antognetti P, Milutinović V (eds) Chapter 7 in neural networks—concepts, applications, and implementations, vol 2, Prentice-Hall Advanced Reference Series, Prentice-Hall, pp 155–206
51. Saaty TL (1980) The analytic hierarchy process. McGraw-Hill, New York
52. Yager RR (1988) On ordered weighted averaging aggregation operators in multi-criteria decision making. IEEE Trans SMC 18:183–190
53. Block J (ed) (1978) Bowen Island: a resource analysis for land use planning. The Islands Trust and Ministry of Municipal Affairs and Housing, Victoria
54. Bowen Island Municipality (2013) http://www.bimbc.ca/

An Algorithmic Approach for Simulating Realistic Irregular Lattices

Juan C. Duque, Alejandro Betancourt, and Freddy H. Marin

Introduction

The complexity of computational experimentation in regional science has drastically increased in recent decades. Regional scientists are constantly developing more efficient methods, taking advantage of modern computational resources and geocomputational tools, to solve larger problem instances, generate faster solutions or approach asymptotics. The first formulation of the p-median problem provides a numerical example that required 1.51 min to optimally locate four facilities in a 10-node network [52]; three decades later, Church [16] located five facilities in a 500-node network in 1.68 min. As noted by Anselin et al. [7], spatial econometrics has also benefited from computational advances; the computation of the determinant required for maximum likelihood estimation of a spatial autoregressive model proposed by Ord [47] was feasible to apply for data sets not larger than 1000 observations. Later, Pace and LeSage [48] introduced a Chebyshev matrix determinant approximation that allows the computation of this determinant for over a million observations in less than a second. According to Blommestein and Koper [11], one of the first algorithms for constructing higher-order spatial lag operators, which was devised by Ross and Harary [54], required 8000 s (approximate computation time) to calculate the sixth-order contiguity matrix in a 100×100 regular lattice. Anselin

J.C. Duque (✉) • A. Betancourt
Research in Spatial Economics (RISE-group), Department of Economics, Universidad EAFIT, Carrera 49 7 Sur - 50, Medellin, Colombia
e-mail: jduquec1@eafit.edu.co; juanca.duque@gmail.com

F.H. Marin
Mathematical Science Department, Universidad EAFIT, Carrera 49 7 Sur - 50, Medellin, Colombia

J.-C. Thill, S. Dragicevic (eds.), *GeoComputational Analysis and Modeling of Regional Systems*, Advances in Geographic Information Science,
DOI 10.1007/978-3-319-59511-5_15

and Smirnov [5] proposes new algorithms that are capable of computing a sixth-order contiguity matrix for the 3111 U.S. contiguous counties in less than a second.

An important aspect when conducting computational experiments in regional science is the selection of the way that the spatial phenomena are represented or conceptualized. This aspect is of special relevance when using a discrete representation of continuous space, such as polygons [34]. This representation can be accomplished through regular and irregular lattices; the use of one or the other could cause important differences in the computational times, solution qualities or statistical properties. We suggest four examples, as follows: (1) The method proposed by Duque et al. [21] for running the AMOEBA algorithm [1] requires an average time of 109 s to delimit four spatial clusters on a regular lattice with 1849 polygons. This time rises to 229 s on an irregular lattice with the same number of polygons. (2) For the location set covering problem, Murray and O'Kelly [46] concluded that the spatial configuration, number of needed facilities, computational requirements and coverage error all varied significantly as the spatial representation was modified. (3) Elhorst [24] warns that the parameters of the random effects spatial error and spatial lag model might not be an appropriate specification when the observations are taken from irregular lattices.[1] (4) Anselin and Moreno [4] finds that the use of regular or irregular lattice affects the performance of test statistics against alternatives of the spatial error components form.

However, returning to the tendency toward the design of computational experiments with large instances, there is an important difference between generating large instances of regular and irregular lattices. On the one hand, regular lattices are easy to generate, and there is no restriction on the maximum number of polygons. On the other hand, instances of irregular lattices are usually made by sampling real maps. Table 1 shows some examples of this practice.

The generation of large instances of irregular lattices has several complications that are of special interest in this paper. First, the size of an instance is limited to the number of polygons of the available real lattices. Second, the possibility of generating a large number of different instances of a given size is also limited (e.g., generate 1000 instances of irregular lattices with 3000 polygons). Third, as shown in Fig. 1, the topological characteristics of irregular lattices built from real maps change drastically, depending on the region from where they are sampled, which could bias the results of the computational experiments.[2]

This paper seeks to contribute to the field of computational experiment design in regional science by proposing a scalable recursive algorithm (*RI-Maps*), which combines concepts from stochastic calculus (mean reversing processes), fractal theory and computational geometry to generate instances of irregular lattices with large number of polygons. The resulting instances have topological characteristics that are a good representation of the irregular lattices sampled from around the

[1]See also Anselin [3], p. 51.

[2]Later in this paper, we show that the topological characteristics of Voronoi diagrams are far from those for an "average" map sampled in different parts of the world.

Table 1 Annotated chronological listing of studies that use irregular lattices generated by sampling real maps

Study	Purpose	Source of irregular lattices
Mur Lacambra [45]	Compares different methods to detect spatial autocorrelation	Spain provinces in 1985 (sizes 14 and 48 polygons)
Anselin et al. [6]	Performance of a diagnostic test for spatial dependence	"COROP" and "economic geographic" regions in The Netherlands (sizes 40 and 81, respectively)
Smirnov and Anselin [55]	Performance of a new method for evaluating the Jacobian term	921 counties (Kreise) for Germany; 3107 U.S. continental counties; 3140 U.S. counties and 29,762 U.S. postal zip codes
Anselin and Moreno [4]	Extend the knowledge about the properties of spatial correlation tests, especially in empirical applications	Spatial grouping of Western U.S. counties for dimensions 46, 80, 124, 264, 413 and 1013
Duque et al. [23]	Performance of an algorithm for spatial clustering (the max-p-regions model)	Sacramento census tracks (403), Colombian municipalities (1068) and U.S. census tracks (3085)

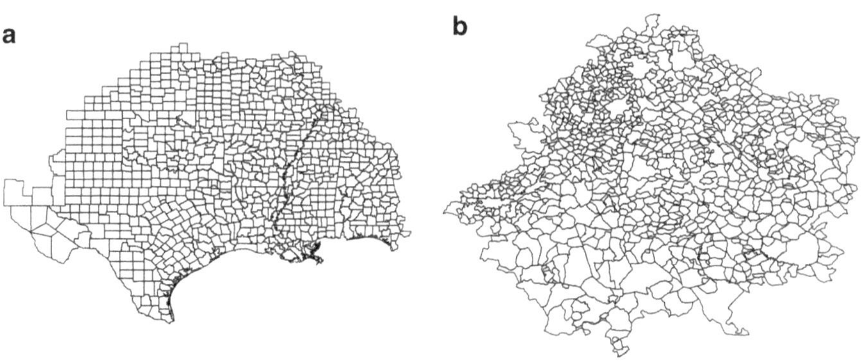

Fig. 1 Examples of two instances of 900 irregular polygons. (**a**) United States. (**b**) Spain

world. Last, the use of these instances guarantee that the difference in the results of computational experiments are not consequence of differences in the topological characteristics of the used lattices.

The remainder of this paper is organized as follows: Section "Conceptualizing Polygons and Lattices" introduces the basic definitions of the polygons and lattices and proposes a consensus taxonomy of the lattices. Section "Topological Characteristics of Regular and Irregular Lattices" presents a set of indicators that are used to characterize the topological characteristics of a lattice and shows the topological differences between regular and irregular lattices. Section "RI-Maps:

An Algorithm for Generating Realistic Irregular Lattices" presents the algorithm for generating irregular lattices. Section "Results" evaluates the capacity of the algorithm to generate realistic irregular lattices. Finally, Section "Application of *RI-Maps*" presents the conclusions.

Conceptualizing Polygons and Lattices

A polygon is a plane figure enclosed by a set of finite straight line segments. Polygons can be categorized according to their boundaries, convexity and symmetry properties, as follows:

(i) Boundary: A polygon is *simple* when it is formed by a single plain figure with no holes, and it is *complex* when it contains holes or multiple parts.[3]
(ii) Convexity: In a *convex* polygon, every pair of points can be connected by a straight line without crossing its boundary. A *concave* polygon is simple and non-convex.
(iii) Symmetry: A *regular* polygon has all of its angles of equal magnitude and all of its sides of equal length. A non-regular polygon is also called *irregular* [19, 38].

A lattice is a set of polygons of any type, with no gaps and no overlaps, that covers a subspace or the entire space. Next, a more formal definition: A lattice is the division of a subspace $S \subseteq R^n$ into k subsets $i \subseteq S$ such that $\cup_i = S$ and $\cap_i = \phi$, where ϕ is the empty set of R^n [32].[4] There exist different taxonomies of lattices depending on the field of study. In an attempt to unify these taxonomies, a consensus lattice taxonomy is presented in Fig. 2. This taxonomy classifies lattices according to the shapes of their polygons, their spatial relationship and the use, or not, of symmetric relationships to construct the lattice[5]:

(i) According to the variety of the shapes of the polygons that form the lattice: *Homomorphisms* are lattices that are formed by polygons that have the same shape, and *polymorphisms* are lattices that are formed by polygons that have different shapes.
(ii) According to the regularity of the polygons that form the lattice and the way in which they intersect, each vertex[6]: *Regular*, lattices formed by regular polygons in which all of the vertexes join the same arrangement of polygons [57]; *semi-regular*, when the polygons are regular but there are different configurations of vertexes; and *irregular* otherwise [28].

---

[3]Complex polygons do not refer to polygons that exist in the Hilbert plane [19].

[4]This paper focuses exclusively on bidimensional lattices (i.e., $n = 2$).

[5]An alternative category is proposed for lattices formed by fractal polygons that are informally defined by Mandelbrot [42] as rough fragmented geometric shapes that could be infinitely divided into scalable parts.

[6]Considering the vertexes to be all of the points of the lattice that intersect three or more polygons.

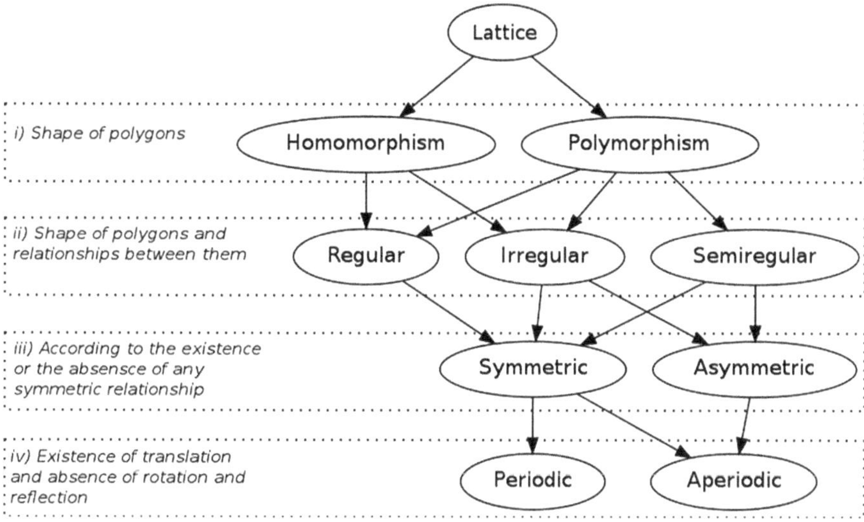

Fig. 2 Consensus taxonomy of lattices

(iii) According to the existence of symmetric relationships within the lattice[7]: *Symmetric*, when the lattice implies the presence of at least one symmetric relationship; and *asymmetric* otherwise.

(iv) According to the symmetric relationship of translation: A lattice is *periodic* if and only if it implies the use of translation without rotation or reflection; it is *aperiodic* otherwise [57].

Table 2 shows an example of each category of this consensus taxonomy.

The topological characteristics of lattices are usually summarized through the properties of the sparse matrix that represent the neighboring relationships between the polygons in the map, the so-called W matrix [8, 12, 30, 41, 50].[8] This paper uses six indicators of which the first three are self-explanatory: The maximum ($\mathbf{M_n}$), minimum ($\mathbf{m_n}$) and average number of neighbors per polygon ($\boldsymbol{\mu}_1$). The fourth indicator, the sparseness (S), see Eq. (1), is defined as the percentage of ones entries with respect to the total number of entries in a binary W matrix (k^2, where k is the number of polygons in the lattice). The fifth indicator is the first eigenvalue of the W matrix ($\boldsymbol{\lambda}_1$). It is an algebraic construct commonly used in graph theory [26, 58] and regional science [12–14, 30] to summarize different aspects of the W matrix. The first eigenvalue, λ_1, is the maximum real value, λ, that solves the system given by Eq. (2), where I_k is the identity matrix of order $k \times k$. The last indicator, ($\boldsymbol{\mu}_2$), is the

[7]There are three types of symmetrical relationships: *Translation*, when the lattice is formed by translating a subset of polygons; *reflection*, when there are axes of reflection in the lattice; and *rotation*, when it is possible to obtain the same lattice after a rotation process of less than 2π [51].

[8] See Anselin [3] for more information about this matrix.

Table 2 Example lattices

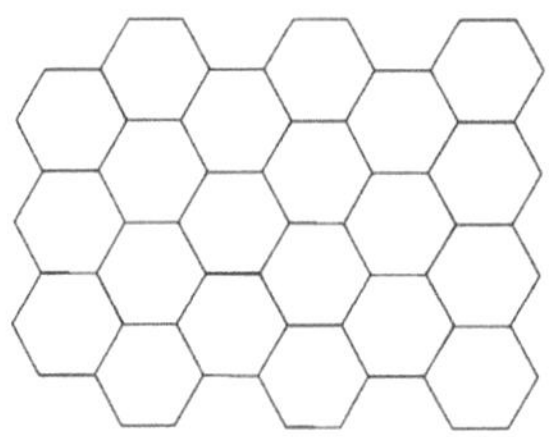

(a) Homorphism
Regular
Periodic Symmetric

(b) Homorphism
Irregular
Periodic Symmetric

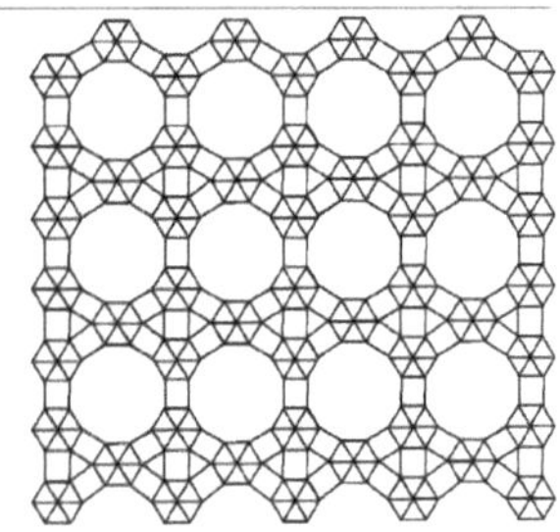

(c) Polymorphism
Semiregular
Periodic Symmetric.
(Ghyka, 2004)

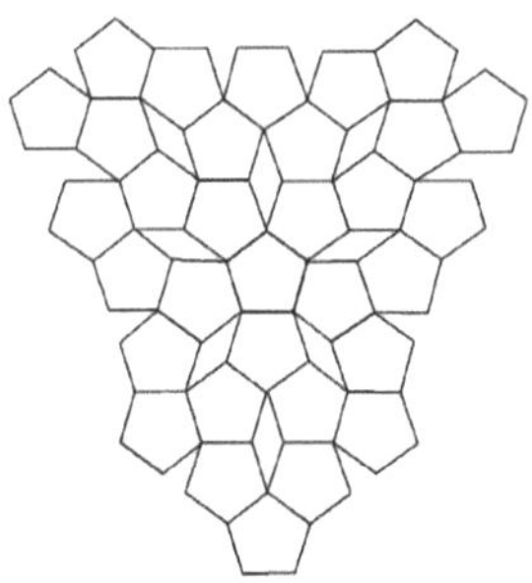

(d) Polymorphism
Semiregular
Aperiodic Symmetric

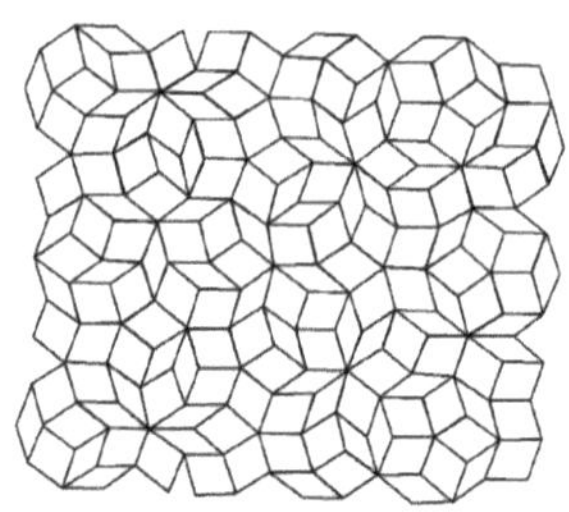

(e) Polymorphism
Semiregular
Aperiodic Asymmetric
(Penrose, 1974)

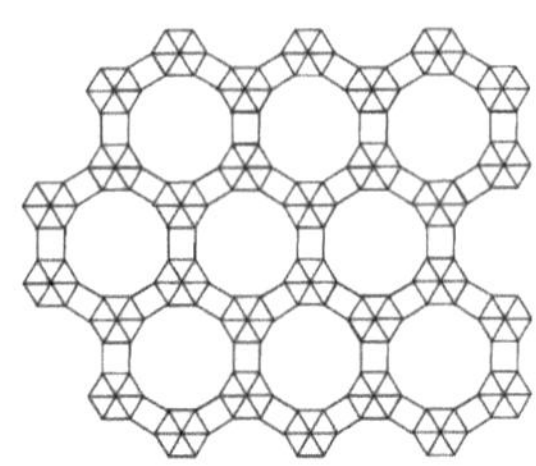

(f) Polymorphism
Regular
Periodic Symmetric
(Ghyka, 2004)

(g) Polymorphism
Irregular
Aperiodic Symmetric

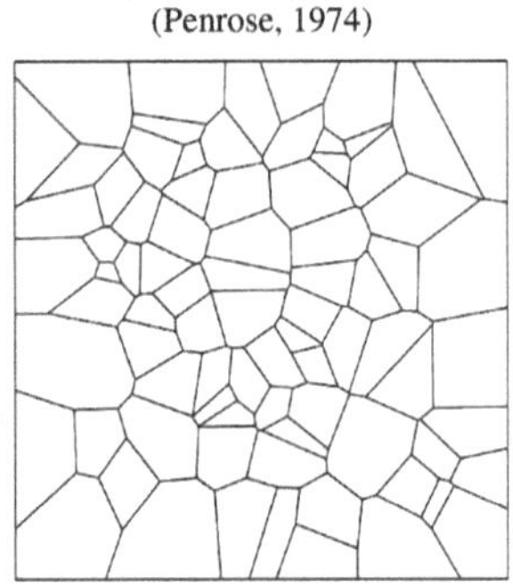

(h) Polymorphism
Irregular
Aperiodic Asymmetric

(i) Polymorphism
Irregular
Periodic Symmetric

variance of the number of neighbors per polygon. It measures the spatial disorder of a lattice, and is given by Eq. (3), where W_{ij} denotes the value of W in the row i and column j.

$$S = \frac{\sum W}{k^2} \tag{1}$$

$$(W - \lambda I_k)v = 0 \tag{2}$$

$$\mu_2 = \frac{\sum_{i=1}^{k} \left(\sum_{j=1}^{k} W_{ij} - \mu_1 \right)^2}{k - 1} \tag{3}$$

Within the field of regional science, lattices are frequently used with two purposes: First, real lattices can be used to study real phenomena, e.g., to analyze spatial patterns, confirm spatial relationships between variables and detect spatio-temporal regimes within a spatial panel, among others. Second, lattices can be used to evaluate the behavior of statistical tests [4, 45], algorithms [21] and topological characteristics of lattices [8, 40, 41]. In these cases, it is necessary to use sets of lattices that satisfy some requirements imposed by the regional scientist, e.g., the number of polygons, regularity or irregularity of the polygons and the number of instances. To accomplish this goal, it is a common approach to use a geographical base for real or simulated data *polymorphism irregular aperiodic asymmetric* (e.g., real lattices and Voronoi diagrams) or *homomorphism regular periodic symmetric lattices* (e.g., regular lattices). The following sections are restricted to the second use of lattices.

Topological Characteristics of Regular and Irregular Lattices

As stated above, regional scientists have the option of using regular or irregular lattices in their computational experiments. However, this section will show that there are important topological differences between these types of lattices.

Real lattices have topological characteristics that vary substantially from location to location. As an example, Fig. 3 presents the topological characteristics of lattices of different sizes (100, 400 and 900 polygons) sampled in Spain and the United States. Each box-plot summarizes 1000 instances. Important differences emerge between these two places: Spanish polygons tend to have more neighbors, are more disordered and their first eigenvalues are higher in mean and variance. These differences in the topological characteristics have direct repercussions on the performance of algorithms whose complexity depends on the neighboring structure [1, 21].

Regular lattices and Voronoi diagrams are also commonly used for computational experiments because they are easy to generate, there is no restriction on the size of the instances (the number of polygons in the map) and their over-simplified structure allows for some mathematical simplifications or reductions [9, 31, 61]. However, the topological characteristics of these lattices are substantially different from real,

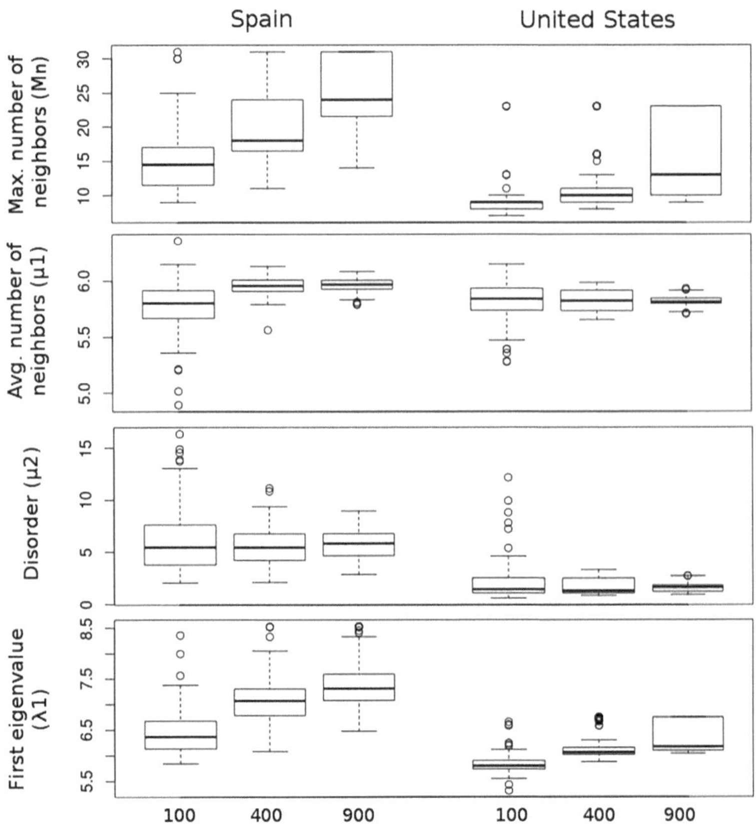

Fig. 3 Topological differences of lattices from Spain and the United States

irregular lattices. These differences can lead to biased results in theoretical and empirical experiments, e.g., spatial stationarity in STARMA models [36], improper conclusions about the properties of the power and sample sizes in hypothesis testing [4, 45] and the over-qualification of the computational efficiency of the algorithms [1, 21], among others. Table 3 shows the topological differences between real maps, two types of regular lattices and Voronoi diagrams.

To illustrate the magnitude of these differences, we calculated the topological indicators (M_n, m_n, μ_1, μ_2, S and λ_1) for six thousand lattices of different sizes (1000 instances each of 100, 400, 900, 1600, 2500 and 3600 polygons) that were sampled around the world at the smallest administrative division available in Hijmans et al. [35]. As an example, Fig. 4 shows seven of those instances. These real instances are then compared to regular lattices that have square and hexagonal polygons and Voronoi diagrams.[9] To avoid the boundary effect on M_n, m_n, μ_1 and

[9]Each one of the six-thousand instances of Voronoi diagrams come from uniformly distributed points.

Table 3 Average topological characteristics for real maps, regular lattices and Voronoi diagrams

		Number of polygons						
		81	100	400	900	1,600	2,500	3,600
Real lattices	M_n	12.28	13.22	23.22	29.55	42.77	48.53	60.64
		±7.52	±9.90	±29.89	±36.27	±50.59	±55.07	±64.58
	m_n	2.33	2.13	1.55	1.23	1.04	1.01	1.00
		±1.11	±1.06	±0.86	±0.60	±0.22	±0.10	±0.00
	μ_1	5.57	5.59	5.67	5.69	5.70	5.72	5.72
		±0.65	±0.61	±0.49	±0.45	±0.46	±0.37	±0.37
	μ_2	5.85	6.72	9.76	7.90	8.85	7.73	8.00
		±13.85	±22.58	±28.79	±15.35	±12.82	±9.39	±8.11
	S	5.98	4.91	1.30	0.58	0.33	0.21	0.15
		±0.51	±0.43	±0.11	±0.046	±0.02	±0.01	±0.01
	λ_1	5.96	6.09	6.89	7.30	8.03	8.33	8.92
		±0.53	±0.65	±1.52	±1.82	±2.42	±2.62	±3.02
Reg. lattice *(squares)*	M_n	4	4	4	4	4	4	4
	m_n	4	4	4	4	4	4	4
	μ_1	4	4	4	4	4	4	4
	μ_2	0	0	0	0	0	0	0
	S	4.44	3.64	0.95	0.43	0.24	0.16	0.11
	λ_1	3.80	3.84	3.96	3.98	3.99	3.99	3.99
Reg. lattice *(hexagons)*	M_n	6	6	6	6	6	6	6
	m_n	6	6	6	6	6	6	6
	μ_1	6	6	6	6	6	6	6
	μ_2	0	0	0	0	0	0	0
	S	6.30	5.19	1.39	0.64	0.36	0.23	0.16
	λ_1	5.55	5.62	5.88	5.94	5.96	5.97	5.98
Voronoi diagrams	M_n	9.15	9.36	10.37	10.90	11.26	11.49	11.71
		±0.77	±0.79	±0.75	±0.74	±0.70	±0.67	±0.68
	m_n	3.36	3.26	3.00	3.00	3.00	3.00	3.00
		±0.48	±0.44	±0.03	±0.00	±0.00	±0.00	±0.03
	μ_1	5.75	5.77	5.88	5.92	5.94	5.95	5.96
		±0.07	±0.05	±0.02	±0.01	±0.00	±0.00	±0.00
	μ_2	1.68	1.70	1.75	1.76	1.76	1.77	1.77
		±0.31	±0.27	±0.13	±0.09	±0.07	±0.05	±0.04
	S	6.67	5.47	1.44	0.65	0.37	0.24	0.17
		±0.08	±0.05	±0.00	±0.00	±0.00	±0.00	±0.00
	λ_1	5.88	5.96	6.20	6.26	6.28	6.29	6.30
		±0.05	±0.05	±0.03	±0.02	±0.02	±0.02	±0.02

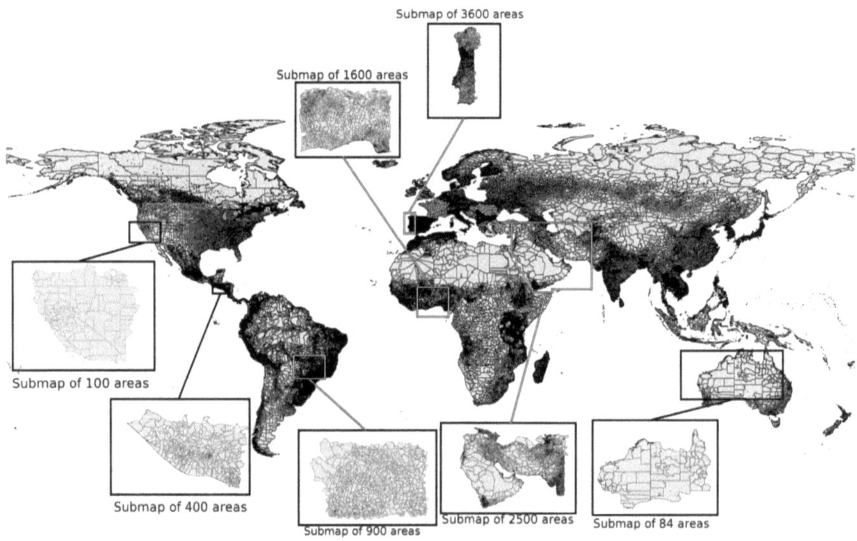

Fig. 4 Base map and example of a random irregular lattice obtained from it

μ_2, the bordering polygons are only considered to be neighbors of interior polygons. Last, S and λ_1 are calculated using all of the polygons. Table 3 shows that regular lattices are not capable of emulating the topological characteristics of real lattices in any of the indicators: $\mu_2 = 0$ and M_n, m_n, $\mu_1 = 4$ and 6 (for squares and hexagons, respectively) are values that are far from those of real lattices. The values obtained for λ_1 and S indicate that regular lattices of hexagons are more connected than real lattices, while regular lattices of squares are less connected than real lattices. With regard to Voronoi diagrams, M_n and m_n indicate that they are not capable of generating atypically connected polygons. The values of μ_1 are close to real lattices. Finally, Voronoi diagrams are more ordered than real lattices, with values of μ_2 close to 1.7, while real lattices report values of μ_2 that are close to 8.

RI-Maps: An Algorithm for Generating Realistic Irregular Lattices

This section is divided into two parts. The first part introduces an algorithm that generates irregular polygons based on a mean reverting process in polar coordinates, and the second part proposes a novel method to create polymorphic irregular aperiodic lattices with topological characteristics that are similar of those from real lattices.

Mean Reverting Polygons (MR-Polygons)

The problem of characterizing the shape of irregular polygons is commonly addressed in two ways, that is, evaluating its similitude with a circle [33] or describing its boundary roughness through its fractal dimension [10, 25].[10] In this paper, we apply both concepts in different stages during the creation of a polygon: The similitude with a circle to guide a mean reverting process in polar coordinates, and the fractal dimension to parameterize the mean reverting process.

Mean Reverting Process in Polar Coordinates

Different indexes are used to compare irregular polygons with a circle: Elongation ratio [60], form ratio [37], circularity ratio [44], compactness ratio [18, 29, 53], ellipticity index [56] and the radial shape index [17]. As Chen [15] states, all of these indexes are based on comparisons between the irregular polygon and its area-equivalent circle. Under this relationship, an irregular polygon can be conceptualized as an irregular boundary with random variations following a circle, which lead us to use a mean reverting process in polar coordinates to create irregular polygons.[11] A mean reverting process is a stochastic process that takes values that follow a long-term tendency in the presence of short-term variations. Formally, the process x at the moment t is the solution of the stochastic differential equation (4), where μ is the long-term tendency, α is the mean reversion speed, σ is the gain in the diffusion term, $x(t_0)$ is the value of the process when $t = 0$ and $\{B_t\}_{t\geq 0}$ is an unidimensional Brownian [43]. Equation (5) shows the general solution; however, for practical purposes, hereafter we use the Euler discretization method, which is given by Eq. (6), where ϵ_t is white noise.

$$dX_t = \alpha(\mu - X_t)d_t + \sigma dB_t \tag{4}$$

$$x(t) = e^{-\alpha(t-t_0)}\left(x(t_0) + \int_{t_0}^{t} e^{\alpha(s-s_0)}\alpha\mu ds + \int_{t_0}^{t} e^{\alpha(s-s_0)}\sigma dB(s)\right), \tag{5}$$

$$X_t = X_{t-1} + \alpha(\mu - X_{t-1})\Delta_t + \sigma\sqrt{\Delta_t}\epsilon_t \tag{6}$$

Algorithm 1 presents the procedure for generating an irregular polygon P in polar coordinates using, as a data generator, a mean reverting process (X_t). This algorithm guarantees that the distance between two points in X_t, following the process X_t, is equal to the distance between the same two points in P when following the process P counterclockwise. The purpose of this equivalence is to preserve the

[10] Chen [15] established a relationship between these two approaches.

[11] Polar coordinates allow us to "wrap" a mean reverting process, with fractal characteristics, around a circle to build a polygon. But, it is important to clarify that once we get those coordinates, we draw them in the Cartesian coordinate system.

Algorithm 1 MR-Polygon: mean reverting polygon.

function MEANREVERTINGPOLYGON($\alpha, \sigma, \mu, X_0, \Delta_t$)
  $X_{t-\Delta_t} = X_0$ ▷ Initial point of the mean reverting process
  $P = [(0, X_0)]$ ▷ Irregular polygon in polar coordinates
  while $\theta < 2\pi$ **do**
    $\epsilon_t \leftarrow \mathit{RandomNormal}(0, 1)$
    $X_t = X_{t-\Delta_t} + \alpha(\mu - X_{t-\Delta_t})\Delta_t + \sigma\sqrt{\Delta_t}\epsilon_t$
    $d \leftarrow \mathit{distance}(X_t, X_{t-\Delta_t})$
    $R_\theta \leftarrow$ Last radius of the irregular polygon
    $\phi_1 = \arccos\left(\frac{2R_\theta - \Delta_t^2}{2R_\theta^2}\right)$
    if $Xt \geq X_{t-\Delta_t}$ **then**
      $\Delta_R = d\left(\cos\left(\arcsin\left(\frac{\Delta_t}{d}\cos\left(\frac{\phi_1}{2}\right)\right)\right) - \sin\left(\frac{\phi_1}{2}\right)\frac{\Delta_t}{d}\right)$
    else
      $\Delta_R = -d\left(\cos\left(\arcsin\left(\frac{\Delta_t}{d}\cos\left(\frac{\phi_1}{2}\right)\right)\right) + \sin\left(\frac{\phi_1}{2}\right)\frac{\Delta_t}{d}\right)$
    end if
    $R_{\theta+\phi_1} = R_\theta + \Delta_R$
    Add $(\theta + \phi_1, R_{\theta+\phi_1})$ to P
    Increase θ in ϕ_1
  end while
  Replace last point of P to $(0, X_0)$
  return P
end function

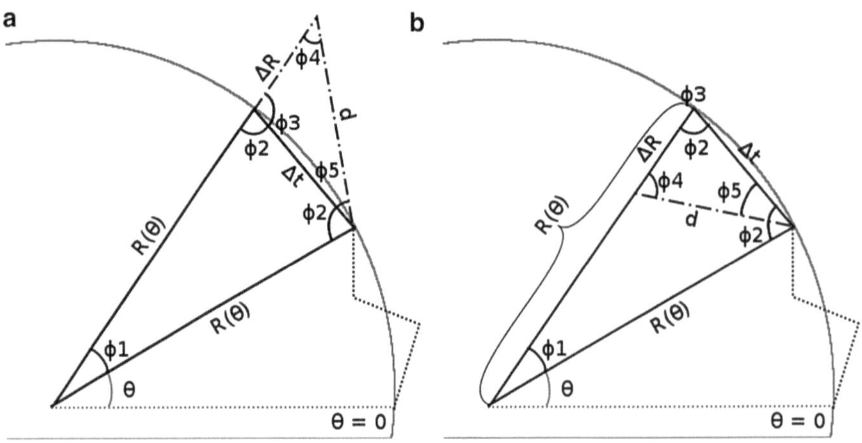

Fig. 5 Geometric problem to preserve the length and the fractal dimension of the mean reverting process when it is used to create an irregular polygon. (**a**) $X_t \geq X_{t-\Delta_t}$. (**b**) $X_t < X_{t-\Delta_t}$

fractal dimension of X_t in P. The angles Δ_R and ϕ_1 in Algorithm 1 are the result of solving the geometric problem presented in Fig. 5. These two angles are used in Eq. (7) to establish the location of the next point in P. The points of P are denoted as P_θ, with θ between 0 and 2π.

$$P_{\theta+\phi_1} = \begin{cases} P_\theta + \Delta_R & \text{if } X_{t+\Delta_t} \geq Xt \\ P_\theta - \Delta_R & \text{if } X_{t+\Delta_t} < Xt. \end{cases} \quad (7)$$

Because the process P depends on the parameters α, μ and σ, it is worthwhile to clarify their effect on the shape of polygon P: α is the speed at which the process reverts to the circle with radius μ and σ is the scaling factor of the irregularity of the polygon. High values of α and low values of σ generate polygons that have shapes that are close to a circle with radius μ. Finally, Δ_t is utilized to preserve the fractal dimension of both processes, X and P, and determines the angular step, ϕ_1 (see Fig. 5).

MR-Polygon Parameterization

The process of establishing the values for α, μ, σ, Δ_t and X_0 is not an easy task, and their values must be set in such a way that the shape of P is similar to a real irregular polygon. However, how do we determine whether a polygon P satisfies this condition? In this case, the fractal dimension appears to be a tool that offers strong theoretical support to assess the shape of a given polygon.

According to Richardson [53], the fractal dimension D of an irregular polygon (such as a coast) is a number between 1 and 2 (1 for smooth boundaries and 2 for rough boundaries) that measures the way in which the length of an irregular boundary L (Eq. (8)) changes when the length of the measurement instrument (ϵ) changes. The fractal dimension is given by Eq. (9), where $\hat{C}$ is a constant.

In general, an object is considered to be a fractal if it is endowed with irregular characteristics that are present at different scales of study [42]. For practical purposes, D is obtained using Eq. (9) and is given by 1 minus the slope of $\log(L(\epsilon))$. This procedure is commonly known as the Richardson plot.

$$L(\epsilon) = \hat{C}\epsilon^{1-D} \quad (8)$$

$$\log(L(\epsilon)) = (1 - D)\log(\epsilon) - \log(\hat{C}) \quad (9)$$

In almost all cases, the Richardson plot can be explained with two line segments that have different slopes; then, two fractal dimensions can be obtained: textural, for small scales, and structural, for large scales [39]. As illustrated, Fig. 6 shows a segment of the United States east coast taken from Google maps in two resolutions. Note that as the resolution increases, some irregularities that were imperceptible at low resolution become visible. In this sense, it can be said that irregularities at low resolution define the general shape and are related to the structural dimension, while irregularities at high resolution capture the noise and are related to the textural dimension. Regional scientists tend to use highly sampled maps, which preserve the general shape but remove the small variations. This simplification does not change the topological configuration of the maps [20]. Figure 7 presents the Richardson plot of the external boundary of the United States and its textural and structural fractal dimension.

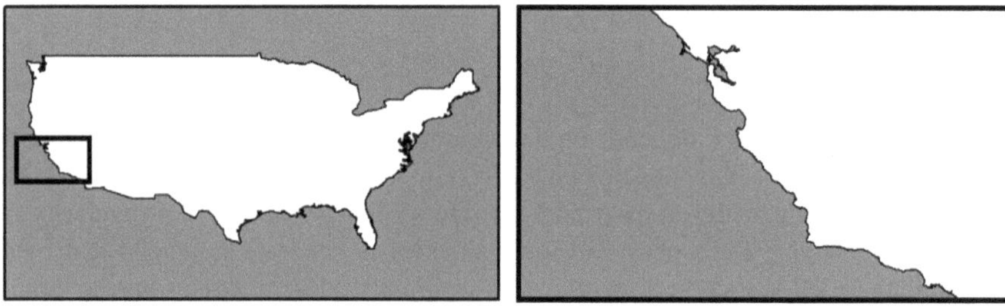

Fig. 6 Illustrative example of irregularities explained by the structural and textural dimension

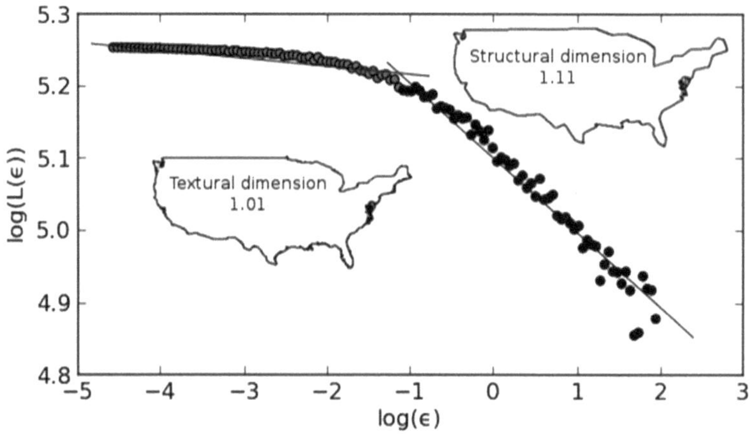

Fig. 7 Richarson plot to estimate the textural and structural dimension of the external boundary of the United States

In the field of stochastic processes, some approaches, which are based on different estimations of the length, have been made to characterize them through their fractal dimension. In our case, an experimental approach based on the fractal dimension of real polygons is proposed to select an appropriate combination of the parameters α and σ to generate realistic irregular polygons. Because our interest is on general shape rather than small variations, we account only for the structural dimension.[12] The parameterization process is divided into two parts: In the first part, the frequency histogram of the fractal dimensions of the real polygons is constructed. In the second part, we propose a range of possible values for α and σ, given μ, X_0, Δ_t, which generates fractal dimensions that are close to those obtained in the first part. Because the level of the long-term tendency μ does not affect the length of X and because Algorithm 1 guarantees that the length is preserved, μ

[12]To calculate the structural dimension, we use the EXACT procedure, which is devised by Allen et al. [2], with a small value for Δ_t. Next, both of the dimensions were determined by using a k-means clustering algorithm over the cloud of points on the Richardson plot.

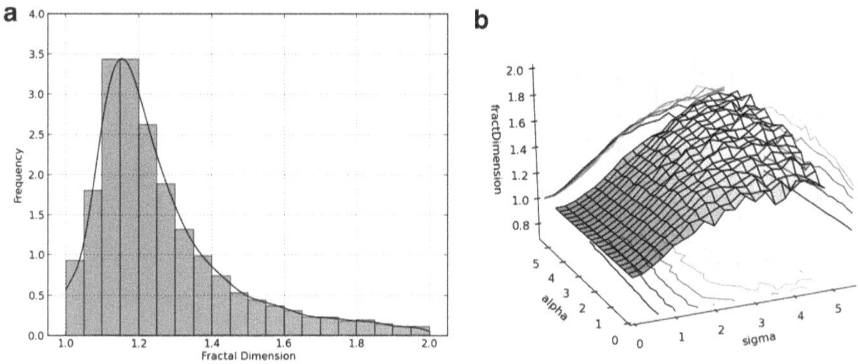

Fig. 8 Stages to find the values of α and σ. (**a**) Fractal dimensions of real polygons. (**b**) Fractal dimension of simulated polygons as a function of α and σ

can be defined as a constant without affecting the fractal dimension. Hereafter, it is assumed that $\mu = X_0 = 10$. The value of Δ_t is set to be 0.001 to properly infer both of the fractal dimensions.

The empirical distribution of the fractal dimension of the irregular polygons is calculated over a random sample of 10,000 polygons from the world map used in Section "Topological Characteristics of Regular and Irregular Lattices". The result of this empirical distribution is presented in Fig. 8a. To find the fractal dimension of the *MR-Polygons*, we generate a surface of the average dimensions as a function of the values of α and σ, which range from 0.01 to 5 with steps of 0.1 (Fig 8b). The resulting surface indicates that the fractal dimension is mainly affected by σ, especially when looking at small dimensions. Additionally, it is found that fractal dimensions close to 1.23 are obtained when σ takes on values between 1.2 and 1.5, regardless of the value of α.

Figure 9 presents some examples of polygons using different values of α and σ. The polygons in the second row, which correspond to $\sigma = 1.5$, produce irregular polygons that have a realistic structural fractal dimension. Additionally, in the same figure, both the original (gray line) and sampled (black line) polygons reinforce the fact that sampling a polygon does not affect the structural dimension. From now on, we will use sampled polygons to improve the computational efficiency.

Recursive Irregular Maps (RI-Maps)

Up to this point, we were able to generate irregular polygons with fractal dimensions that are similar to those from real maps. The next step is to use these polygons to create irregular lattices of any size whose topological characteristics are close to the average values obtained for these characteristics in real lattices around the world. For this step, we formulate a recursive algorithm on which an irregular

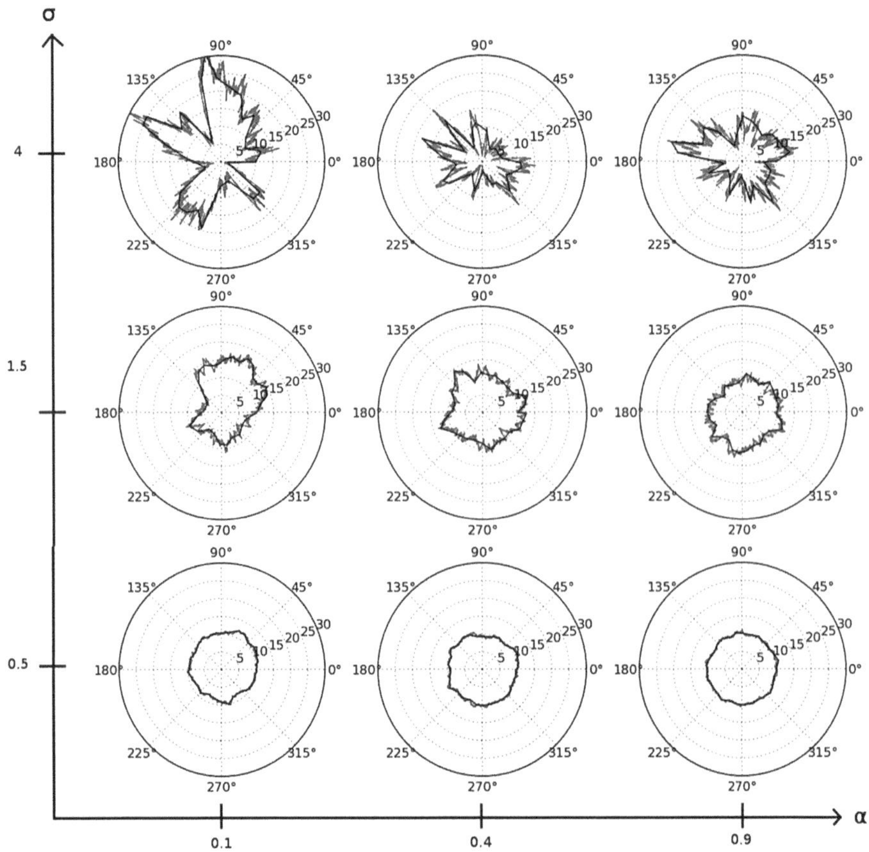

Fig. 9 Examples of stochastic polygons generated using Algorithm 1 with different values of σ and α

frontier is divided into a predefined number of polygons using *MR-Polygons*. Our conceptualization of the algorithm was made under three principles: (1) *Scalability:* Preserving the computational complexity of the algorithm when the number of polygons increases; (2) *Fractality:* Preserving the fractal characteristics of the map at any scale; and (3) *Correlativity:* Encouraging the presence of spatial agglomerations of polygons with similar sizes, which is commonly present in real maps in which there are clusters of small polygons that correspond to urban areas.

Algorithm 2 presents the *RI-Maps* algorithm to create polymorphic irregular aperiodic asymmetric lattices with realistic topological characteristics. This algorithm starts with an initial empty irregular polygon, *pol*, (the outer border of the *RI-Map*) and the number of polygons, *n*, to fit inside. In a recursive manner, a portion of the initial polygon *pol* starts being divided following a depth-first strategy until that portion is divided into small polygons.[13] This process is repeated for a new uncovered portion of *pol* until the whole area of *pol* is covered. Because the recursive partitions are made by using *MR-Polygons*, we take the values of α from a uniform distribution between 0.1 and 0.5, and the values of σ from a uniform distribution between 1.2 and 1.5. Regarding μ, X_0 and Δ_t, we use values proposed in Section "Mean Reverting Polygons (MR-Polygons)". Finally, to guarantee the computational treatability of the geometrical operations, each polygon comes from a sampling process of 30 points. The main steps of the *RI-Maps* algorithm are summarized in Fig. 10.

The *RI-Maps* algorithm has three unknown parameters:

- p_1: Because each polygon is created by the *MR-Polygons* using a polar coordinate system that is unrelated to the map being constructed with *RI-Maps*, it is necessary to apply a scaling factor, $\sqrt{\frac{p_1 \times area(pol)}{n \times \pi \times \mu^2}}$, that adjusts the size of the *MR-Polygon* before being included into the *RI-Map*.
- p_2: When a new polygon is used to divide its predecessor, its capacity to contain new polygons (measured by the number of polygons) is proportional to its share of the unused area of its predecessor. However, to enforce the appearance of spatial agglomerations of small polygons, the number of polygons that the new polygon can hold is increased with a probability of p_2.
- p_3: When p_2 indicates that a new polygon will hold more polygons, the number of extra polygons is calculated as the p_3 percent of the number of missing polygons that are expected to fit into the unused area of its predecessor polygon. The number of extra polygons is subtracted from the unused area to keep constant the final number of polygons (n).

Table 4 illustrates the effect of the parameters p_2 and p_3 on the topological characteristics of *RI-Maps*. In the first row, p_2 and p_3 equal 0, which generates highly ordered lattices without spatial agglomerations. The second and third rows are more disordered than the first row and have spatial agglomerations, with those in the second row less frequent and evident than those in the third row. As will be shown in the next section, lattices in the third row are more realistic in terms of their topological characteristics.

To find a combination of p_1, p_2 and p_3 that generates realistic *RI-Maps* in terms of their topological characteristics, we use a standard genetic algorithm, where the population γ at iteration i, denoted as γ^i, is formed by the genomes $\gamma_j^i = [p_{j1}^i, p_{j2}^i, p_{j3}^i]$, where p_{j1}^i, p_{j2}^i and p_{j3}^i are real numbers between 0 and 1, representing instances of p_1, p_2, p_3, which are denoted as phenomes. In this case, $i \in \mathbb{N}$ between 0 and 20 and $j \in \mathbb{N}$ between 0 and 100. To evaluate the quality of each genome's fitness function, $F(\gamma_j^i)$ is defined in Eq. (10), where θ is a set of polygons, ϕ_k is

[13]There is not a proven computational advantage or theoretical reason behind the decision of implementing a depth-first strategy. We follow this strategy because it simplified the coding structure.

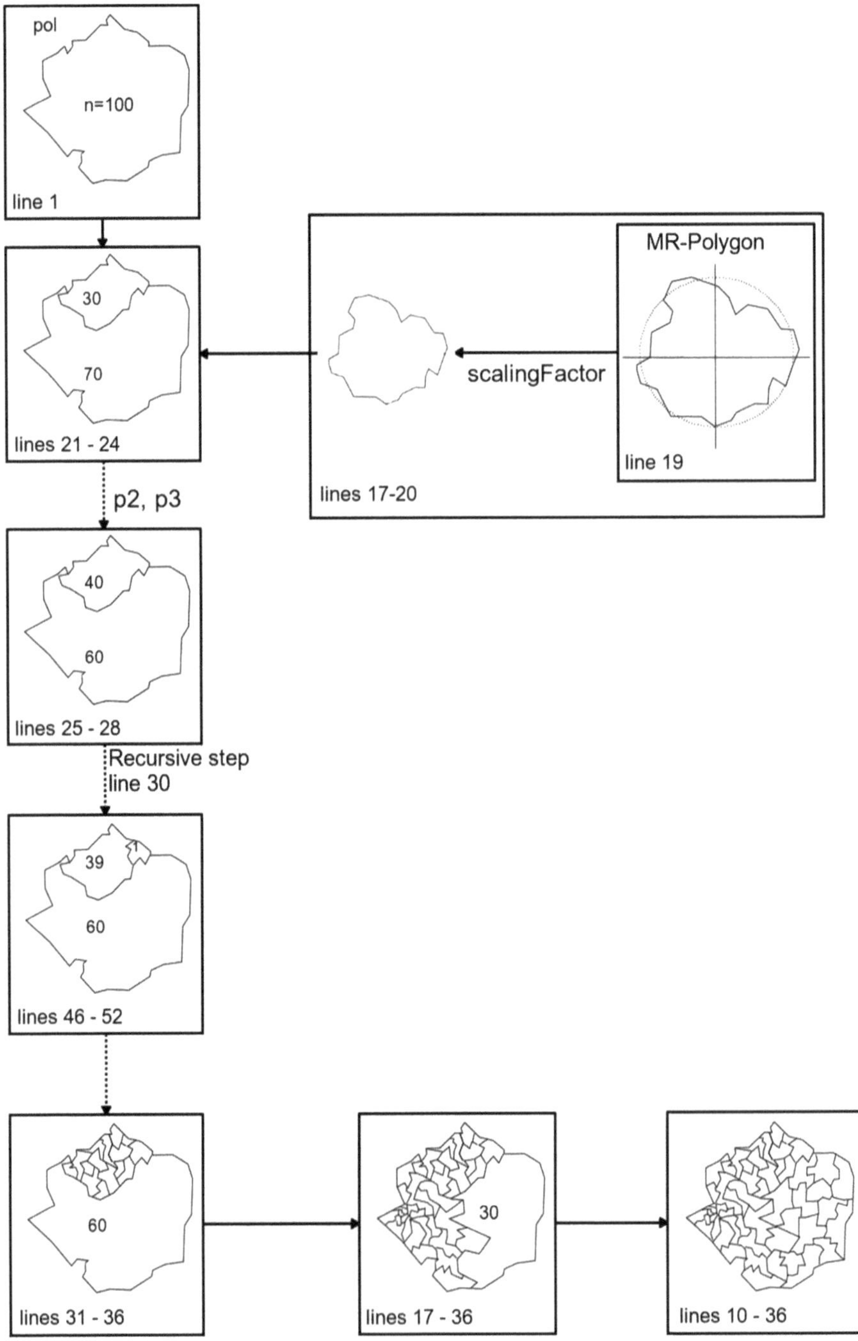

Fig. 10 Diagram of the main steps of the *RI-Maps* algorithm

Algorithm 2 RI-Map: recursive irregular map.

```
function RecursiveIrregularMap(n, pol)
    (α_min, α_max, σ_min, σ_max, μ, X_0, Δ_t) = (0.1, 0.5, 1.2, 1.5, 10, 10, 0.001)
    p_1 ∈ ℝ, p_2 ∈ ℝ, p_3 ∈ ℝ
    if n > 2 then
        missingPolygons ← n
        uncoveredPolygon ← pol
        coveredPolygon ← ϕ
        polygons ← []
        scalingFactor ← sqrt( (p_1 × area(pol)) / (n × π × μ^2) )
        while area(uncoveredPolygon) / area(pol) >= 0.03 do
            uncovered2select ← Bigger part of uncoveredPolygon
            if missingPolygons × area(uncovered2select) / area(uncoveredPolygon) <= 1.5 then
                polygons.put(uncovered2select)
                coveredPolygon ← coveredPolygon ∪ uncovered2slect
                missingPolygons ← missingPolygons − 1
            else
                α ← RandomUniform(α_min, α_max)
                σ ← RandomUniform(σ_min, σ_max)
                pol_i ← MeanRevertingPolygon(α, σ, μ, X_0, Δ_t)
                pol_i ← Multiply each ratio of pol_i by scalingFactor
                pol_i ← Center pol_i randomly into uncovered2select
                pol_i ← (pol_i − coveredPolygon) ∩ pol
                pol_i ← Bigger part of pol_i
                n_i ← missingPolygons × area(pol_i) / area(uncoveredPolygon)
                if Uniform(0, 1) < p_2 then
                    n_i = n_i + missingPolygons × p_3
                end if
                n_i ← Round(n_1)
                if n_i ≥ 1 then
                    polygons_i ← RecursiveIrregularMap(n_i, pol_i)      ▷ Recursive step
                    polygons ← polygons ∪ polygons_i
                    coveredPolygon ← coveredPolygon ∪ polygons_i
                    missingPolygons ← missingPolygons − n_i
                end if
            end if
            uncoveredPolygon ← pol − coveredPolygon
        end while
        Append interior holes of coveredPolygon to polygons
        coveredArea ← ⋃ polygons
        while length(polygons) < n do
            Append the smaller polygon to its larger neighbor
        end while
        while length(polygons) > n do
            Divide the larger polygon
        end while
    else if n = 1 then                                              ▷ Terminating case
        polygons ← [pol]
    else                                                            ▷ Terminating case
        pol_1, pol_2 ← Divide pol in 2
        polygons ← [pol_1, pol_2]
    end if
    return polygons
end function
```

Table 4 Examples of *RI-Maps* of 400, 1600 and 3600 polygons using different combinations of parameters

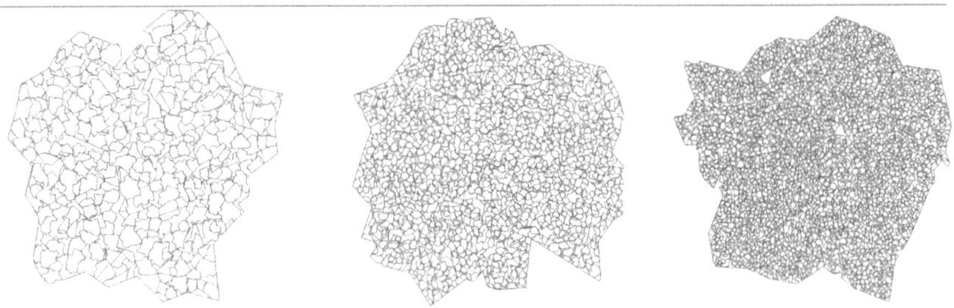

i) $p_1 = 0.1$, $p_2 = 0.0$, $p_3 = 0.0$

ii) $p_1 = 0.1$, $p_2 = 0.1$, $p_3 = 0.5$

iii) $p_1 = 0.010$, $p_2 = 0.050$, $p_3 = 0.315$

the relative importance for a map of k polygons and $f_k(\gamma_j^i)$ is a function given by Eq. (11) that measures the average difference between the values of the topological indicators of real lattices and those values of *RI-Maps* formed by k polygons using the phenome γ_j^i. For the sake of simplicity, in Eq. (11), $\Psi_k = [M_n, m_n, \mu_1, \mu_2, S, \lambda_1]$ denotes the vector of real indicators and $\Psi_k(\gamma_j^i)$ denotes the vector for the mean values of *RI-Maps* with k polygons using γ_j^i. The superindex l is used in Ψ_k^l and $\Psi_k^l(\gamma_j^i)$ to refer to the lth indicator in the real and simulated values, respectively. Finally, ns is the number of simulations to be generated with each genome.

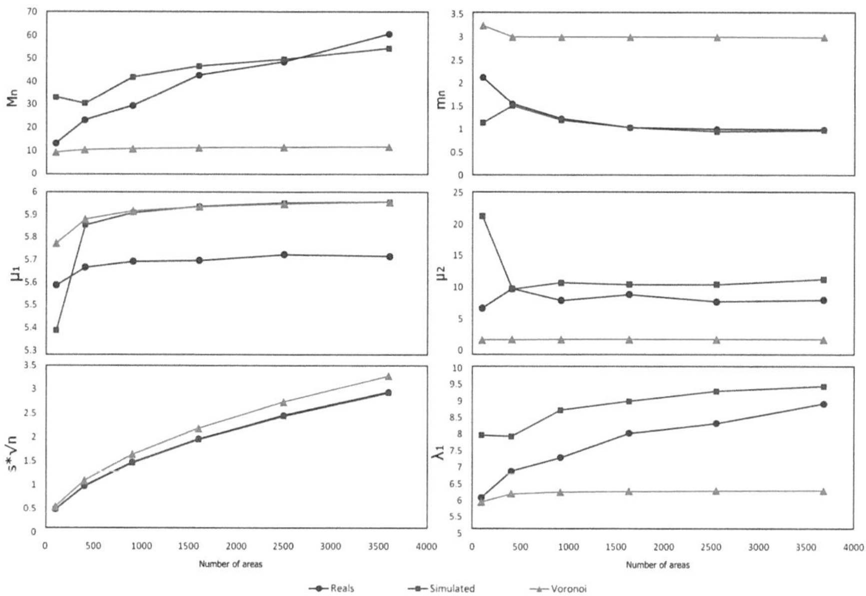

Fig. 11 Comparison of the topological characteristics of real lattices, *RI-Maps* and Voronoi diagrams

$$F(\gamma_j^i) = \frac{\left(\sum_{k\in\theta} \phi_k f_k(\gamma_j^i)\right)}{\sum_{k\in\theta} \phi_k} \quad (10)$$

$$f_k(\gamma_j^i) = \frac{\sum_{l=1}^{6} \frac{\left(\sum_{s=i}^{ns} \Psi_k^l(\gamma_j^i)\right) - ns\Psi_k^l}{ns\Psi_k^l}}{6} \quad (11)$$

The algorithm starts with an initial random population of 100 genomes to obtain the best four genomes. The subsequent populations are composed of two parts. The first 64 genomes are all of the possible combinations of the last best 4 genomes, and the other 36 genomes are random modifications of those 64 genomes. Because of the computational time required to evaluate Eq. (10), only lattices of 400 and 1600 were used, with an importance of $\phi_{400} = 1$ and $\phi_{1.600} = 2$, respectively. The algorithm reached the optimal value after 13 iterations with $p_1 = 0.010, p_2 = 0.050$ and $p_3 = 0.315$.

Results

Figure 11 presents a graphical comparison of the topological characteristics of real *RI-Maps* and Voronoi diagrams. The values for the *RI-Maps* were obtained from 100 instances.[14] The results show that *RI-Maps* have a maximum (M_n) and a minimum

[14]The code to generate *RI-Maps* is available to the academic community as a utility within the module "inputs" in clusterPy V.0.10.0, an open source cross-platform library of spatial clustering

(m_n) number of neighbors that are very close to the values found in the real lattices. Regarding the average number of neighbors, both *RI-Maps* and Voronoi diagrams show similar values that are slightly higher than those observed in real lattices. However, because the number of neighbors is an integer value, it can be concluded for all three cases that the average number of neighbors is 6, which verifies the findings by Weaire and Rivier [59] in irregular lattices. Regarding μ_2, *RI-Maps* are a better approach to simulate the level of disorder found in real lattices. To facilitate the visualization, the values of S are reported as $S * \sqrt{n}$. The results show that *RI-Maps* replicate the values of real lattices at any size, while Voronoi diagrams report higher values that tend to increase with the number of polygons. Last, *RI-Maps* have values of λ_1 that are closer to the values of real lattices, especially for large instances.

Table 5 presents the average and standard deviation of *RI-Maps* under the optimal parameters ($p_1 = 0.010$, $p_2 = 0.050$, $p_3 = 0.315$) found in the previous section. This table completes the topological information on lattices presented in Table 3. Figure 12 shows the running times for different instance sizes using a HP ProLiant DL140 Generation 3 computer running the Linux Rocks 6.0 operating system equipped with 8 GB RAM and a 2.33 GHz Intel Xeon Processor 5140. The dotted line shows the $x = y$ values, but its non-linear appearance is due to the quadratic scale used in the x-axis to improve the visualization of the plot. Although the reported times correspond to a non-optimized code, the plot shows an almost linear relationship between the problem size and the running time.[15]

Table 5 Topological characteristics (mean and standard deviation) for *RI-Maps*

	Number of polygons						
	81	100	400	900	1600	2500	3600
M_n	26.500	33.100	30.460	41.870	46.760	49.730	54.396
	± 12.765	±14.751	±13.664	±16.035	±16.429	±16.886	±15.156
m_n	1.260	1.140	1.510	1.200	1.040	0.950	0.979
	±0.691	±0.513	±0.847	±0.550	±0.374	±0.261	±0.204
μ_1	5.347	5.388	5.855	5.909	5.937	5.952	5.957
	±0.333	±0.304	±0.093	±0.052	±0.036	±0.032	±0.027
μ_2	17.397	21.313	9.722	10.708	10.426	10.443	11.276
	±13.734	±14.729	±6.189	±3.831	±2.277	±1.506	±1.229
S	5.974	4.879	1.372	0.620	0.353	0.226	0.157
	±0.324	±0.219	±0.020	±0.006	±0.002	±0.001	±0.001
λ_1	7.431	7.969	7.931	8.724	8.993	9.299	9.449
	±0.804	±0.808	±0.979	±0.962	±0.960	±0.957	±0.798

algorithms written in Python [22]. To access the repository go to: https://code.google.com/p/clusterpy.

[15]Future research will be devoted to reduce computational time and exploit the possibilities of parallelization.

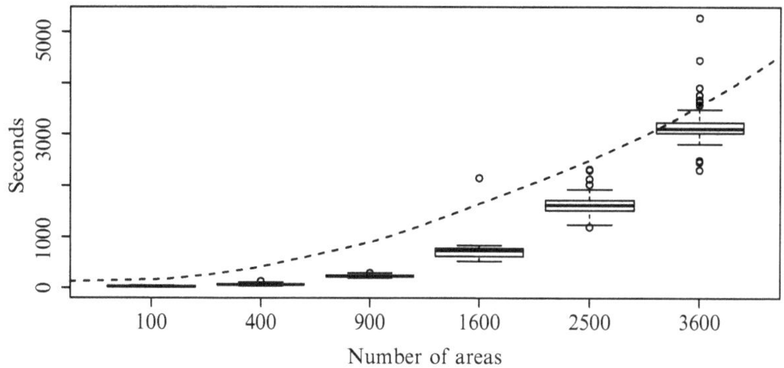

Fig. 12 Running times of RI-Maps while the number of areas increases

Table 6 Kolmogorov-Smirnov test to compare the distributions of AMOEBA execution times using different lattices

	Regular lattices	RI-Maps	Real maps
Regular lattices	0.00 (p=1)	0.51 (p=0.0e^{-4})	0.61 (p=2.8e^{-5})
RI-Maps	0.51 (p=0.0e^{-4})	0.00 (p=1)	0.19 (p=0.607)
Real maps	0.61 (p=2.8e^{-5})	0.19 (p=0.607)	0.00 (p=1)

Application of *RI-Maps*

In this section, we present an example of the use of *RI-Maps* based on the computational experiments designed by Duque et al. [21] to compare the efficiency of the improved AMOEBA algorithm. To present the results, Duque et al. [21] proposed three computational experiments; one of them reports the running time of AMOEBA as the number of polygons of regular lattices increases. In this paper, we will run the same algorithm not only for regular lattices but also for real irregular and simulated irregular lattices (*RI-Maps*). First, we want to see whether the conclusions that are obtained for regular lattices can be extrapolated to irregular lattices. Second, we want to see if the results obtained with *RI-Maps* are also valid for real irregular maps. This experiment was executed with a HP ProLiant DL140 Generation 3 computer running the Linux Rocks 6.0 operating system equipped with 8 GB RAM and a 2.33 GHz Intel Xeon Processor 5140.

In the generated experiment, for each type of lattice, there were 30 instances with 1600 polygons. For each instance, we generated a spatial process that had four clusters, each using the methodology proposed by Duque et al. [21]. Last, the instances for real maps were obtained from sampling the same world map that was used in previous sections. Figure 13 presents the distribution of the running times obtained for each type of lattice, and Table 6 compares the distributions with the two-sided Kolmogorov-Smirnov test [27]. The null hypothesis of the Kolmogorov-Smirnov test is that the two samples come from the same probability distribution. Regarding the first question, it is clear that using a regular lattice for testing the

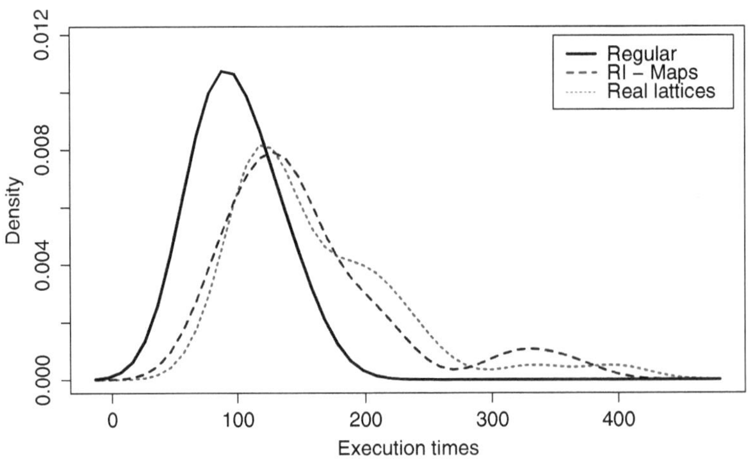

Fig. 13 Execution times of AMOEBA over regular lattices and *RI-Maps* of 1600

AMOEBA underestimates the execution times. On the other hand, the distribution of the running times obtained for real maps and *RI-Maps* is statistically equal, which shows the benefits of using *RI-Maps* because it can automatically generate instances without limiting the maximum number of polygons.

Conclusions

This paper introduces an algorithm that combines fractal theory, the theory of stochastic processes and computational geometry for simulating realistic irregular lattices with a predefined number of polygons. The main goal of this contribution is to provide a tool that can be used for geocomputational experiments in the fields of exploratory spatial data analysis, spatial statistics and spatial econometrics. This tool will allow theoretical and empirical researchers to create irregular lattices of any size and with topological characteristics that are close to the average characteristics found in irregular lattices around the world.

As shown in the last section, the performance of some geocomputational algorithms can be affected by the topological characteristics of the lattices in which these algorithms are tested. This situation can lead to an unfair comparison of algorithm performances in the literature. With the algorithm proposed in this paper, the differences in the computational performances will not be affected by the topological characteristics of the lattices.

This paper also shows that the topological characteristics of regular lattices (with squared and hexagonal polygons) and Voronoi diagrams (commonly used to emulate irregular lattices) are far from the topological characteristics that are found in real lattices.

Acknowledgements The authors wish to thank Colciencias (Departamento Administrativo de Ciencia y Tecnología e Inovación) for their financial support under the program "Jovenes Investigadores." The authors also thank the Cyberinfrastructure Service for High Performance Computing, "Apolo," at Universidad EAFIT, for allowing us to run our computational experiments in their supercomputer. The usual disclaimer applies.

References

1. Aldstadt J, Getis A (2006) Using AMOEBA to create a apatial weights matrix and identify spatial clusters. Geogr Anal 38(4):327–343
2. Allen M, Brown G, Miles N (1995) Measurement of boundary fractal dimensions: review of current techniques. Powder Technol 84(1):1–14
3. Anselin L (1988) Spatial economtrics: methods and models, 1st edn. Kluwer Academic, Dordrecht
4. Anselin L, Moreno R (2003) Properties of tests for spatial error components. Reg Sci Urban Econ 33(5):595–618
5. Anselin L, Smirnov O (1996) Efficient algorithms for constructing proper higher order spatial lag operators. J Reg Sci 36(1):67–89
6. Anselin L, Bera A, Florax R, Yoon M (1996) Simple diagnostic tests for spatial dependence. Reg Sci Urban Econ 26(1):77–104
7. Anselin L, Florax R, Rey SJ (2004) Advances in spatial econometrics: methodology, tools and applications. Springer, Berlin
8. Aste T, Szeto K, Tam W (1996) Statistical properties and shell analysis in random cellular structures. Phys Rev E Stat Phys Plasmas Fluids Relat Interdiscip Topics 54(5):5482–5492
9. Bartlett MS (1975) Probability, statistics, and time: a collection of essays, 1st edn. Chapman and Hall, New York
10. Batty M, Longley P (1994) Fractal Cities: A Geometry of Form and Function. Harcout Brace & Company, London
11. Blommestein H, Koper N (2006) Recursive algorithms for the elimination of redundant paths in spatial lag operators. J Reg Sci 32(1):91–111
12. Boots B (1982) Comments on the use of eigenfunctions to measure structural properties of geographic networks. Environ Plan A 14:1063–1072
13. Boots B (1984) Evaluating principal eigenvalues as measures of network structure. Geogr Anal 16(3):270–275
14. Boots B (1985) Size effects in the spatial patterning of nonprincipal eigenvectors of planar networks. Geogr Anal 17(1):74–81
15. Chen Y (2011) Derivation of the functional relations between fractal dimension of and shape indices of urban form. Comput Environ Urban Syst 35:442–451
16. Church RL (2008) BEAMR: an exact and approximate model for the p-median problem. Comput Oper Res 35(2):417–426
17. Clark W (1964) The concept of shape in geography. Am Geogr Soc 54(4):561–572
18. Cole J (1964) Study of major and minor civil divisions in political geography. In: 20th International geographical congress, Sheffield, University of Nottingham
19. Coxeter HSM (1974) Regular complex polytopes. CUP Archive, Cambridge
20. Douglas D (1973) Algorithms for the reduction of the number of points required to represent a digitized line or its caricature. Cartographica: Int J Geogr Inf Geovisualization 10(2):112–122
21. Duque J, Aldstadt J, Velasquez E, Franco J, Betancourt A (2011) A computationally efficient method for delineating irregularly shaped spatial clusters. J Geogr Syst 13:355–372
22. Duque JC, Dev B, Betancourt A, Franco JL (2011) ClusterPy: {Library} of spatially constrained clustering algorithms, {Version} 0.9.9.
23. Duque JC, Anselin L, Rey SJ (2012) the max-p-regions problem. J Reg Sci 52(3):397–419

24. Elhorst JP (2003) Specification and estimation of spatial panel data models. Int Reg Sci Rev 26(3):244–268
25. Frankhauser P (1998) The fractal approach: a new tool for the spatial analysis of urban agglomerations. Popul Engl Sel 10(1):205–240; New Methodological Approaches in the Social Sciences. www.persee.fr/doc/pop_0032-4663_1998_hos_10_1_6828
26. Garrison WL, Marble DF (1964) Factor-analytic study of the connectivity of a transportation network. Pap Reg Sci Assoc 12(1):231–238. doi:10.1007/BF01941256
27. George Marsaglia WWT, Wang J (2003) Evaluating Kolmogorov's distribution. J Stat Softw 8(18):1–4
28. Ghyka M (2004) The geometry of art and life. Kessinger Publishing, Whitefish, MT
29. Gibbs J (1961) A method for comparing the spatial shapes of urban units. In: Urban research methods. D. Van Nostrand Company, Inc, Princeton, pp 96–106
30. Gould P (1967) On the geographical interpretation of eigenvalues. Trans Inst Br Geogr 42(42):53–86
31. Griffith D (1987) Toward a theory of spatial statistics: another step forward. Geogr Anal 19(1):69–82
32. Grunbaum B, Shephard GC (2011) Tilings and patterns. Dover Publications, New York
33. Haggett P (1977) Locational analysis in human geography. Wiley, New York
34. Haining R (2010) The nature of georeferenced data. In: Fischer MM, Getis A (eds) Handbook of applied spatial analysis. Springer, Berlin, pp 197–217
35. Hijmans R, Guarino L, Jarvis A, O'Brien R, Mathur P, Bussink C, Cruz M, Barrantes I, Rojas E (2005) DIVA-GIS 7.1.7. Available in http://www.diva-gis.org/
36. Hooper P, Hewings G (1981) Some properties of space-time processes. Geogr Anal 13(3):203–223
37. Horton R (1932) Drainage basin characteristics. Trans. AGU 13(1):350–361
38. Johnson DL (2001) Symmetries, 1 edn. Springer, London
39. Kindratenko V, Treiger B (1996) Chemometrical approach to the determination of the fractal dimension (s) of real objects. Chemom Intell Lab Syst 34:103–108
40. Le Caer G, Delannay R (1993) The administrative divisions of mainland France as 2D random cellular structures. J Phys I 3(8):1777–1800
41. Le Caer G, Delannay R (1995) Topological models of 2D fractal cellular structures. J Phys I 5(11):1417–1429
42. Mandelbrot BB (1982) The fractal geometry of nature. W.H. Freeman, San Francisco
43. Mao X (1997) Stochastic differential equations and applications, 1 edn. Horwood Publishing, Chichester
44. Miller V (1953) A quantitative geomorphic study of drainage basin characteristics in the clinch mountain area Virginia and Tenesse. Department of Geology, Columbia University
45. Mur Lacambra J (1992) Contrastes de autocorrelación espacial: Un estudio de Monte Carlo. Estadística Española 34(130):285–308
46. Murray AT, O'Kelly ME (2002) Assessing representation error in point-based coverage modeling. J Geogr Syst 4(2):171–191
47. Ord K (1975) Estimation methods for models of spatial interaction. J Am Stat Assoc 70(349):120–126
48. Pace R, LeSage, JP (2004) Chebyshev approximation of log-determinants of spatial weight matrices. Comput Stat Data Anal 45(2):179–196
49. Penrose R (1974) The Rôle of Aesthetics in Pure and Applied Mathematical Research. J Inst Math Appl 10:266–271
50. Peshkin M, Strandburg K, Rivier N (1991) Entropic predictions for cellular networks. Phys Rev Lett 67(13):1803–1806
51. Radin C (1993) Symmetry of tilings of the plane. Bull Am Math Soc 29(2):213–217
52. ReVelle C, Swain R (1970) Central facilities location. Geogr Anal 2(1):30–42
53. Richardson L (1961) The problem of contiguity: an appendix of statistics of deadly quarrels. General Systems Yearbook 6(13):139–187

54. Ross IC, Harary F (1952) On the determination of redundancies in socioeconometric chains. Psychometrika 17(2):195–208
55. Smirnov O, Anselin L (2001) Fast maximum likelihood estimation of very large spatial autoregressive models: a characteristic polynomial approach. Comput Stat Data Anal 35(3):301–319
56. Stoddart D (1965) The shape of atolls. Mar Geol 3(5):269–283
57. Tilley R (2006) Crystals and crystal structures, 1st edn. Wiley, Chichester
58. Tinkler K (1972) The physical interpretation of eigenfunctions of dichotomous matrices. Trans Inst Br Geogr 55(55):17–46
59. Weaire D, Rivier N (2009) Soap, cells and statistics—random patterns in two dimensions. Contemp Phys 50(1):199–239
60. Weeitty A (1969) On the form of drainage basins, 1st edn. Department of Geography, Pennsylvania State University, Pennsylvania
61. Whittle P (1954) On stationary processes in the plane. Biometrika 41(3):434–449

A Robust Heuristic Approach for Regionalization Problems

Kamyoung Kim, Yongwan Chun, and Hyun Kim

Introduction

Geographical districting is a process of partitioning a large areal unit into a fixed number of sub-regions. In practice, sub-regions are constructed by aggregating smaller areal units within a larger areal unit. This process is implemented to achieve a solution that produces an optimal outcome based on preset criteria. These criteria are generally set to achieve a maximum balance among resulting sub-regions or the maximum homogenous characteristic of each sub-region. Hence, an optimization model is often used for geographical districting problems. Geographical districting has been utilized in various fields, such as school attendance zone design [1], functional region delineation [2], census region redistricting [3], police district design [4], and hazards and disasters management [5]. Specifically, one well-known application is political districting [6].

Political districting (also known as electoral districting) is a process to define political boundaries with a fixed number of districts in which an election is

K. Kim (✉)
Department of Geography Education, Kyungpook National University, 80 Daehak-ro, Buk-gu, Daegu, 702-701, South Korea
e-mail: kamyoungkim@knu.ac.kr

Y. Chun
School of Economic, Political and Policy Sciences, University of Texas at Dallas, 800 West Campbell Road, Mail Station GR31, Richardson, TX, 75080-3021, USA
e-mail: ywchun@utdallas.edu

H. Kim
Department of Geography, University of Tennessee at Knoxville, 311 Burchfiel Geography Building, 1000 Phillip Fulmer Way, Knoxville, TN, 37996-0925, USA
e-mail: hyunkim@utk.edu

J.-C. Thill, S. Dragicevic (eds.), *GeoComputational Analysis and Modeling of Regional Systems*, Advances in Geographic Information Science,
DOI 10.1007/978-3-319-59511-5_16

performed. "Fairness" and "balance" are well emphasized in political districting because political districting can favor a specific political party or group (e.g., Gerrymandering). Although various factors are well-recognized, three fundamental factors commonly identified in the literature are equal population, spatial contiguity, and compactness (e.g., [7, 8]). Equal population (or population balance) ensures the comparable sizes of political districts, while spatial contiguity ensures that each individual district does not have an isolated portion or hole within its boundary. The compactness, which refers to how contorted a political district is, can be used to avoid intentional manipulations, such as the 12th congress district of North Carolina in 1992 (see [9]). Another factor recognized in the literature is the respect of pre-existing boundaries (e.g., [10]). These pre-existing boundaries can include natural boundaries, such as major water bodies (e.g., rivers) and administrative or political subdivisions. Because physical and social barriers can represent social and geographical integrity, incorporating these pre-existing boundaries may lead to better solutions for political districting problems.

Spatial optimization approaches are generally utilized for political districting. Although an exact approach can be applied (e.g., [11]), heuristic methods have been often involved due to the complex and multi-criteria nature of political districting. For example, heuristic approaches include not only a simple greedy method (e.g., [12]), but also meta-heuristic methods such as tabu [10], simulated annealing [13], and genetic algorithm [14, 15]. However, the performance of heuristic and meta-heuristic methods may vary considerably depending on how to deal with constraints, calling into question the robustness of these methods in many cases.

This paper focuses on the development of a robust heuristic method for political districting problems. Specifically, this paper investigates the robustness of the heuristic method with a physical barrier in the model in addition to the common three criteria. Although real physical barriers exist in many regions, the incorporation of physical barriers in heuristic modeling is scarce (e.g., [10]). Many exact solutions and heuristic approaches do not incorporate the presence of physical barriers when they are developed because of the difficulty in explicitly prescribing the barriers' characteristic form in the model. This paper proposes a new heuristic method to effectively search an optimal solution with physical barriers as constraint. The rest of this paper is organized as follows. Section "Literature Review" briefly reviews related studies for political districting and related methods. Section "Problem Statement" describes the proposed heuristic algorithms, and section "Application Results" presents an application for the city of Seoul, South Korea with the Han River as a physical barrier. Conclusions and implications are provided in section "Conclusions".

Literature Review

There are several important conditions in political districting problems. The main goal of political districting problems is to group similar spatial units into a fixed number of political boundaries while satisfying criteria. The districted boundaries are created for legislative purposes, including elections, taxes, and governance. From a modeling perspective, the objective function of a political districting problem is either to maximize social and geographical *equity* or minimize the difference between them among districted regions. The most common attribute in the objective function is the population size in each district. Some essential constraints for spatial characteristics include spatial contiguity, compactness, and physical barriers. *Spatial contiguity* means that the spatial extent of a constructed district should not be fragmented. *Compactness* can be formulated as the district's area to perimeter ratio. *Physical barriers*, such as rivers and mountains, are recognized as pre-boundary conditions because they affect the integrity of districted regions directly. In other words, the resulting districts by pre-defined physical barriers greatly influence the direction of political decisions, the consolidation of residents, and the integrity of the community within a district [8, 16, 17]. Although physical barriers have not been well-addressed, notable research contributions exist in literature. Mills [18] utilized a permanent assignment method to determine electoral boundaries to ensure that a particular population group is assigned to a specific unit. Segal and Weinberger [19] used an adjacency graph where adjacency among spatial units is represented as link in a graph form. This method represents the contiguity of spatial units involved in natural barriers. As a districting method, they employed a shortest algorithm to identify districts by calculating the shortest paths between each center and all the candidate units in a pre-defined set. As a way to avoid unnecessary splitting of communities of common interests and crossing natural barriers, George et al. [20] proposed algorithms which solve the districting problem using the a piecewise-linear objective function and a prespecified radius constraints (i.e., previously existing electoral districts) to avoid the undesirable splits of geographic units. To ensure a similar population size with a tolerance among districts, the algorithm employed a penalty function that limits the sum of population flows below the target value. As an application for the sales territory alignment problem, Zoltners and Sinha [21] introduced an adjacency tree structure which explicitly takes into account nontraversable geographic obstacles such as mountains, lakes, and rivers as well as traversable road network.

In response to the need for realistic political redistricting problems with various criteria, developing an appropriate automated process based on mathematical programming is necessary. Unfortunately, as criteria become more complicated and the applications expand, most exact-solution seeking models quickly become intractable for formulating optimal solutions, such that only small instances are solvable. Hence, many political districting problems are solved by taking algorithmic approaches rather than mathematical programming to compromise between the quality and solvability of solutions. However, algorithmic approaches do not always guarantee optimal solutions.

According to Duque et al. [22], regionalization problems are classified into two categories according to the design of their models: the model with explicit contiguity constraints or implicit contiguity constraints. Because of the difficulty to incorporate spatial contiguity constraints into models, many algorithmic solution approaches have been proposed to improve the solvability of the problems without explicit contiguity conditions [22]. Hess et al. [12] developed a political districting problem that was formulated using the structure of a location-allocation model. Their mathematical formulation is known as an effective regionalization approach. However, this approach does not guarantee the spatial contiguity requirement for resulting districts; therefore, they provided a method to resolve the contiguity problem using a greedy-based heuristic method where the solution is completed only after non-contiguous spatial units are reassigned, which assures contiguity. Garfinkel and Nemhauser [11] proposed a multi-step algorithm to ensure the contiguity of boundaries in districting regions. They used a set of pre-defined districts to improve solvability, and the algorithm was designed to check for spatial contiguity requirements while districting. However, this study implies that the results are not consistently generated and lead to a large variation in district size because the districts generated by the solution are considerably dependent on what pre-defined feasible solution is used.

In contrast, many exact solution approaches using mixed integer programming (MIP) have also been proposed in regionalization problems. The most difficult part of the solution is prescribing the spatial contiguity requirement with explicit mathematical forms. Zoltners and Sinha [21] proposed an efficient formulation to maintain the contiguity of a region using the concept of the connectedness of the tree that is formed among the units within each district. Shirabe [23] proposed flow-based contiguity constraints to avoid fragmented regions. The constraints use the concept of p-partitioned tree networks for a given number of areal units. Spatial units are represented as network nodes where the flow from a seed unit should be connected to other units in the region. As discussed by Duque et al. [24], the various strategies for contiguity requirements in exact solution approaches, such as tree-associated, ordered-area assignment, and flow-based constraints, help to solve regionalization problems, but the MIP models are inherently *NP*-hard, where the solution is only valid for a limited instance. Alternatively, a stepwise MIP solution approach, *Analytical Target Reduction* (ATR) by Kim et al. [2], was proposed to improve the capability of determining the optimal solution for a large instance. The main concept of the ATR is focused on reducing the solution space of hard problems by using the set of known solutions. For example, the optimal solutions for the instances of low computational complexity can be a pre-solution in other hard instances. However, the method does not directly reduce the complexity of the model itself.

As non-exact solution search methods, heuristic approaches are preferred in non-tractable instances. Heuristic approaches for districting or regionalization problems are capable of generating near-optimal or optimal solutions, although the quality of solution is not always consistently maintained due to the characteristics of the constraints that are required in districting problems. A simple approach includes clustering or partitioning methods to quickly determine a solution (see [25–28]). Such methods often require additional processes to ensure the spatial contiguity of the resulting regions.

Meta-heuristic approaches have been applied to districting problems since the 1990s. The goal of a meta-heuristic approach is to provide a higher level of frameworks when a certain type of heuristic algorithm is implemented for particular optimization problems [29]. The meta-heuristic strategies consist of diversification and/or intensification to avoid solutions entrapped in local optima. *Diversification* refers to a way that the solution space is explored, while *intensification* is an accumulation process for improving solutions based on the search experience. Both strategies expand the set of starting points in solution space by prohibiting particular *moves* to produce a new best solution. Classical regionalization problems employed greedy frameworks or simple hill-climbing techniques for the diversification of solution space [30–32]. Tabu search (hereafter TABU) uses a set of *tabu* lists that suppress units in short- and long-term memory to explore a new local solution space. The main principle of TABU is to overcome local optima using tabu lists when the procedure moves from one solution to another. An element in the tabu lists is defined as a *move* that is used in a current solution but cannot be released as a solution element until the objective function reaches a certain expiration level to avoid entrapment into sub-optima (see [29] in detail). This meta-heuristic is also commonly found in the literature for districting problems (for example, [3, 10, 33–36]).

Simulated annealing (SA) is used as an effective solution method for redistricting problems that consider multi-attribute criteria. The SA allows deteriorated solutions with a certain probability to avoid local optima [37]. Since its introduction by Browdy [16], this method has become popular in political districting problems with several variants [13, 35, 38].

The SA method is focused on the strategy of diversification, particularly how to effectively swap the basic spatial units among districts and build the local search method to reduce computational time, while improving the objective function. In the context of the districting problems, the strategy is basically performed on the *move* of a basic spatial unit from one region to one of the *neighboring* regions because the quality of the solutions depends on the handling of the move among adjacent districts [2, 34]. In detail, because the complexity of the MIP approach is due to the constraints of spatial contiguity, any greedy or meta-heuristic approach suffers from computational burden when checking for spatial contiguity among spatial units. To address this issue, Openshaw and Rao [3] and MacMillan [38] proposed spatial contiguity checking procedures using matrix operations. These methods are simple and very effective in identifying the adjacent spatial units that need to be swapped between adjacent districts. However, the procedures require considerable computational time if the adjacent matrix becomes large, which degrades the performance for finding good solutions. In addition, the performance of meta-heuristic approaches is contingent upon establishing parameters to control random components. In other words, many trials and errors may occur before the best set of parameters for a given instance is determined.

Problem Statement

In this section, we present the problem statement for political redistricting that motivates the development of a new robust heuristic algorithm, named the Dissolving/Splitting heuristic algorithm (DS). The performance of this algorithm will be compared with two generic meta-heuristic approaches, TABU and SA. As mentioned in the previous section, heuristic algorithms should establish the rule needed to move a basic spatial unit from its current district to an adjacent district, leading to the creation of a new neighboring solution. The DS algorithm is expandable to the other regionalization problems because the main purpose of the algorithm is the enhancement of the regrouping of districts (or regions) with neighborhood spatial units, rather than providing a general framework at a higher level. We consider three main criteria for the political districting problem: population equality, spatial contiguity, and compactness. These criteria are considered in the three heuristics presented in this chapter (e.g., [7, 8]). Additionally, we hypothesize that a physical barrier is a critical factor that influences the performance of heuristic approaches, which may cause the entrapment of solutions in local solution space. For this purpose, we add a physical barrier constraint to the model formulation and implement it in the case study of Seoul, Korea, where the Han River is recognized as a natural barrier for delineating political districts.

We first use the objective function that is generally adopted in existing political districting research, namely minimizing population deviation among districts [3, 35, 36]. The objective function is formulated as follows:

$$\text{Minimize } Z = \sum_{j=1}^{k} \left| p_j - \overline{p} \right| \tag{1}$$

where

k = the number of districts to be delineated
p_j = population of district j
$\overline{p}$ = average district population, $\overline{p} = \sum_j p_j / k$.

The objective function minimizes the difference between the average population ($\overline{p}$) and the population of each district. The $\overline{p}$ value is calculated *a priori*.

Second, compactness is considered as a constraint. As found by Young [39], there are different versions of compactness measures for a district, but we adopt a simple measure of compactness that has previously been used by Bozkaya et al. [10], Wei and Chai [36], and Ricca and Simeone [35]. The area and perimeter of each basic spatial unit are calculated using the geometry function in geographic information systems (GIS) software as a part of data preprocessing. Based on these input data, the area of a district can be obtained by summing the area of basic spatial units when assigned to the district. In detail, the compactness measure S_k is defined as:

$$S_k = \frac{Perimeter_k}{2\sqrt{\pi}\sqrt{Area_k}} \tag{2}$$

where the perimeter of district k is computed as:

$$Perimeter_k = \sum_{j \in N_k} Perimeter_j - 2\sum_{i \in N_k}\sum_{j \in N_k} l_{ij} \tag{3}$$

Here l_{ij} is the length of shared boundaries between spatial units i and j, and N_k is set of spatial units assigned to district k. Note that the equation of $Perimeter_k$ of district k should consider only the outer boundaries of spatial units of district k, and the lengths of all shared boundaries among spatial units within district k should be excluded.

Third, in this study, we consider the additional spatial constraint that a linear spatial feature such as a river may entail as physical barrier, when the algorithm couples two spatial units in a district, k. The principle is as follows. If spatial units i and j share a boundary, but the physical barrier is on the boundary between them, the algorithm manages the spatial units as separate while delineating the districts. Three algorithms, SA, TABU and DS, are structured accordingly with these model components.

Simulated Annealing (SA)

We construct an SA algorithm for comparison purposes. The SA algorithm for the political districting problem is formulated on the basis of the structure suggested by previous research, such as Kirkpatrick et al. [40], Openshaw and Rao [3], and Duque et al. [34]. The common structure of their algorithms is as follows. After generating a feasible initial solution by grouping n basic spatial units into p regions, the algorithm randomly selects a region and moves its basic spatial unit into a neighboring region under the contiguity constraint. If the *move* improves the objective value, then the move is accepted. Moves that do not improve the objective value, i.e., bad moves, are accepted to explore more of the solution space and to escape from local optima. Such moves are allowed if the acceptance probability $R(0, 1) < e^{-\Delta H/T}$, where $R(0, 1)$ is a random number in the interval [0, 1]. $\Delta H$is the change in the objective value caused by the move, and T is the current temperature. The temperature is lowered with a predefined cooling schedule. The acceptance probability of bad moves decreases as T is lowered. The SA is terminated when topping criteria are satisfied, that is, when T reaches a predefined tolerance or the algorithm iterates a predefined number of times without an improvement in the objective value. In detail, the procedure of the SA algorithm is as follows:

- Step 1: Set initial temperature T, cooling rate α, tolerance ε, and the maximum number of non-improving moves *max_it*.

- Step 2: Randomly generate an initial set of p districts from n basic spatial units under the contiguity requirement ($p < n$), and calculate the objective value of the initial solution.
- Step 3: Randomly select a district and generate a list of its edge spatial units that share a boundary with the adjacent districts without violating the contiguity requirement.
- Step 4: Randomly select an edge spatial unit from the list and move it to an adjacent district. Then, calculate the objective value of the new solution.
  - Step 4.1: If the objective value is improved, then accept the move, $k = 1$.
  - Step 4.2: If the move does not improve the objective value, accept it only if $R(0, 1) < e^{-\Delta H/T}$ and reduce the temperature with a cooling rate and increase the iteration count, k: $T = \alpha T$, $k = k + 1$.
- Step 5: Update the list of edge spatial units and return to step 4 until all spatial units in the list have been processed.
- Step 6: Repeat Steps 3–5 until $T < \varepsilon$ or $k > max\_it$.

TABU

In this section, we describe a simple TABU algorithm that has a structure based on the work by Openshaw and Rao [3] and Duque et al. [34]. Our TABU algorithm uses a tabu restriction and an aspiration criterion as the means for constraining and guiding the search process to the best districting set. To generate an environment similar to previous studies, only short-term memory is considered in our algorithm. The algorithm starts with a feasible initial solution and then identifies all possible moves (i.e., all possible alternative solutions) in the current solution. The best move among all possible moves is accepted if it improves the current objective value. In the case that no improving moves exist, the algorithm selects moves randomly or selects the best move and accepts them, even if it results in a worse objective value. In our experiment, random selection provides better results than accepting the best move because non-best moves allow the algorithm to explore a new solution space. To prevent cycles, the reverse move is prohibited during tabu tenure (i.e., tabu restriction). In particular, the tabu tenure of our algorithm is changed adaptively with the improvement of the current solution. A move in the tabu list (i.e., a tabu move) is allowed only if it produces a better solution than the local best solution, which is the aspiration criterion. The TABU algorithm is terminated when it reaches the maximum number of iterations (i.e., *max_it*) without improving the aspiration criterion. The procedure of TABU search is described as follows:

- Step 1: Set the length of tabu list *tabulist* and the maximum number of non-improving moves *max_it*.

- Step 2: Randomly generate an initial set of p districts from n basic spatial units under the contiguity constraint ($p < n$), and calculate the objective value of the initial solution. The initial solution becomes the current and local best solution.
- Step 3: For the current solution, find all possible moves M between two neighboring districts and identify the best move (the move with the biggest improvement in the objective value) among them.
- Step 4: Accept the best move if it yields a solution better than the local best solution. The solution accepting the best move becomes the current and local best solution. Set $k = 1$ and proceed to step 3.
- Step 5: If no move improves the solution, then consider a tabu move. If a tabu move yields a solution better than the local best solution, then accept it and delete it from the tabu list. The solution accepting the tabu move becomes the current and local best solution. Set $k = 1$ and proceed to step 3.
- Step 6: If there is no improving move in M and the tabu list, randomly select a move from M and allow it, even if a worse objective value is returned. Add the move to the tabu list, set $k = k + 1$, and proceed to step 3.
- Step 7: Stop the tabu search algorithm when $k > max\_it$.

A Robust Heuristic Algorithm: The Dissolving–Splitting (DS) Algorithm

In general, the meta-heuristics, such as SA and TABU, can produce higher quality solutions than traditional greedy algorithms because meta-heuristics explicitly implement strategies for widening the solution space. In the context of districting problems, the search capability by meta-heuristics may be restricted because most diversification strategies regarding local moves are made by the handover of a basic spatial unit from its region to a *neighboring* region or the exchange of basic spatial units among regions under spatial contiguity. In particular, meta-heuristic algorithms based on a local move become more inherently restricted when physical barrier constraints are involved in districting procedures. For example, it is reasonable that physical barriers, such as rivers and mountain ranges, become the outer boundaries of a region rather than the interior boundaries among the basic spatial units of a district (or region). Therefore, switching the basic spatial units of a district with other neighboring or adjacent districts can entail inefficient procedures, resulting in increased computational time, and performance degradation towards improving the quality of solution.

In contrast, often in districting problems, a greedy-based algorithm is more straightforward for finding a higher quality solution compared to meta-heuristics. Our robust heuristic algorithm, called the dissolving-splitting algorithm (DS), is also based on the framework of a greedy-search method rather than a meta-heuristic approach. The DS is designed to effectively explore the solution space and avoid deteriorated solutions without parameter settings, which are required in meta-heuristics. The DS consists of two sub-procedures, the DS procedure of districts and the intensive local search, where the objective value is gradually improved by moving a basic spatial unit from its current district to a neighboring region. As implied by the term dissolving-splitting, the procedure focuses on randomly dissolving *small* districts into a larger district while simultaneously splitting a larger population district into two districts to achieve the equality of the population, thus satisfying the objective of the districting problem. Spatial contiguity is checked during the dissolving and splitting procedures. The DS procedure is specified as follows:

- Step 1: Randomly generate an initial set of p districts from n basic spatial units under the contiguity constraint ($p < n$). The initial solution becomes the current solution and global best solution.
- Step 2: Dissolving-splitting procedure
  - Step 2.1: Randomly select a district from the current solution and dissolve it. That is, the spatial units consisting of the selected district are randomly assigned to neighboring districts.
  - Step 2.2: Select a district with the largest population and split it into two districts. That is, randomly select two seeds (spatial units) among the spatial units of the selected district. Then, assign the remaining spatial units to the seeds with the contiguity constraint. The result of this procedure becomes the DS solution.
- Step 3: Local search procedure
  - Step 3.1: Calculate the objective value (*local objective*) of the DS solution.
  - Step 3.2: Randomly select a district and generate a list of its edge spatial units that share a boundary with their adjacent districts and can be assigned to adjacent district(s) without violating the contiguity requirement.
  - Step 3.3: Randomly select an edge spatial unit from the list and move it to an adjacent district. Then, calculate the objective value of the new solution.
  - Step 3.4: If the objective value is improved compared with *local objective*, then accept the move and update *local objective*.
  - Step 3.5: Update the list of edge spatial units and return to step 3.3 until all spatial units in the list have been examined.
  - Step 3.6: Repeat Steps 3.2–3.5 until there is no improvement in *local objective*. The result of this local search procedure becomes the local best solution.
- Step 4: Comparison of global and local best solutions
  - Step 4.1: If the local best solution is better than the global best solution, then the local best solution becomes the current solution and the global best solution. Proceed to step 2.
  - Step 4.2: Otherwise, proceed to step 2.
- Step 5: Repeat steps 2–4 until there is no improvement in the objective value.

The procedure of DS is similar to the merge–split methods devised by Sammons [41] and Openshaw [42]. Sammons's algorithm considers particular districts with a considerably large population to be divided into two sub-districts, and only two neighboring districts with small populations are merged together into a district. In Openshaw's [42] approach, merging and splitting are allowed only when the value of the objective function is improved. However, these predefined conditions of the merging and splitting procedures often prevent the algorithms from exploring other solution spaces, resulting in the entrapment of solution at local space. Compared to these previous methods, the DS algorithm allows any district to be dissolved and split without any constraint.

Application Results

Data and Design for the Experiments

The proposed method is applied to a political districting problem in the city of Seoul, South Korea. Redistricting is an important issue, as South Korea elects the National Assembly every 4 years. In the 2012 South Korea legislative election, 48 electoral districts were defined in the city, with each district represented by one seat in the National Assembly. We investigate the performance of the DS algorithm in comparison to the SA and TABU algorithms by constructing 48 electoral districts based on the level of *dong*, the lowest administrative unit in Korea. Figure 1 displays 424 *dongs*, with their population sizes based on the 2010 Population and Housing Census of Korea: the total population is approximately ten million (9,804,065). Specifically, the population equality and compactness (or conformity) of districts are the most critical factors in redistricting for the legislative election [43]. A physical barrier is incorporated in addition to the three common factors in this political districting problem. The Han River, which passes through Seoul, is widely accepted as a boundary of economic and cultural disparities between the River North (or *Gangbuk*) and River South (or *Gangnam*) regions in Seoul. Additionally, as a natural barrier, the river is used to construct the subdivisions of Seoul. Hence, it is not desired for any single district to be drawn across the river in the political districting problem.

The performance of the DS method is compared with those of two other heuristic methods, SA and TABU. Each method produces results with 100 random initial feasible solutions for each of four scenarios with different numbers of districts ($p = 30$, 40, 48, and 50). Regarding population equality, the ideal population sizes for each district are 326,802 ($p = 30$), 245,102 ($p = 40$), 204,251 ($p = 48$), and 196,081($p = 50$). In each solution, compactness is maintained less than or equal to 2.0 of the compactness measure (S_k). To compare the performance among the methods, we set parameter values to make the computation times comparable. The parameter values for the SA are set as follows: initial temperature $T = 100{,}000$,

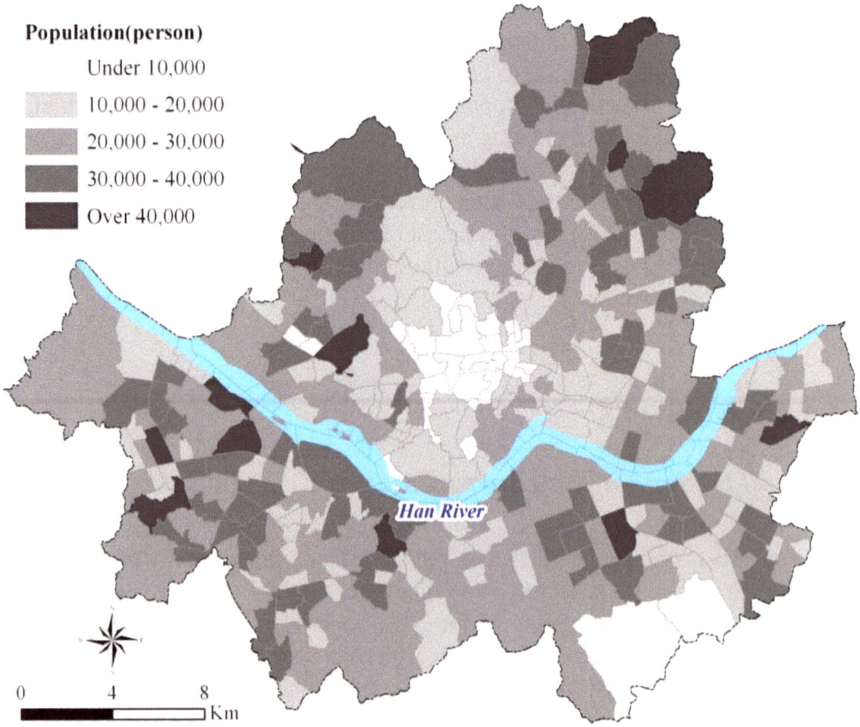

Fig. 1 Population distribution of Seoul Metropolitan City (2010)

cooling rate $\alpha = 0.997$, tolerance $\varepsilon = 5$ and the maximum number of non-improving moves *max_it* = 2000. The length of *tabulist* is 30 and the maximum number of non-improving moves *max_it* is set to 2000 for the TABU. A wide range of parameter values are explored for these models through a number of solutions with a different parameter set each time. The parameter values that produced the best results are chosen. In addition, computational time balance among three heuristics is considered. These three algorithms are programmed in Microsoft Visual Studio 2010 Express with Visual Basic.NET. The problem instances are solved on a machine with an Intel Core(TM) *i*5–3.40 GHz processor and 4 GB RAM on the Windows 7 operating system.

Comparison of Performance

Table 1 summarizes the computational results of three algorithms considering no physical barrier in the model. Descriptive statistics for the 100 solutions are also provided. The column labelled improved rate (%) measures the degree of

Table 1 Summary of computation results for Seoul (without physical barrier constraint)

		Initial solutions		SA				TABU				DS			
p	Descriptive statistics	Objective	S_k	Objective	S_k	Improved (%)	Time (s)	Objective	S_k	Improved (%)	Time (s)	Objective	S_k	Improved (%)	Time (s)
30	Mean	4,748,336	1.70	404,077	1.86	91.51	201.29	372,293	1.86	92.09	232.92	137,737	1.83	97.03	135.92
	Std. dev.	681,507	0.04	547,830	0.02	11.57	77.76	470,382	0.02	10.00	103.68	25,840	0.03	0.73	16.41
	Range	3,259,432	0.23	2,235,400	0.11	48.00	317.04	2,064,752	0.12	48.44	621.1	182,552	0.11	3.74	150.00
40	Mean	4,806,513	1.66	310,744	1.84	93.74	164.64	323,238	1.84	93.48	229.00	170,822	1.80	96.34	173.98
	Std. dev.	774,486	0.03	377,740	0.02	6.87	52.62	404,205	0.02	7.35	79.56	19,751	0.02	0.80	9.85
	Range	3,440,422	0.19	1,786,696	0.10	30.83	228.58	2,551,522	0.11	43.29	404.15	102,458	0.12	4.04	52.11
48	Mean	4,866,935	1.65	343,469	1.81	93.10	158.93	367,006	1.81	92.60	251.57	196,804	1.78	95.89	223.36
	Std. dev.	617,844	0.03	380,907	0.02	6.97	51.11	342,919	0.02	6.37	85.13	18,487	0.02	0.68	13.51
	Range	2,998,292	0.15	2,494,908	0.09	47.7	246.19	2,053,890	0.11	39.27	406.76	91,166	0.08	3.61	71.48
50	Mean	6,239,467	1.64	314,204	1.81	93.73	152.53	314,654	1.82	93.67	223.09	208,387	1.78	95.71	223.27
	Std. dev.	4,927,075	0.03	284,367	0.02	4.66	40.99	288,794	0.02	4.85	66.28	18,404	0.02	0.61	19.55
	Range	3,156,774	0.16	2,171,632	0.12	35.96	220.44	2,075,802	0.09	34.44	297.62	91,558	0.11	2.78	52.84

improvement on the objective value from that of each initial solution for each method. There is a noticeable difference in the mean and standard deviation of the results among the three methods with DS clearly outperforming the other methods. In general, the DS produces better solutions than the SA and the TABU. For example, the mean of the DS for $p = 30$ is 137,737 which is much lower than the TABU (372,293) and SA (404,077). This tendency is observed in other p levels. Considering the standard deviation and range of the 100 solutions, the solution quality of the DS algorithm appears very consistent compared to the other methods. Although the SA and TABU may produce better solutions than the DS in some instances, our experiments show that the DS clearly outperforms the meta-heuristics in most cases. The compactness values (S_k) of the solutions by the three heuristics get larger than those of the initial solutions. This indicates that compactness deteriorates, which is a result of controlling only the upper bound of the compactness measure in the three methods Note that the solution time of the DS is more sensitive to the number of districts than the SA and the TABU, because the frequency of the DS is a function of p.

To compare the consistency of the solutions, for an initial solution, we also examine the histograms of the objective values. As shown in Fig. 2, the three

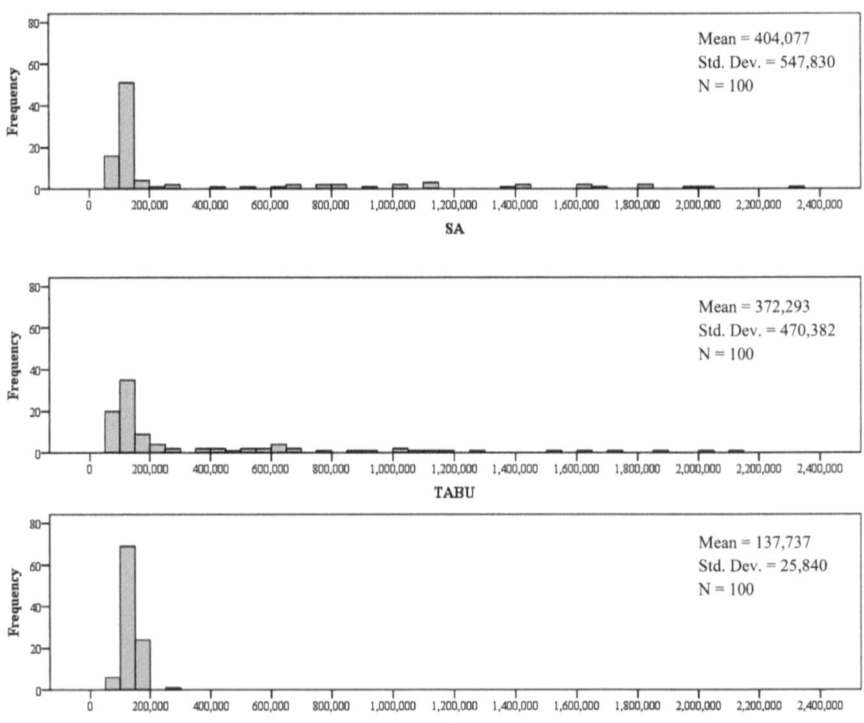

Fig. 2 Histogram of objective values of 100 simulations for an initial solution ($p = 30$)

algorithms produce different distributions of 100 objective functions due to their random components. The SA and TABU form a positively skewed distribution with the largest range between the best and worst solutions. This skewed distribution indicates that the SA and TABU may produce good but also poor solutions over a wide range. The modes of the meta-heuristics range from 100,000 to 150,000 and at a range of greater than 2,000,000 for the worst solutions. In contrast to those two meta-heuristics, the objective values of the DS are intensively concentrated between 50,000 and 200,000, with very small variances. For example, even the worst solution is smaller than 300,000. From this observation, the consistency in generating solutions is well maintained in the DS, while the SA and the TABU are considerably dependent on the random components in their algorithms. The DS process ensures a better quality of solutions with a less variation and is less affected by the quality of initial solutions. One notable fact from Table 1 and Fig. 2 is that the DS algorithm can be improved further by allowing the moves that give rise to a deteriorated result like the SA or the TABU.

Table 2 summarizes the results of the three algorithms when a physical barrier, the Han River, is considered. Notice that the solution times of the three methods decrease compared to the results of the instances without a physical barrier. This can be explained by the decreased number of move cases among districts because the number of neighbor units is reduced due to the presence of the physical barrier. The quality of the solutions from the SA and the TABU decreases for the range, standard deviation, and mean in problems with physical barriers than those with non-physical barriers. However, the DS solutions are not negatively influenced by physical barriers. In terms of the quality of the solutions, the DS clearly outperforms the two meta-heuristics. In many instances, the worst solutions of the DS are better than the best solutions of the SA and TABU. This is because the dissolving and splitting procedure is very effective in handling spatial contiguity and districting spatial units compared to the traditional meta-heuristics.

Figure 3 shows the spatial configuration of 48 districts derived from three heuristics for an initial solution. In these maps, red and green symbolize districts with larger and smaller populations than the average district population (204,251), respectively. In the SA and the TABU districting problems without physical barrier constraints, random components heavily influenced the results, generating poor objective values. For example, Fig. 3a and b show that districts with positive or negative population deviations are spatially clustered. In this situation, any local move could not improve the objective value even when allowing deteriorated solutions. Comparatively, in the case of the spatial solution of the DS without the physical barrier constraint (Fig. 3c), the population deviation of districts is greatly reduced, and districts with a larger or smaller population deviation are spatially dispersed. When considering the physical barrier constraint, the computational results of the SA and the TABU depend heavily on initial solutions. For example, the number of districts in the South of Han River at the initial solution is 20, but this number of districts is kept in the solutions of the SA and the TABU (see Fig. 3d and e). As a result, while districts with positive deviation are distributed in the south of the Han River, districts with negative deviation are distributed in

Table 2 Summary of computation results for Seoul (with physical barrier constraint)

p	Descriptive statistics	Initial solutions Objective	Initial solutions S_k	SA Objective	SA S_k	SA Improved (%)	SA Time (s)	TABU Objective	TABU S_k	TABU Improved (%)	TABU Time (s)	DS Objective	DS S_k	DS Improved (%)	DS Time (s)
30	Mean	5,021,442	1.70	1,980,877	1.84	61.36	127.41	1,909,820	1.83	62.68	118.92	290,493	1.81	94.06	115.39
	Std. dev.	808,360	0.04	1,190,192	0.03	20.13	46.07	1,182,787	0.03	20.19	66.80	35,822	0.03	1.24	16.25
	Range	4,077,018	0.23	5,737,598	0.13	78.67	276.02	5,228,832	0.13	76.66	402.01	146,000	0.17	5.78	87.78
40	Mean	4,921,490	1.66	1,493,986	1.81	70.22	120.05	1,455,982	1.82	70.90	119.50	263,969	1.78	94.54	143.09
	Std. dev.	642,167	0.03	1,005,285	0.03	18.83	30.13	997,137	0.03	18.75	62.02	19,736	0.03	0.82	11.32
	Range	3,087,808	0.19	3,944,760	0.13	72.61	155.89	3,960,218	0.13	72.59	323.09	97,072	0.15	4.25	67.04
48	Mean	5,103,082	1.65	1,558,598	1.80	69.80	119.23	1,514,494	1.79	70.66	120.62	236,836	1.77	95.28	176.81
	Std. dev.	663,141	0.03	932,065	0.02	17.01	31.36	916,112	0.03	16.78	60.80	31,958	0.02	0.89	10.51
	Range	3,026,238	0.15	3,740,780	0.11	62.08	161.6	3,313,742	0.15	62.30	289.33	130,102	0.12	4.12	56.05
50	Mean	4,158,749	1.64	1,425,384	1.79	72.98	117.06	1,421,023	1.79	73.05	127.08	232,013	1.77	95.38	184.10
	Std. dev.	5,113,475	0.03	979,077	0.03	16.58	32.30	969,021	0.02	16.49	70.96	29,090	0.02	0.85	9.87
	Range	3,519,782	0.16	4,019,840	0.13	66.84	159.45	3,620,706	0.12	66.86	260.84	134,666	0.10	4.84	52.84

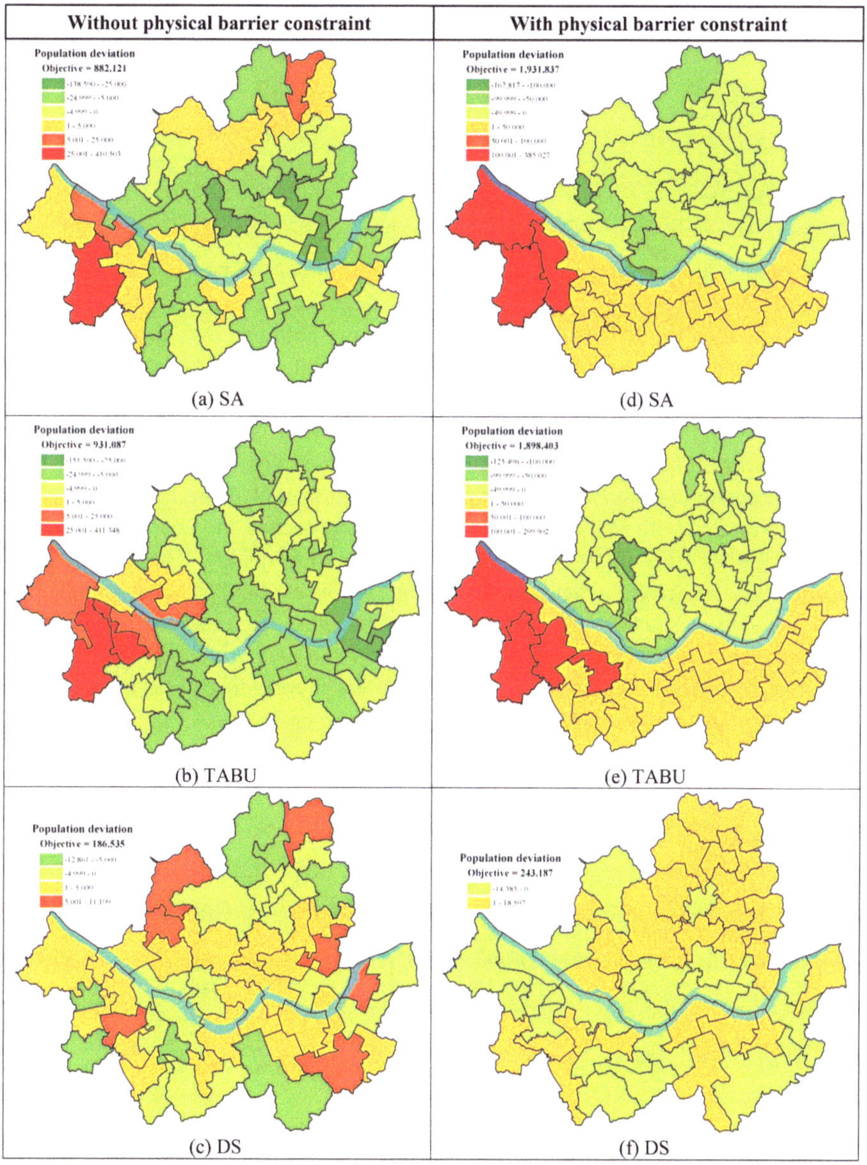

Fig. 3 Spatial solutions of three heuristic algorithm for an initial solution (average district population = 204,251)

the North of it, producing large objective values. The main reason for this result is that local moves were restricted by the physical barrier. In the case of the DS, due to dissolving/splitting procedures, the algorithm can overcome the effect of the physical barrier. As a result, the number of districts in the South of the Han

River changed from 20, at the initial solution, to 25 at the final solution, producing a greatly improved objective value compared to the other two meta-heuristics (Fig. 3f).

Conclusions

Many exact solution approaches using spatial optimization have been developed for political districting problems. The major drawback of the exact solution approach is the limited solvability for large instances that are common in real-world applications. In particular, the complexity in handling spatial contiguity and compactness is a major concern. To make heuristic approaches more promising, producing consistently high-quality solutions with little variation is a key component. This study developed a robust heuristic method for political districting problems using a simple procedure, dissolving/splitting (DS) algorithm, for districting spatial units. When compared with traditional meta-heuristic approaches, the DS outperforms the SA and TABU in terms of solution time and solution quality. The DS algorithm is effective, especially when a physical barrier is explicitly incorporated along with population equality, contiguity, and compactness. Because the proposed DS is simple and the structure does not rely on the meta-heuristic frameworks that are sensitive to random components, we believe this algorithm can be used to solve other regionalization problems as well as districting problems.

References

1. Armstrong MP, Lolonis P, Honey R (1993) A spatial decision support system for school redistricting. J Urban Reg Inf Syst Assoc 5:40–52
2. Kim H, Chun Y, Kim K (2015) Delimitation of functional regions using a p-regions problem approach. Int Reg Sci Rev 38(3):235–263. doi:10.1177/0160017613484929
3. Openshaw S, Rao L (1995) Algorithms for reengineering 1991 census geography. Environ Plan A 27:425–446
4. D'Amico SJ, Wang S-J, Batta R, Rump CM (2002) A simulated annealing approach to police district design. Comput Oper Res 29:667–684
5. Cutter SL (ed) (2001) American hazardscapes: the regionalization of hazards and disasters. Joseph Henry Press, Washington, DC
6. Barkan JD, Densham PJ, Rushton G (2006) Space matters: designing better electoral systems for emerging democracies. Am J Polit Sci 50:926–939
7. Fryer RG Jr, Holden R (2011) Measuring the compactness of political districting plans. J Law Econ 54(3):493–535
8. Ricca F, Scozzari A, Simeone B (2011) Political districting: from classical models to recent approaches. 4OR 9:223–254
9. Webster GR (1997) The potential impact of recent supreme court decisions on the use of race and ethnicity in the redistricting process. Cities 14(1):13–19

10. Bozkaya B, Erkut E, Laporte G (2003) A tabu search heuristic and adaptive memory procedure for political districting. Eur J Oper Res 144:12–26
11. Garfinkel RS, Nemhauser GL (1970) Optimal political districting by implicit enumeration techniques. Manag Sci 16:495–508
12. Hess SW, Weaver JB, Siegfelat HJ, Whelan JN, Zitlau PA (1965) Nonpartisan political redistricting by computer. Oper Res 13(6):998–1006
13. MacMillan W, Pierce T (1994) Optimization modeling in a GIS framework: the problem of political redistricting. In: Fotheringham S, Rogerson P (eds) Spatial analysis and GIS. Taylor and Francis, London, pp 221–246
14. Bação F, Lobo V, Painho M (2005) Applying genetic algorithms to zone-design. Soft Comput 9:341–348
15. Forman SL, Yue Y (2003) Congressional districting using a TSP-based genetic algorithm. Lect Notes Comput Sci 2724:2072–2083
16. Browdy M (1990) Simulated annealing: an improved computer model for political redistricting. Yale Law Policy Rev 8(1):163–179
17. Williams JC (1995) Political redistricting: a review. Pap Reg Sci 74:13–40
18. Mills G (1967) The determination of local government electoral boundaries. Oper Res Q 18:243–255
19. Segal M, Weinberger DB (1977) Turfing. Oper Res 25(3):367–386
20. George JA, Lamar BW, Wallace CA (1997) Political district determination using large-scale network optimization. Socio Econ Plan Sci 31(1):11–28
21. Zoltners AA, Sinha P (1983) Sales territory alignment: a review and model. Manag Sci 29:1237–1256
22. Duque JC, Ramos R, Surinach J (2007) Supervised regionalization methods: a survey. Int Reg Sci Rev 30(3):195–220
23. Shirabe T (2005) A model of contiguity for spatial unit allocation. Geogr Anal 37(1):2–16
24. Duque JC, Church RL, Middleton RS (2011) The p-regions problem. Geogr Anal 43:104–126
25. Fischer MM (1980) Regional taxonomy: a comparison of some hierarchic and non-hierarchic strategies. Reg Sci Urban Econ 10:503–537
26. Masser I, Scheurwater J (1980) Functional regionalisation of spatial interaction data: an evaluation of some suggested strategies. Environ Plan A 12(12):1357–1382
27. Openshaw S, Wymer C (1995) Classifying and regionalizing census data. In: Openshaw S (ed) Census users handbook. GeoInformation International, Cambridge, pp 239–270
28. Openshaw S (1973) A regionalisation program for large data sets. Comput Appl 3(4):136–147
29. Glover F (1989) Tabu search-part I. ORSA J Comp 1(3):190–206
30. Liittschwager J (1973) The Iowa redistricting system. Ann N Y Acad Sci 219:221–235
31. Nagel S (1965) Simplified bipartisan computer redistricting. Stanford Law Rev 17(5):863–899
32. Openshaw S (1977) A geographical solution to scale and aggregation problems in region-building, partition and spatial modeling. Trans Inst Br Geogr 2(4):459–472
33. Blais M, Lapierre S, Laporte G (2003) Solving a home-care districting problem in an urban setting. J Oper Res Soc 54(11):1141–1147
34. Duque JC, Anselin L, Rey SJ (2012) The max-p-regions problem. J Reg Sci 52(3):397–419
35. Ricca F, Simeone B (2008) Local search algorithms for political districting. Eur J Oper Res 189:1409–1426
36. Wei BC, Chai WY (2004) A multiobjective hybrid metaheuristic approach for GIS-based spatial zoning model. JMMA 3:245–261
37. Goldern B, Skiscim C (1986) Using simulated annealing to solve routing and location problems. Nav Res Logist Q 33:264–280
38. MacMillan W (2001) Redistricting in a GIS environment: an optimization algorithm using switching-points. J Geogr Syst 3:167–180

39. Young HP (1988) Measuring the compactness of legislative districts. Legis Stud Quart 13(1):105–115.
40. Kirkpatrick S, Gelatt DC, Vecchi MP (1983) Optimization by simulated annealing. Science 220(4598):671–680
41. Sammons R (1978) A simplistic approach to the redistricting problem. In: Masser I, Brown P (eds) Spatial representation and spatial interaction. Martinus Nijhoff, Leiden, pp 71–94
42. Openshaw S (1978) An optimal zoning approach to the study of spatially aggregated data. In: Masser I, Brown P (eds) Spatial representation and spatial interaction. Martinus Nijhoff, Leiden, pp 95–113
43. Kim MJ, Kim K (2013) Spatial optimization approaches to redistricting for National Assembly election: a case study on Yongin city. J Korean Geogr Soc 48(3):1–15

iGLASS: An Open Source SDSS for Public School Location-Allocation

Min Chen, Jean-Claude Thill, and Eric Delmelle

Introduction

Every year a large budget is spent on public education in the United States. Optimal public school location and assignment are essential for school operation to keep budgets under control and deliver the school service to population consistently with the core property of public good. Although these planning activities have been studied for many years, solving them is not easy, especially when the capacity of schools must be respected (i.e. how much demand a facility can accommodate) and assignment of pupils to their closest school is intended. Previous work has focused on the location phase of the problem, taking the closest assignment in the allocation phase for granted. However, the enforcement of school capacity constraints may render the closest assignment difficult or even impossible to implement. In such case simple solution strategies applied in the allocation phase fall short of meeting the needs of the problem at hand and more contextualized approaches need to be adopted because demand allocation directly influences the objective function and changes the location selection accordingly. For planners and decision-makers, an interactive Spatial Decision Support System (SDSS) integrating Geographic Information Systems (GIS) and optimal location-allocation models is also desired to solve the problem and display the solutions in a more interpretable way. In such SDSS, various objectives can be considered by school planners as part of the planning process, such as minimizing the pupils' travel time, minimizing their travel distance, maximizing the number of students assigned to their closest school, and minimizing the number of students who would travel over a distance deemed so

M. Chen • J.-C. Thill (✉) • E. Delmelle
Department of Geography and Earth Sciences, University of North Carolina at Charlotte, Charlotte, NC 28223, USA
e-mail: jfthill@uncc.edu

J.-C. Thill, S. Dragicevic (eds.), *GeoComputational Analysis and Modeling of Regional Systems*, Advances in Geographic Information Science,
DOI 10.1007/978-3-319-59511-5_17

long that their parents would wage complaints against the school district. In this research, we will keep these three objectives in mind as functional requirements of the proposed system.

The contribution of this research is threefold. First, we formulate a generalized model for the school location-allocation problem incorporating minimum and maximum school capacity constraints. In this model, the main objective is to minimize the total travel time or distance for all the students attending public schools. Additionally, to make the model more practical, maximizing the number of students who would be assigned to their closest schools or minimizing the number of students sent to remote schools is taken into consideration.

The second contribution is the development of a new operational approach integrating Tabu Search to solve the school location selection problem and Greedy Algorithm/Genetic Algorithm for the student allocation problem with the aim of overcoming the limitations of exact algorithms. The proposed heuristic algorithm considers both the location and the allocation phases jointly.

The third main contribution of this chapter is the implementation of a SDSS for school location and allocation based on an open source GIS, called the "interactive Graphical Location-Allocation System for Schools" (iGLASS). The SDSS provides planners with interactive ways to select the location of new schools, for closing existing schools, and to decide the modalities of allocating students to schools.

The remainder of this chapter is organized as follows. In the second section, the literature regarding location-allocation modeling and the contributions of geospatial analysis to location science is reviewed. In the third section, we present in detail the proposed capacitated school location-allocation problem, while the algorithm proposed to solve this optimization problem is outlined in section "Solution Algorithms". Then the iGLASS decision tool will be discussed in section "iGLASS Implementation". Finally, a case study of Charlotte-Mecklenburg Schools is studied as a demonstration of the school location problem and its implementation. Concluding remarks and directions for future research are given in section "Conclusions".

Literature Review

Location-Allocation Problems

Location-allocation problems are typical combinational problems and have widely been studied in different fields, including economics, industrial engineering, operations research, regional science, urban planning, geography, computer science and mathematics [1, 2]. Location-allocation models have benefited much from development in computing technologies and from the interest of scholars in diverse areas, who have contributed diverse perspectives towards the analytics of facility location and facility service areas. For example, GIS has contributed greatly to location analysis in terms of data input and management, visualization, problem solution and theoretical advances [1].

Location problems deal with locating one or more facilities in such a way to optimize one or multiple objectives [3]. Location-allocation problems form an important class of location problems, rooted in location theory and whose purpose is to "*locate a multiple number of facilities and allocate the demands served by those facilities so that the system service is as efficient as possible*" [3]. Location theory was not formalized until Weber's seminal 1909 [4] treatise on industrial location [5]. Most discussions about location-allocation problems were triggered by Hakimi [6], after he developed a formulation for finding one or more facilities on a graph to minimize the sum of the distances or the maximum distances between facilities and points on the graph. Since his work, application of location-allocation models has blossomed and a number of different location-allocation models have been identified.

Francis, McGinnis, and White [7] suggested that the literature on location-allocation modeling is organized in four classes according to the discrete versus continuous characteristics of the space where facilities are sited. For the discrete scenarios, the facilities can only be built at a restricted number of discrete candidate locations, while for the continuous scenarios, the facilities can be built anywhere in the region they are designed to serve. These classes are the continuous space, discrete space, mixed space and discrete network location-allocation problems. Brandeau and Chiu [5] presented a survey of over 50 location problems and gave a broad overview of location problems studied before 1989 by providing a framework of classification based on their objective functions, system parameters and decision variables. This paper can be viewed as an excellent starting point to get an overview of the research work in the location and location-allocation area before the 1990s. About the same time, Current, Min, and Schilling [8] pointed out that there are often multiple objectives to implement in location (including location-allocation) problems, rendering the search for solutions more complex. Four main categories of objectives were uncovered in their work: minimizing the operating cost, minimizing the travel impedance (e.g., distance or time), maximizing the coverage and maximizing the demand assignments.

In the class of discrete location-allocation problems, there are four typical problems: the p-median problem, the p-center problem, the uncapacitated and capacitated facility location problem (UFLP/CFLP) and the quadratic assignment problem (QAP) [9]. The p-median problem is known as a *minisum* location-allocation problem, which means that the objective is to minimize the total distances or costs between demands (customers) and providers (facilities). The p-median problem was first studied by Hakimi [6, 10]. ReVelle and Swain [11] mathematically formulated the p-median problem, presenting it as an integer programming problem.

Location-allocation problems are combinational optimization problems, for which a solution contains the information of where facilities can be located, given the spatial distribution of demand within the area, to minimize the total cost (travel time, distance or other impedances) or to minimize the operating cost or some other overall objective. It is non-trivial to solve in that it has a very large solution space, and it is NP-hard, and there is no way to find the exact optimal solution within polynomial time. Traditional exact algorithms of linear programming provided by commercial software like CPLEX or Lingo [3] are not efficient for large-scale problem. Consequently, many heuristic algorithms have been proposed to solve location problems. Among them, greedy [12], alternate [13] and vertex substitution [14] were applied earlier on. Later, especially in the last decade, more advanced heuristics and metaheuristics were proposed to solve the p-median problem, just to list a few: Genetic Algorithms [15], Tabu Search [16], simulated annealing [17] and ant colony optimization algorithms [18]. Bischoff and Dächert [19] used different methods to solve the generalized class of location-allocation problems, in which N new uncapacitated facilities are to be located in the plane with respect to M objects, and their performances are compared. More specifically, they compared the multi-start, (variable) neighborhood search, tabu search, simulated annealing, and an evolutionary algorithm. Their numerical results show all the methods perform very similarly to each other, while the termination criterion may change the computation time and quality of the objectives.

Planning for Public Services and School Locations

Teitz [20] advocated for the systematic study of location of public facilities as a system [2]. He stressed the necessity to balance efficiency (similar to the optimization objective) and equity in public facility locations [21]. Thus, public facilities should be distributed and established mainly by government guided by governmental welfare criteria within budgetary constraints, rather than determined by profit-making, which governs private sector operations [21]. After Teitz's advocacy for location modeling in the domain of public service planning and delivery, location-allocation models became more widely used for various public facilities such as schools, hospitals, libraries, fire stations and police stations. Various objectives have been identified as pertinent when designing the distribution of public facilities [22]: (1) minimize the total travel time or distance (p-median, see Hakimi [6], Hillsman [23]), (2) minimize the maximum travel time or distance that separates a user from his/her closest public facility [24], (3) minimize the number of necessary facilities while keeping a certain level of coverage (Location Covering Set Problem) [24]. All these models have been extensively studied and applied under different scenarios. Their general goal is to minimize travel impedance from an agglomerative or individual perspective; hence they require all the demand nodes to be assigned to their closest facility, or a facility to be located within an acceptable distance from each demand node. However, they inherently lack the functionality to address capacity constraints (i.e. how much demand should be reached to open a facility and how much demand a facility can accommodate at most), so that Ellwein and Gray [25] and others have proposed an alternative model to handle the capacity and regional constraints to make sure an acceptable level of service will be provided by these facilities.

Education is one of the most expensive public services provided by government entities in the United States. The State of North Carolina spent 38.5% of its general state funds on public schools in 2011–2012 [26]. It is essential to optimize the location of public schools not just to minimize the system efficiencies from a user perspective but also to make sure that a high level of equity is offered. Public school location-allocation problems have caught much attention since the beginning of location analysis and especially in recent years. We discuss hereafter some of the challenges associated with school location analysis that have been highlighted in the literature.

(i) Capacitated Models. In early location models, no capacity constraints were taken into consideration. Following Ellwein and Gray [25] and Mirchandani and Francis [9], Murray and Gerrard [27] proposed to add capacity constraints for each school. These authors also tried to address the efficient and effective provision of services, regional or zonal requirements, aside from preventing excessive facility workload. A solution approach based on Lagrangian relaxation was developed. Capacitated location problems reflect the capacity constraints that must be considered by location planners or decision-makers. Given the existence of maximum capacity constraints, the sum of the demand assigned to a particular facility cannot exceed the upper bound of the capacity; in the case of schools location-allocation problems, minimum capacity constraints are also regarded as essential, and when these constraints are added [17], students may be forced to cross school district boundaries in order to meet minimum capacity requirements. Also under-utilized schools (schools with fewer students than the minimum capacity constraint) may be suggested as candidates for closing because of the excessive operating costs per student, while over-utilization (overcrowding) can only be addressed by increasing the capacity of the school or adding new schools. When the school authority wishes to reduce class sizes, capacity adjustment is also required. A recent study by Delmelle et al. [28] accounts for adjustable school capacity in a longitudinally dynamic environment.

(ii) Assignments to Closest Facilities. In location-allocation problems, it is ideal if all the demands can be sent to their closest facilities, and in many studies closest assignments have been taken for granted. However, this ideal situation may be violated in practice due to the inherent constraints on facility capacity or because the closest facility is not available to be used. In the case of school location-allocation problem, although in certain circumstances, a student may be willing to travel a longer distance to attend a school that offers special programs [29], it is always desirable to send as many students as possible to their closest school [30] or minimize the number of non-closest assignments. Generally speaking, the non-closest assignment in school allocation problems is mainly due to the maximum capacity constraints; hence, increasing the capacity of schools will generally result in a higher rate of closest assignments. Gerrard and Church [31] addressed closest assignment constraints in integer-linear location-allocation models, and they also gave a summary of the applications with closest assignments, and presented some improvements. Delmelle et al. [28] proposed a mathematical formulation that alleviates the incidence of non-closest assignments.

(iii) Modifying an Existing Facilities Network. In urban regions with high population growth, adding new facilities or augmenting the capacities of existing facilities may be required, while closure and capacity reduction may be deemed necessary in areas with population decline. In a location-allocation modeling context for public schools, Antunes and Peeters [17, 32] and Antunes et al. [33] introduced a p-median problem with minimum-maximum constraints capable of handling the opening of new schools and closing of existing schools to address the variations of enrollment. By examining the school capacity utilization after consolidation, Church and Murray [34] addressed the analytical issues in school closure and consolidation. However, their proposed model cannot force facilities to remain open until they reach a certain age in order to reach scheduled amortization. This functionality is an integral part of the model proposed by Delmelle et al. [28].

(iv) Multi-periods Planning Problem. Due to the dynamic characteristics of school systems, some scholars have proposed to incorporate temporal issues into school location-allocation models. Wesolowsky [35] extended the static location-allocation problem into a multi-period one. A model was provided in their paper, and a possible solution was also discussed. The dynamic modeling of a school network is particularly challenging for several reasons: (1) demand is likely to fluctuate over time, requiring the opening of new schools and the closing of existing ones [17, 28, 32]; (2) capacity regulates how much demand can be served at a particular time [32]; (3) the uneven quality of enrollment forecasts degrades the reliability and efficiency of planning decisions [28, 36]; and (4) social costs arise with the reassignment of students to a different school over successive periods over the planning horizon.

(v) Ethnic and Racial Balance. In societies with significant ethnic and racial diversity, social inclusion policies may dictate that a certain level of social and ethnic balance must be maintained in each school [37, 38], which may increase overall travel time in the student population. Clarke and Surkis [39] developed a system called "MINTRAN" to alleviate the racial segregation problems in public schools by assigning students to schools in an efficient way, given the racial distribution of students and locations and capacities of schools.

(vi) School Choice. An increasingly popular practice in public education is to allow parents and students some choice of the school to attend. This serves to alleviate political backlash associated with other policies and restore confidence in decision makers and managers. Thus assignment is modeled through multivariate choice modeling modules. Such approaches are presented by Church and Schoepfle [40] and Müller, Haase, and Kless [41].

In this research we address the first three considerations in our model, leaving the remaining three issues for future research.

GIS, SDSS, and Location Science

Although the development of location theory is independent of the development of GIS and major advances in location science come from the development of mathematical models [1], the integration of location and location-allocation models in GIS provides new insights and visualization of solution results [3]. In the early stage of location science, most location-allocation models were operations research based rather than GIS-based, but later, commercial GIS software like ArcGIS started to provide functionalities to meet the application requirements of the locational planning process. For example, ArcGIS 10.4 [42] provides toolboxes to solve six different types of location-allocation problems: minimize impedance, maximize coverage, minimize facilities, maximize attendance maximize market share and target market share, respectively.

Church [43] and Murray [1] summarized the contributions of GIS to location science developments in terms of data input, visualization, problem solution and theoretical advances, and argued that the role of GIS in location models should not be confined as a "*mere spatial data input mechanism*", which is commonly held by researchers in location sciences. GIS as a data input tool have been extensively used, and the simplest application is to identify the set of potential sites of public facilities based on basic requirements (e.g., distance from existing facilities, population density and topological requirements). These potential locations can be retrieved by basic functionalities (buffering and overlay, map algebra) provided by GIS. Regarding the visualization, GIS can be used as a powerful tool to display solutions interpretable to even inexperienced user, and more meaningful patterns can be discovered from the location-allocation maps. Armstrong and Densham [44] suggested a new cartographic framework to visualize network-based location-allocation solutions with a goal to support collaborative group decision-making. In this framework, the synthetic maps were created by decomposing the location-allocation solution map into atomic elements, and were accessible to group members. In turn, group members can discover the similarities and dissimilarities in alternative solutions and work collaboratively.

A few operational systems have integrated location-allocation models and GIS along the lines of a full-fledge decision-support environment aiming at providing semi-structured or unstructured spatial decisions. SDSS is a computer-based system providing interactive ways for decision-makers to assist them to solve complex spatial problems [45]. A typical SDSS contains three main components: a database management system, a library of models used to solve the problems and an interface to aid users to modify the parameters and analyze outcome of different decisions. In particular, research on SDSS for resource allocation applications has caught on by integrating Artificial Intelligence (AI) techniques with GIS (e.g., [46]). In the field of location analysis, efforts dedicated to developing such a system where GIS plays an integral part remain weak. A majority of research has used a loose coupling approach: exporting data and displaying results in GIS software and solving the location-allocation problem by optimization software like CPLEX or Lingo. Ribeiro and Antunes [47] developed such a system by using MapObjects components of ArcGIS.

Planners and decision makers need a flexible and portable SDSS integrating GIS and location-allocation modeling functionality that provides them with an interactive way to design new school distributions and student allocations. In particular, it is critical to be able to selected different objectives on the fly and evaluate associated solution outcomes to better inform the decision process.

Problem Formulation

We present a min-max capacitated school location-allocation model with multiple objectives. First, we aim to assign all the students to their closest school. However, schools with an excessive number of students will provide very poor services to them; as a result, it is necessary to place maximum capacity constraints on the schools. Minimum capacity constraints are significant when considering limited budgets. In most cases, it is easier to expand the capacity of an existing school than opening a new one regarding both the money required at the startup stage and maintenance phase. So when it is possible, we would like to recommend expanding the maximum capacity of an existing school rather than opening a new school. When capacity constraints are placed on the problem, it may become impossible to assign all the students to their closest school. Consequently, the objective morphs from single (minimize the total travel time or distance) to multiple, which include minimizing the total time or distance travelled by students, maximizing the number of students sent to their closest school, and minimizing the number of students sent to a school that is much further than their closest one.

Our capacitated school location-allocations problem can be formulated as an integer-linear programming model, subject to a fixed number of facilities and max-min-capacity constraint with the objectives described above. We use the following notation.

Indexation and Sets

i, I = index and set of demand nodes (student).
j, J = index and set of school locations.

Parameters

a_i= demand at node *i*.
d_{ij}= travel impedance (distance, time, or cost) between locations *i* and *j*.
p= number of schools that can be opened.
C_j^-= minimum capacity of school site *j*.
C_j^+= maximum capacity of school site *j*.

Decision Variables

$$X_{ij} = \begin{cases} 1, & \text{if we assign a student } i \text{ to school } j \\ 0, & \text{otherwise} \end{cases}$$

$$Y_j = \begin{cases} 1, & \text{if we locate a school at } j \\ 0, & \text{otherwise} \end{cases}$$

Derived Variables

$$I_{i0} = \begin{cases} 1, & \text{if student } i \text{ assigned to their closest school} \\ 0, & \text{otherwise} \end{cases}$$

$$I_{ic} = \begin{cases} 1, & \text{if student } i \text{ assigned much further than their closest school,} \\ & \text{which would arouse complaints} \\ 0, & \text{otherwise} \end{cases}$$

Formulation

Extending the *p*-median, a generic formulation of the school location problem is as follows.

$$\text{Minimize Z} = \sum\nolimits_{i \in I} \sum\nolimits_{j \in J} a_i * d_{ij} * X_{ij} \tag{1}$$

$$\text{Maximize A} = \sum\nolimits_{i \in I} a_i * I_{i0} \tag{2}$$

$$\text{Minimize C} = \sum\nolimits_{i \in I} a_i * I_{iC} \tag{3}$$

Subject to:

$$\sum\nolimits_{j \in J} X_{ij} = 1 \qquad \forall i \in I \tag{4}$$

$$X_{ij} \leq Y_j \qquad \forall i \in I, \forall j \in J \tag{5}$$

$$\sum\nolimits_{j \in J} Y_j = p \tag{6}$$

$$C_j^- \leq \sum\nolimits_{i \in I} a_i * X_{ij} \qquad \forall j \in J \tag{7}$$

$$C_j^+ \geq \sum\nolimits_{i \in I} a_i * X_{ij} \qquad \forall j \in J \tag{8}$$

$$X_{ij} \in \{0, 1\} \qquad \forall i \in I \tag{9}$$

$$Y_j \in \{0, 1\} \qquad \forall j \in J \tag{10}$$

$$Y_j = 1 \qquad \forall j \in J^o \tag{11}$$

$$Y_j = 1 \qquad \forall j \in J^c \tag{12}$$

where:

$Y_j = 1$ means that the j-th school will be open, otherwise, it will be closed,

$X_{ij} = 1$ means the student in the i-th demand will be assigned to the j-th school.

Objective (1) is to minimize the aggregated travel impedance (time or distance) for all the students in the school system; objective (2) is to maximize the number of students sent to their closest school, while (3) is to minimize the number of students sent so far away that they will file complaints with the school district. Constraint (4) ensures that each student is assigned to a school, while constraint (5) stipulates that a student can be allocated to a school only if that school is currently open. Constraint (6) restricts the number of open schools to a certain number (p). Minimum and maximum capacity constraints (7) and (8) limit the flow of students to each facility. Integer constraints (9) prevent fractional assignments. If certain schools must remain open (for instance due to political pressure, or if a school is a neighborhood landmark or if a school was just opened), this can be enforced by introducing a new set J^O which is the set of schools that must remain open. Similarly, J^C is the set of schools that must close.

In order to combine the three objectives from Eqs. (1)–(3), a standardized objective function can be formulated as Eq. (13), where ($\gamma_1, \gamma_2, \gamma_3$) reflects the importance imputed to each sub-objective by the community and decision makers:

$$\text{Minimize } Z = \gamma_1 * \sum_i \sum_j a_i d_{ij} X_{ij} - \gamma_2 * \sum_i a_i I_{i0} + \gamma_3 * \sum_i a_i I_{ic} \tag{13}$$

It should be noted that the model in Eqs. (4)–(13) can be made dynamic by including a temporal component. However, the number of time periods over which the problem is optimized will increase the number of location-allocation variables dramatically, thus affecting the run time and making heuristic methods a more appealing solution approach.

Solution Algorithms

Location-allocation is an NP-hard problem, rendering it impossible to solve in polynomial time when an exact optimal solution is intended. Heuristic algorithms have been proposed to efficiently solve these location-allocation problems, especially when the problem becomes very large. In this section, a two + one-phase approach consisting of Tabu Search (TS) and Greedy Algorithm/Genetic Algorithm (GA) will be presented. TS and Greedy Algorithm/GA will be used to solve the location of schools and the allocation of students to schools, respectively. The heuristic ends with a local re-optimization of the solution achieved through the earlier two-phase iterative process.

Overview of the Two + One-Phase TS-Greedy/GA Approach

We propose an algorithm with two main phases, location and allocation. The location phase is guided by TS, while the allocation phase is solved by greedy algorithms or GA. For the school location phase, the sequence is as follows. First, a fixed number of (p) schools are selected randomly from the set of school candidates. Then, demand nodes are assigned to each school according to some priority score assigned to the demand nodes; the priority score is devised on the basis of some exogenous considerations relevant to the student assignment problem at hand. The process ends with the best solution when the solution found has not changed for a certain number of iterations. The flowchart of the algorithm is illustrated in Fig. 1.

The best solution comprises the set of schools and associated student assignments that minimizes the objective function; it is called the incumbent solution. At each iteration, the incumbent solution becomes the new, initial solution, and is

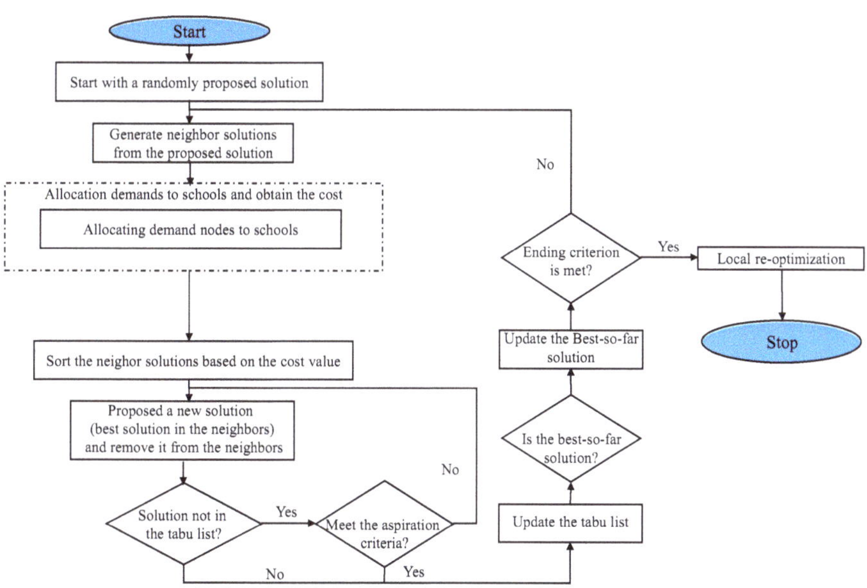

Fig. 1 Flowchart of public school location-allocation heuristic (with location selection as framework)

changed by generating a new set of schools (swapping opened and closed schools, respectively), and students are reassigned to schools that are open. The process is repeated until the best solution found cannot be improved over a preset number of consecutive iterations.

The two phases of the heuristic are discussed in more detail in sections "Phase 1: Tabu Search Approach for School Location" and "Phase 2: Greedy Algorithms and GA for Allocating Students to Schools", respectively, while local re-optimization is presented in section "Local Re-optimization".

Tabu, Greedy, and Genetic Algorithms

In the search for solution to location-allocation problems, an array of heuristic algorithms have been applied. Among them, the Genetic Algorithms [15], Tabu Search [16], and simulated annealing [17] have become rather commonplace. Compared with exact solution methods, heuristics provide an efficient way to find the solution. A heuristic algorithm is an ad hoc and rule-of-thumb way [48] to find the approximate optimal solution rather than exact best solution for an optimization problem. Among these heuristic algorithms, Tabu Search and Genetic Algorithms belong to the most efficient heuristic techniques in that they can find high-quality solutions in relatively short run time.

TS was first proposed by Glover and McMillan [49] and formulated by Glover [50]. It was introduced as a local solution searching strategy addressing combinatorial optimization problems, and was initially applied in fields like scheduling, computer channel balancing, cluster analysis, space planning, travelling salesman, and graph coloring [50]. For the TS algorithm, a random solution is initially generated, and its direct neighbors in the solution space are examined to find a better solution for the problem; then the neighbors of the last selected solution are examined again; the searching loop continues until the stopping criteria are reached. In order to prevent the solution from falling into sub-optimal regions or on plateaus where most of the solutions have the same objective function, memory structures are used as forbidden neighbors, and in this way a global optimal (or approximate optimal) solution can be obtained.

Greedy algorithms [51] are trying to find the best solution by choosing the next optimal step which will provide an immediate benefit to the problem. It is myopic in that locally optimal choice in every step does not guarantee a global optimal solution. However, greedy algorithms are attractive due to their simplicity and easy implementation, and in most cases, they can provide a solution that is close to the global optimum.

Genetic algorithm (GA) was first introduced by Holland [52]. It mimics the natural evolution theory and can be used to find approximate optimal solutions to a wide range of problems. GA is capable of escaping a local optimal solution and finding the global optimum. Hosage and Goodchild [53] is among the first researchers to apply GA to solve location-allocation problems. Although their result showed that GA is unlikely to have the same efficiency as other heuristics, they proved the applicability of GA for this class of optimization problems. Furthermore, GA can be used to find optimal solutions with multiple objectives, providing alternative solutions for decision-makers. For instance, Zhang and Armstrong [54] proposed a multi-objective GA for corridor selection problems, and a large set of Pareto-optimal and near-optimal solutions were provided to the decision-makers for evaluation. Evolutionary Algorithms (EAs), as an extension of GA, have also been used by researchers when they faced multiple objectives during site searches [55].

Phase 1: Tabu Search Approach for School Location

The following components need to be given careful consideration when we the TS algorithm is implemented: the representation of a solution and the definition of its neighbors; the contents of the tabu list, including its length and the aspiration rules; and the stopping criteria. Intuitively, the neighbors should be rather similar to each other, yet different; here we define neighbors as those solutions consisting of identical schools, except for one. More detail regarding these components are reported hereunder.

(1) Let us assume there are n candidate locations from which p school sites will be selected. Then, to represent a solution, an array of n binary values is constructed, where each digit represents the status of a site. A value of 1 means that a school will be built at that site, and 0 otherwise (Fig. 2 shows an example of 5 school sites out of 7 candidates).

(2) Neighbors are generated by swapping the status (open or close) of any two candidates whose status differs (see Fig. 2).

(3) A tabu list records the swaps that occurred in the previous steps, while the tabu length is the maximum number of swaps the list is able to keep in memory; solutions generated by the swaps in the tabu list cannot be selected as the next solution unless an aspiration rule is satisfied.

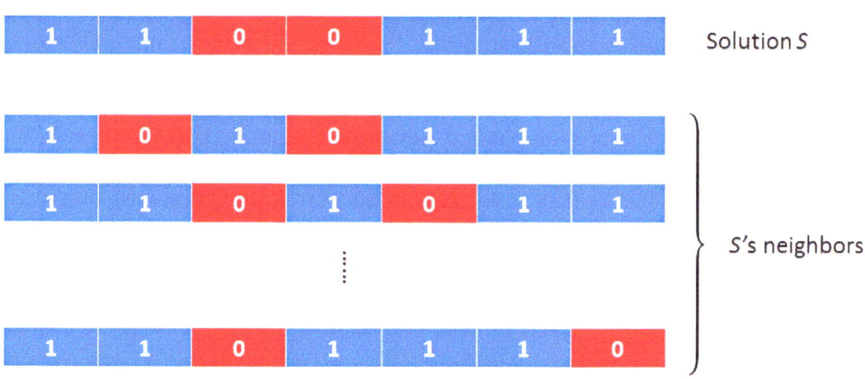

Fig. 2 Representation of a location solution and its neighbors in Tabu Search

(4) An aspiration rule is defined here as when the solution generated by the swap in the tabu list is the best-so-far solution, which means that if the swap for a new solution that resides in the tabu list but the solution is better than all the previously visited solutions, then this swap can be taken and the solution can be selected as the next solution.

(5) The stopping criterion states that when the best-so-far solution does not change for a certain number of steps, it will be treated as the optimal solution and the process ends.

To be more specific, TS starts with an initial location solution of fixed schools. The set N of neighboring solutions is generated by swapping one currently open school with one that is closed and vice versa. This yields a N-number of potential solutions. A tabu list keeps track of these location swaps. The algorithm will not revisit solutions that have already been explored, unless an aspiration criterion is met. The aspiration criterion allows the algorithm to visit a move currently in the tabu list provided it is better than the previously visited solutions. The use of a tabu list and neighboring solutions allows the algorithm to escape from a local optimum.

Phase 2: Greedy Algorithms and GA for Allocating Students to Schools

In the allocation phase, students are allocated to one of the p school sites in the solution set following a greedy (myopic) approach (either based on the magnitude of the demand, proximity, or regret) or a GA approach (the fitness function can be given by eqs. (1), (2) or (3)).

Capacity constraints make school location-allocation problems much harder to solve than regular p-median problems. In order to provide a high level of service to students, the number of students sent to each school cannot exceed a certain quota dictated by its design (maximum capacity constraints), while to make more efficient use of the investments in building a new school, the number of students should be higher than a certain threshold number (minimum capacity constraints). In capacitated location-allocation problems, the strategy to implement the allocation step is assumed to be the closest assignment; however, the capacity constraints make the closest assignment difficult or even impossible to implement. In this algorithm, we will consider the allocation phase explicitly and treat the maximization of the number of students sent to their closest school or minimization of the number of students sent much further than their closest school as an additional objective, aside of the objective of minimizing the total travel time or distance. This is handled through assignment priority lists: a priority value is associated with each demand node; the node with a higher priority will be considered ahead of others in the assignment decisions; thus it has a higher possibility to be sent to its closest school.

Greedy Algorithms

In many situations, a greedy algorithm cannot yield the optimal solution; however, it may produce a locally optimal solution that is close to a global optimum in a relatively shorter time frame. When using greedy algorithms to allocate demand nodes to schools, two kinds of priority lists of the demand nodes are maintained. One priority list is for all the demand nodes; the priority is calculated by their distance/population/regret value, which will be described later. The other kind of priority list is constructed specifically for each site on the solution list, with the priority score determined by geographic proximity.

From Eq. (1), we can identify that two variables (number of demands and distance) contribute to the objective function. It would be easy to conceive of three different greedy ways of minimizing the objective by:

(1) minimizing the weights on population, that is to say we set the priority based on the population. For the entire school district, we sort the demand values from highest to lowest (nodes with larger demand have higher priority), in this way we are trying to assign as many students as possible to their closest school at every step with a global goal of minimizing the total cost (impedance). In a sequential fashion, the demand nodes with the highest value are assigned to their closest schools. When a demand node cannot be assigned to its closest school, due to capacity constraints, the demand is allocated to the second-closest school, or third closest, and this continues in this fashion until a school can accommodate the student demand.

(2) minimizing the weights on distance, that is to say we calculate the priority based on distance. A list of demand nodes sorted by distance is generated for each school in the solution set. Demand is assigned based on its proximity, until the school has reached its maximum capacity. Once that school capacity level is reached, the algorithm moves to the next open school (in a random order) and assigns students in a similar fashion. At the end of the iteration, the remaining school demand that could not be assigned within the user-defined bound is pooled together into one set, and allocated to remaining schools according to the priority list. Figure 3 shows how to assign demand nodes to its closest school until the school is fully utilized; first, we need to find all the closest demands to that school, and then try to assign as many members in the closest demands list as possible to the school; a common list of unassigned demands was maintained as well.

(3) minimizing the product of the two variables (demands of node by distance to its assigned school) at every step. Priority assignment lists are designed, where the priority score incorporates the concept of regret, or excess travel time that would be incurred if a demand node was not assigned to its closest school. The regret R_j is defined as the difference in travel distance (or time) between the closest school (distance d_i*) and the school (distance d_{ij}) where the demand is assigned to, weighted by the population at that demand node. Nodes are assigned to the schools based on this regret value. Specifically:

$$R_j = a_j * \left(d_{ij} - d_i *\right) \quad (14)$$

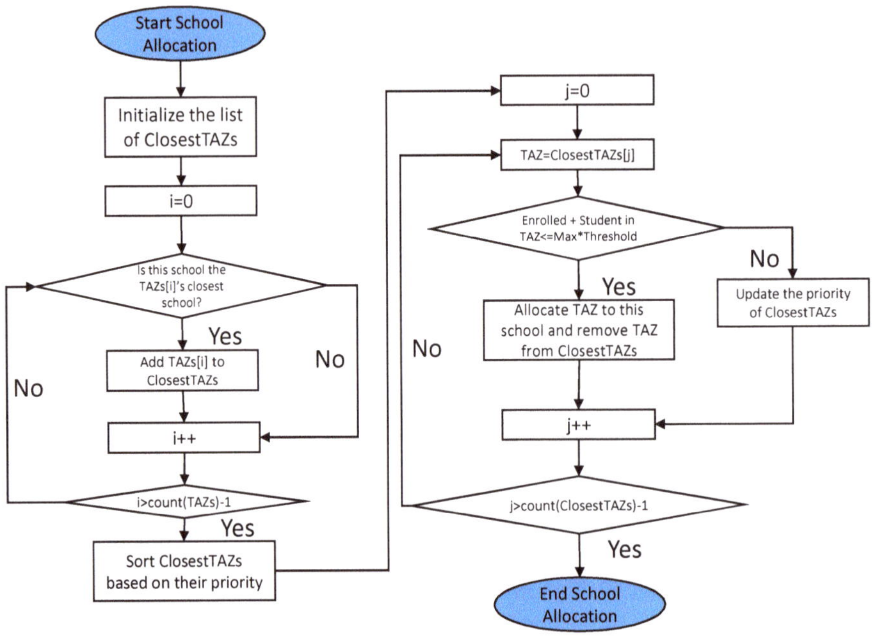

Fig. 3 Proximity-based assignment in greedy heuristic. *TAZ*, traffic analysis zone

For those demands that were not allocated in the first step, a priority list for all the unallocated demands will be generated, and they will be assigned in sequence to the remaining capacity of the schools (see Fig. 4). The details on how to do Allocate(Schools[j]) is illustrated in Fig. 3; after all schools have accepted as many closest demands as they can, a common list of unassigned demands (TAZs in Fig. 4) will be generated; the list of TAZs will then be sorted based on their priority (e.g. regret value between the closest school (not able to accept all the students from this demand) and the closest school which may still be able to accept all the students from this demand). Every time the assignment failed, the order of the unassigned TAZs must be updated based on the new priority of the unassigned TAZ we are working on. The program will stop until all TAZs were assigned to some school.

Genetic Algorithm

Genetic algorithms have been widely used in many optimization problems. GA was used to solve location problems as well [56], but so far they have not been used for the allocation phase when the assignment of all demand nodes to their closest facility is impossible. When capacity constraints are enforced, the assignment of all demand nodes to the closest school may become impossible, but the sequence of the assignment of the demand nodes may be exploited towards the overall performance of closest assignments, and eventually contribute to the objective functions (1), (2) and (3) since different assignment sequences will result in different values of the objective functions. Consequently, the sequence of allocation of the demands

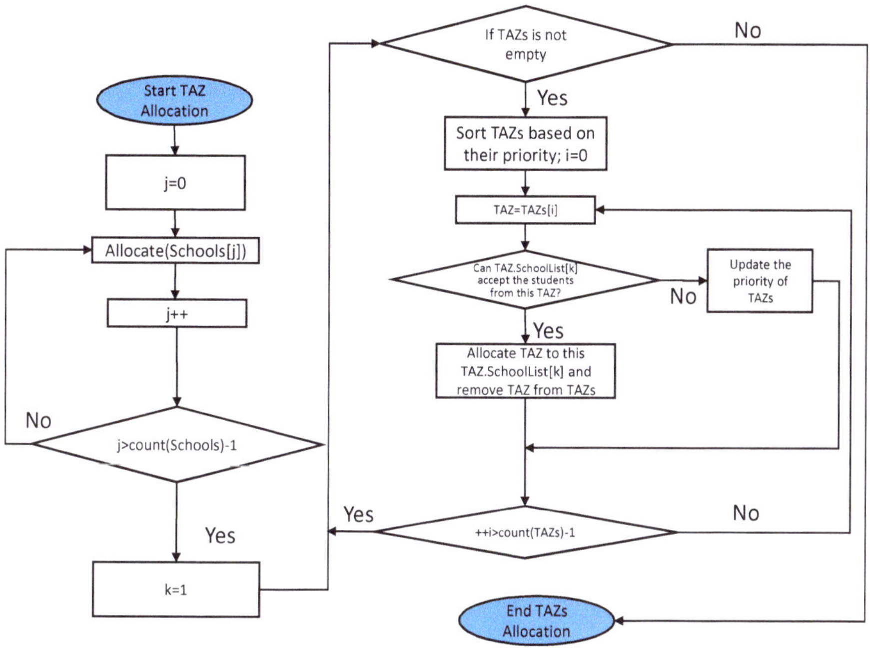

Fig. 4 Assignment of unallocated students to the schools in greedy heuristic

to schools will be used in the design of the representation of GA individuals (assignment instances). Our objective is to find the best assignment sequence that will generate the optimal objective functions. In the design of the GA, every assignment sequence will be treated as an individual and they are coded by their orders. More details on the essential components in the GA design are provided hereunder.

(1) The representation of an individual in the allocation phase is based on the assignment orders of all the demand nodes. Figure 5 is a simple illustration of how the GA operations are designed in our approach. Every demand node has a unique id, and for every individual, it is coded by the order when the demand node is assigned to the schools. For example, if there are seven demand nodes and they are considered in the sequence of 1, 2, 3, 4, 5, 6, 7 (which means the first demand node will be assigned to school first, the second demand node will be assigned after the first demand node has been assigned to a school, and so on, until the seventh demand node, which will be considered only after demands of 1, 2, 3, 4, 5 and 6 have been assigned to a school) and the individual's gene will be coded as 1,234,567.

(2) For the selection operation, a tournament selection strategy is applied, which means two individuals are compared each time, and the one with relatively better fitness will be selected. The fitness can be determined by the total cost (lower total cost has higher fitness), closest assignment percentage (the higher, the better), or further assignments (the lower, the better).

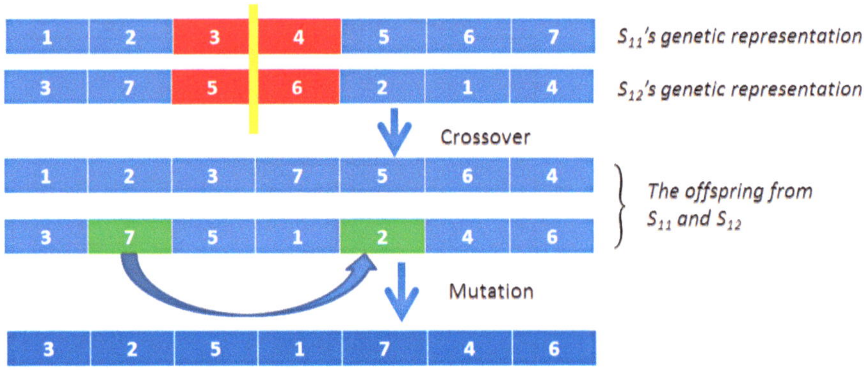

Fig. 5 A simple example of the GA design in the allocation phase

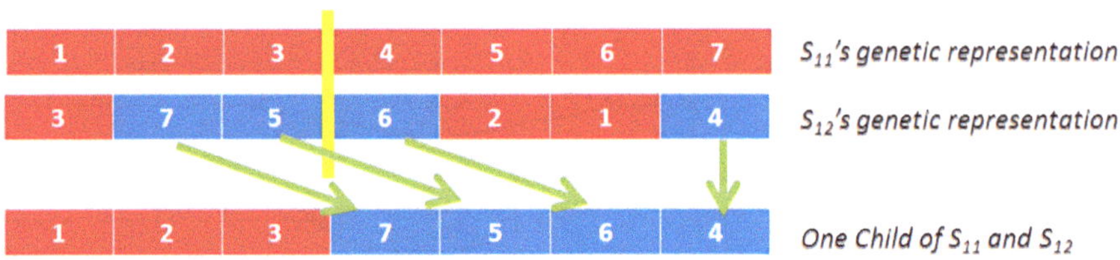

Fig. 6 A demo of how a child is reproduced

(3) The generation of new individuals (allocation solutions) is conducted from their selected parents by crossover and mutations, and order crossover of two parents is used. Crossover is defined as follows: (1) a crossover point is randomly selected, and every parent is partitioned into two sub-sets of genes; (2) the first sub-set of genes (c1) from the first parent (p1) is copied to the child; (3) eliminating the genes in c1 from the second parent (p2); (4) the remaining genes (c2) of p2 after elimination is concatenated to c1 and form the genes of the child. For example, two individuals of 1,234,567 and 3,756,214 crossover between the third and fourth codes (Fig. 5); then two children 1,237,564 (the assignment sequence of demand nodes 1, 2 and 3 were determined by the first parent, and the assignment sequence of remaining demand nodes, that is 4, 5, 6 and 7, were determined by the second parent, and the sequence turns out to be 7, 5, 6 and 4) and 3,751,246 will be generated (see Fig. 5). Figure 6 shows more details regarding how a child is generated from their parents.

(4) With this design, the mutation operation is easy. It is implemented by swapping the order of any two demand nodes or multiple demands nodes, which depends on the mutation probabilities set by the user.

Local Re-optimization

Once the algorithm has iterated through the phases to identify school locations and allocate students to schools to its completion, a local re-optimization is conducted. Local re-optimization is implemented to reduce the incidence of assignment of children to a school other than their closest. This proceeds as follows. We treat each school and its q closest schools as a group, and the students assigned to any of them through the global location and allocation heuristics will be processed as a small-scale allocation problem. The assignment of students will be executed again for these students and schools using the same heuristic option as for the global optimization until the stopping point.

iGLASS Implementation

To implement the "interactive Graphical Location-Allocation System for Schools" (iGLASS), we follow a tight-coupling design approach, where the communication between the DSS and GIS is facilitated by an interface and DSS modules are executed from within the GIS environment. This strategy minimizes data conversion and keeps run time to a minimum, which is an essential consideration for interactive sessions on large scale problems, possibly involving various community stakeholders. Ribeiro and Antunes [47] adopted a similar strategy.

We integrate TS and Greedy/GA algorithms with an open-source GIS environment to provide a SDSS (Fig. 7) for school location-allocation modeling and planning. The stand-alone iGLASS tool is developed based on a scalable open source GIS platform—DotSpatial 1.3, which "*is a geographic information system library written for .NET 4. It allows developers to incorporate spatial data, analysis and mapping functionality into their applications or to contribute GIS extensions to the community. DotSpatial provides a map control for .NET*".[1] We implement the extension of location-allocation modeling with the C# programming language. The model parameters and input data can be altered on the fly through the graphical user interface (GUI) (such as modification of their capacities). Visualization components pertain to student assignments to school and school utilization.

iGLASS is designed to provide great flexibility to users who can customize the system to the specific needs of a case study. It allows users to modify the demand attributes and school capacities on the fly, which enables rapid sensitivity analysis (Fig. 8). Second, demand nodes can be fractioned into smaller demand nodes, with a threshold defined by the user. This property allows to allocate demand associated to a single node (such as a neighborhood) to multiple schools when capacities are very tight, thus achieving a greater school utilization value. Third, parameters for

[1] http://dotspatial.codeplex.com/.

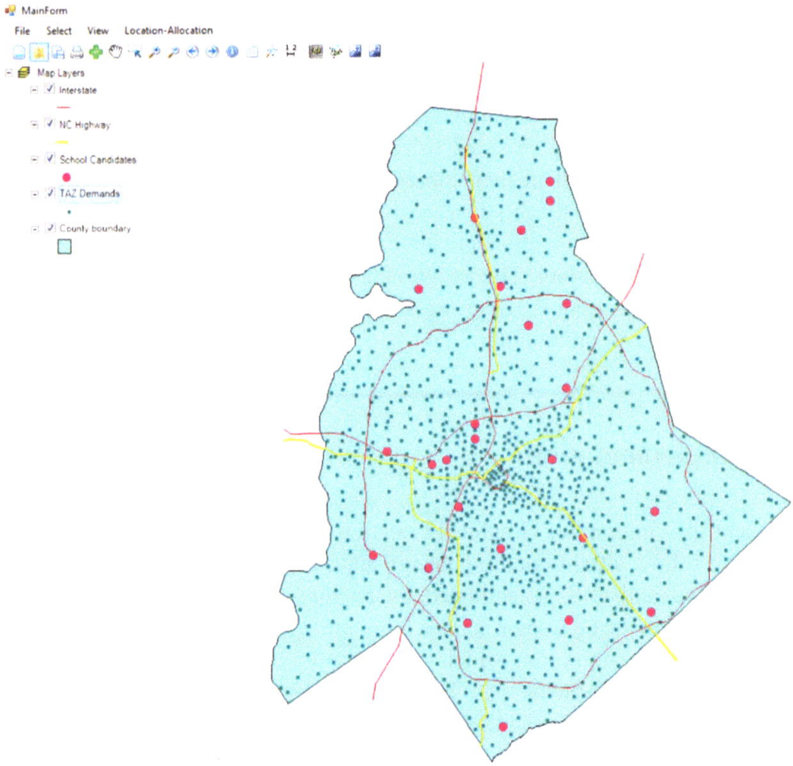

Fig. 7 The iGLASS Interface

the TS, Greedy and GA heuristics can be modified as an option by the users. Fourth, at the end of the main location-allocation processes, the user can choose to re-optimize the best solution found so far to reduce the incidence of non-closest assignments, as discussed in section "Local Re-optimization". Re-optimization is conducted among every set of *q* nearby schools at a time, but only for schools with non-closest assignments. For each school, the algorithm identifies its q-1 closest schools; a smaller instance of the allocation programming model is run then. If the new solution improves the objective value, it is adopted.

Visual outputs include spider maps that reflect the assignment of students to a school (and the magnitude of this allocation); different symbolism is used for the school status (open or close). Schools can also be visualized based on their utilization. Additionally, iGLASS is flexible to add ancillary geographic information in the form of shapefiles to the map display, such as school district boundaries, and highlight the assignments to user-selected schools (Fig. 9).

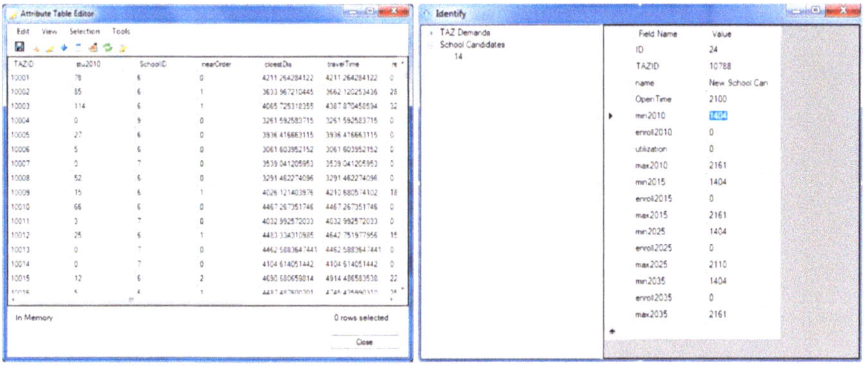

Fig. 8 Demand attribute (*left*) and school capacities (*right*) can be modified on the fly

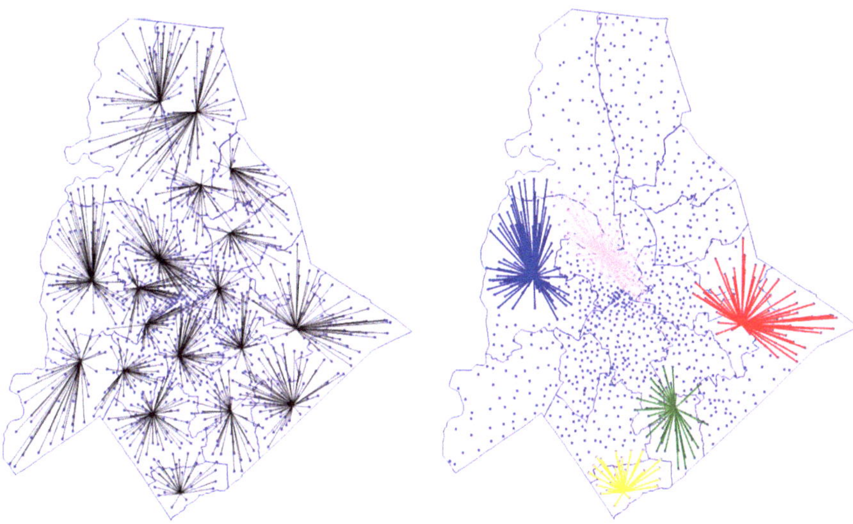

Fig. 9 School district boundaries associated with each school and optimal student assignment. The allocation to each school can be highlighted interactively and displayed by different color

Case Study

Case Study Area

We illustrate the behavior of iGLASS powered by the heuristic algorithms of location-allocation on the case study of high schools in the Charlotte Mecklenburg Schools (CMS) system, which serves the city of Charlotte (North Carolina, USA) and its surrounding county. The case study involves the optimization of the location of public high schools in the CMS system, given the demands and sites that are candidates for schools. The school system has experienced rapidly expanding enrollment over the past 30 years, as a result of one of the highest population

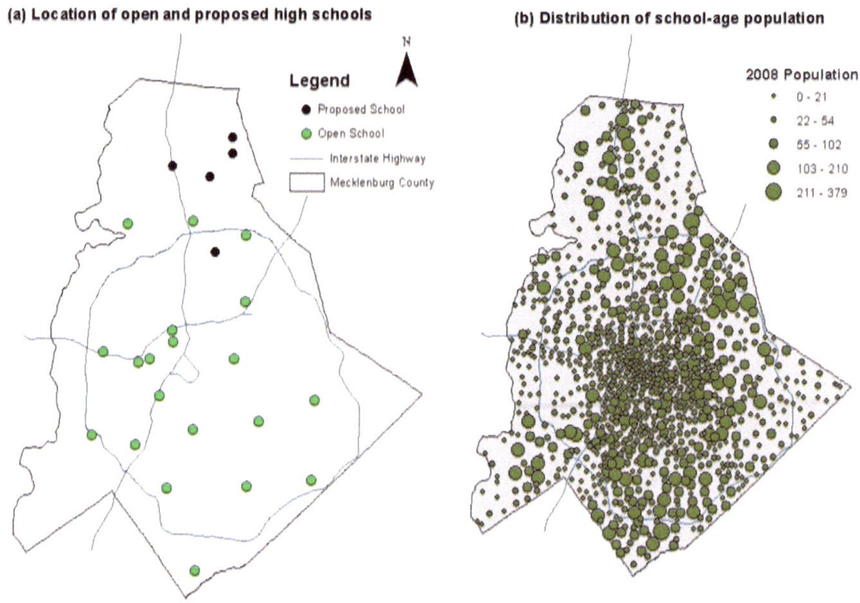

Fig. 10 Open and proposed schools in the CMS system (**a**) and spatial distribution of school demand at the TAZ level (**b**)

growth rates in the United States. In North Carolina alone, the state had a projected enrollment growth of 37% from 1999 to 2009, with the CMS system one of the fastest growing. Population increase in the Charlotte area has had a direct impact on the opening/closing of new and existing schools and the ability of CMS to increase school capacity in the short run.

In 2008, the CMS system operated 20 public high schools (see Fig. 10a) with several additional high schools under consideration to address increasing school-age population. Actual enrolment as well as minimum and maximum capacities of each school were provided by CMS administration. We use the estimated population of children in age of attending high school (14–17 years old) to estimate the demand to be served by CMS high schools. This total demand consists of 37,851 students. The distributed demands within each traffic analysis zone (TAZ) is aggregated into one demand node (n = 1057 demand nodes) (Fig. 10b).

Location-Allocation Results

In this case study, the travel distance estimated on the generalized multimodal network of the Charlotte Department of Transportation is used as impedance from each TAZ (demand node) to potential school sites. All the experiments with iGLASS are done on a computer configured with Intel Pentium (R) CPU B940 (2.00 GHz) with 4.00 GB RAM. The location-allocation solutions will be compared

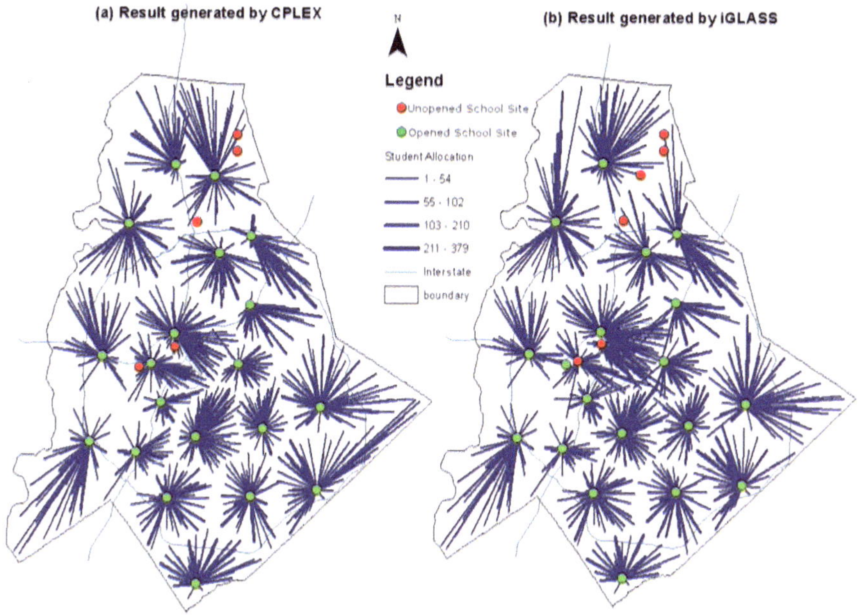

Fig. 11 Location-allocation solution by CPLEX (*left*) and iGLASS (*right*) with the number of facilities p = 20 in both cases (school locations were not fixed). The allocation in iGLASS is based on the minimization of the regret

with each other by using different allocation methods, and a benchmark generated by the mathematical programming solver CPLEX on the mathematical program formulation by objective (1) and constraints (4)–(12) will also be demonstrated.

As a first scenario, we assume that 20 schools should be open out of a feasible set of 25 sites (as shown in Fig. 10). Figure 10 compares the solution obtained by heuristic optimization in iGLASS with the one obtained with CPLEX. The run time for the CPLEX solution is 122.02 s, much more than the iGLASS run time, which is 23.38 s. The objective function obtained by iGLASS is only slightly higher than the CPLEX solution (about 4.5% worse). As far as the school sites in the solution set are concerned, we find that CPLEX and iGLASS locations are identical when the school candidates are broadly dispersed across the service area. However, when there are several candidates bunched up in a small geographic region of the broader study area, CPLEX and iGLASS produce different solutions. This can be seen in the central and northern parts of the study area in Fig. 11. From the maps, we can see that the allocations obtained from iGLASS and CPLEX are similar when the location of schools is dispersed widely, while in regions with multiple options for students (where there are several schools close to each other, that is in the central part of the service area), differences between iGLASS and CPLEX are much greater regarding the allocations.

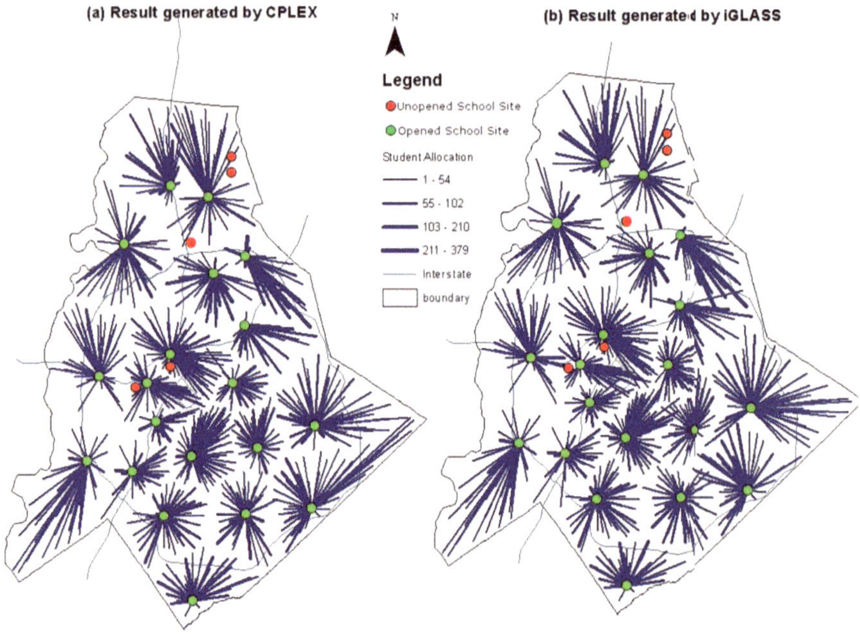

Fig. 12 Location-allocation solution by CPLEX (*left*) and iGLASS (*right*) with the number of facilities p = 20 in both cases (school locations were determined in CPLEX and are fixed in iGLASS). The allocation is based on the minimization of the regret

In a second test scenario, we focus on demand allocations. To this end, we first solve the location-allocation problem with CPLEX. The solution of 20 optimal locations is then used as feasible set in iGLASS to derive demand allocations and compare them to those found with the CPLEX solver. Figure 12 contrasts the CPLEX allocation solution and the iGLASS solution derived with the Greedy algorithm in conjunction with the regret-based priority listing. The difference is in the allocation of pupils to schools, which is influenced by the assignment criterion. We report in Table 1 three performance metrics for each solution method, namely the total travel cost, the percentage of students assigned to their closest school, and the percentage of students assigned to a school that is twice as far as the closest school (labelled as 'further assignment'). By and large, we find that all the heuristic algorithms perform rather similarly to each other as well as to the CPLEX solver. This will be discussed in more detail below. iGLASS's Greedy algorithm with a regret specification and re-optimization is the heuristic with the lowest total travel (only 3% worse than CPLEX), while performance of the others lags behind a little. Clearly, the run time is the critical advantage of the iGLASS heuristic toolbox: iGLASS requires 11.48 s in comparison with 64.29 s for CPLEX. iGLASS is nearly five times faster that CPLEX, which is a significant benefit when a large real-world problem is analyzed.

Table 1 Comparison of different solution strategies applied to the allocation phase of iGLASS. CPLEX is used as a benchmark (data in the parenthesis are the results generated after local re-optimization)

Methods	Total impedance (×10 million)	Closest assignment (%)	Further assignment (%)
CPLEX	1.383	75.45	2.68
iGLASS (greedy, population)	1.469 (1.447)	69.2 (73.3)	7.6 (6.75)
iGLASS (greedy, distance)	1.453 (1.448)	73.9 (73.6)	6.98 (6.7)
iGLASS (greedy, regret)	1.469 (1.424)	80.52 (80.61)	7.46 (5.13)
iGLASS (GA, fitness criterion: Min Total cost)[a]	1.480	82.50	6.50
iGLASS (GA, fitness criterion: Max closest)[a]	1.510	83.00	7.70
iGLASS (GA, fitness criterion: Min further)[a]	1.490	81.50	6.20

[a]Local re-optimization is not applied

Detailed analysis of the results in Table 1 leads to several interesting observations: (1) generally speaking, while heuristic algorithms used in iGLASS have better performance regarding the run time than CPLEX (where a linear programming algorithm is applied), they generate worse values on the total impedance objective function (1.448E8 meters and higher, versus 1.383E8 meters) and the incidence of further assignments (5.13% or higher, versus 2.68%); (2) in different runs, greedy allocation algorithms always generate the same allocation solution when the location of schools are fixed, which indicates this solution is stable; the regret-based greedy method gets the best total impedance, highest closest assignments and lowest further assignments after re-optimization. These results are not consistent with the initial expectations: the population-based greedy algorithm was assumed to get the highest closest assignment percentage because, at every step, we try to send as many students as possible to their closest school, while the greedy method based on the regret value was intended to get the solution with lowest cost in that, at every step, we try to firstly allocate those demands with highest regret value; (3) when the assignment priority is based on regret minimization in iGLASS, the number of closest assignments is slightly higher than with CPLEX; however the same does not apply when the criterion for prioritization is distance or population; (4) the GA algorithm cannot promise a stable solution (there may be some differences according to the fitness function), but it can obtain better rates of closest assignments and of further assignments when these factors are used as fitness functions; actually the final solution generated by GA is highly dependent on how we generate the initial population, here we generate the initial population randomly; (5) in most cases, local re-optimization can improve the solution regarding the total cost, closest assignment and further assignment as a whole.

Conclusions

Location-allocation problems are non-trivial geospatial problems and public school location-allocation is particularly difficult to solve, one reason being that capacity constraints in practice should be incorporated into the model. Moreover, students should be ideally assigned to their closest schools, but the maximum capacity of schools may prevent such outcome from materializing. In this chapter, we addressed several modeling challenges associated with school location-allocation problems when the capacities of the schools are incorporated, and an approach was proposed to solve the location and allocation phases explicitly. It is fair to say that research is lacking on explicitly modeling the allocation phase when the requirement of closest assignment does not hold for all demands. Thus we proposed a generalized multi-objective model of school location that minimizes total travel impedance of all students in the system, while maximizing the number of students assigned to their closest school and minimizing the incidence of "unusually" long trips to school. The model was formulated as a capacitated p-median model. We proposed to solve this complex problem as a two + one-phase heuristic process incorporating Tabu Search in the location phase, and Greedy and Genetic Algorithms in the allocation phase; re-optimization contributes to enhance the optimality of the heuristic solutions. It was implemented as an SDSS based on an open source GIS platform.

The iGLASS executable and stand-alone application gives the user the opportunity to interactively change (1) school capacities; (2) the demand associated with high-school population at each node, (3) local re-optimization to improve the solution. The results from the pilot testing of the interactive and geocomputational iGLASS toolbox presented here are close to those obtained with the CPLEX solver in all respects, but most importantly (1) the run time is significantly reduced, (2) the user has the capacity to change parameters on the fly, and (3) multiple objectives can be effectively handled to support policy and decision making. This is an appealing feature for decision-makers, who may need to weigh different scenarios, such as changes in school capacity, or travel penalties associated with school closing and their objectives when allocation is performed. Furthermore, a range of feasible location-allocation alternatives were provided to the users by iGLASS so they can make decisions accordingly.

We see a number of avenues for further improvement of the work reported in this chapter. First, the initial population for the GA heuristic can be improved by generating a greedy solution rather than by randomly generating it (e.g. initial population would be generated by greedy methods), and different initial populations should be used to evaluate the performance (e.g. stability and improvement of objectives) of GA process in iGLASS. Second, the model can easily be extended to reflect dynamic changes in demand over the time. Third, the impact of the uncertainty in (1) the attribute of the demand and (2) its geographic location need to be evaluated. One way to address the latter is by splitting demand nodes into nodes of smaller demand, which can be redistributed in their respective neighborhood area (geographic perturbation). Finally, further benchmarking should be conducted to fully evaluate the performance of the iGLASS location-allocation toolbox against the full range of solutions that can be generated through trade-offs afforded with the multi-objective formulation embodied in Eq. (13).

Acknowledgements The authors are grateful to the Renaissance Computing Institute of North Carolina for funding this research.

References

1. Murray AT (2010) Advances in location modeling: GIS linkages and contributions. J Geogr Syst 12(3):335–354
2. Scott AJ (1970) Location-allocation systems: a review. Geogr Anal 2(2):95–119
3. Church RL, Murray AT (2009) Business site selection, location analysis, and GIS. Wiley, New York
4. Weber A (1909) Uber den Standort der Industrien, 1909; (translated as Alfred Weber's theory of the location of industries in 1929). University of Chicago, Chicago
5. Brandeau ML, Chiu SS (1989) An overview of representative problems in location research. Manag Sci 35(6):645–674
6. Hakimi SL (1964) Optimum locations of switching centers and the absolute centers and medians of a graph. Oper Res 12(3):450–459
7. Francis RL, McGinnis LF, White JA (1983) Locational analysis. Eur J Oper Res 12(3): 220–252
8. Current J, Min H, Schilling D (1990) Multiobjective analysis of facility location decisions. Eur J Oper Res 49(3):295–307
9. Mirchandani PB, Francis RL (1990) Discrete location theory. Wiley, New York
10. Hakimi SL (1965) Optimum distribution of switching centers in a communication network and some related graph theoretic problems. Oper Res 13(3):462–475
11. ReVelle CS, Swain RW (1970) Central facilities location. Geogr Anal 2(1):30–42
12. Kuehn AA, Hamburger MJ (1963) A heuristic program for locating warehouses. Manag Sci 9(4):643–666
13. Maranzana F (1964) On the location of supply points to minimize transport costs. Oper Res Q 15(3):261–270
14. Teitz MB, Bart P (1968) Heuristic methods for estimating the generalized vertex median of a weighted graph. Oper Res 16(5):955–961
15. Salhi S, Gamal M (2003) A genetic algorithm based approach for the uncapacitated continuous location–allocation problem. Ann Oper Res 123(1):203–222
16. Crainic TG, Gendreau M, Soriano P, Toulouse M (1993) A tabu search procedure for multicommodity location/allocation with balancing requirements. Ann Oper Res 41(4): 359–383
17. Antunes A, Peeters D (2001) On solving complex multi-period location models using simulated annealing. Eur J Oper Res 130(1):190–201
18. Li X, He J, Liu X (2009) Intelligent GIS for solving high-dimensional site selection problems using ant colony optimization techniques. Int J Geogr Inf Sci 23(4):399–416
19. Bischoff M, Dächert K (2009) Allocation search methods for a generalized class of location–allocation problems. Eur J Oper Res 192(3):793–807
20. Teitz MB (1968) Toward a theory of urban public facility location. Pap Reg Sci Assoc 21(1):35–51
21. DeVerteuil G (2000) Reconsidering the legacy of urban public facility location theory in human geography. Prog Hum Geogr 24(1):47–69

22. Church RL, ReVelle C (1974) The maximal covering location problem. Pap Reg Sci Assoc 32(1):101–118
23. Hillsman E (1984) The p-median structure as a unified linear model for location-allocation analysis. Environ Plan A 16:305–318
24. Toregas C, Swain R, ReVelle C, Bergman L (1971) The location of emergency service facilities. Oper Res 19(6):1363–1373
25. Ellwein LB, Gray P (1971) Solving fixed charge location-allocation problems with capacity and configuration constraints. AIIE Trans 3(4):290–298
26. NCDPI (2012) Highlights of the North Carolina Public School Budget [cited]. Available from http://www.ncpublicschools.org/docs/fbs/resources/data/highlights/2012highlights.pdf
27. Murray AT, Gerrard RA (1997) Capacitated service and regional constraints in location-allocation modeling. Locat Sci 5(2):103–118
28. Delmelle EM, Thill J-C, Peeters D, Thomas I (2014) A multi-period capacitated school location problem with modular equipment and closest assignment considerations. J Geogr Syst 16(3):263–286
29. Araya F, Dell R, Donoso P, Marianov V, Martínez F, Weintraub A (2012) Optimizing location and size of rural schools in Chile. Int Trans Oper Res 19(5):695–710
30. Teixeira JC, Antunes AP (2008) A hierarchical location model for public facility planning. Eur J Oper Res 185(1):92–104
31. Gerrard RA, Church RL (1996) Closest assignment constraints and location models: properties and structure. Locat Sci 4(4):251–270
32. Antunes A, Peeters D (2000) A dynamic optimization model for school network planning. Socio Econ Plan Sci 34(2):101–120
33. Antunes A, Berman O, Bigotte J, Krass D (2009) A location model for urban hierarchy planning with population dynamics. Environ Plan A 41(4):996–1016
34. Church RL, Murray AT (1993) Modeling school utilization and consolidation. J Urban Plan Dev 119(1):23–38
35. Wesolowsky GO (1973) Dynamic facility location. Manag Sci 19(11):1241–1248
36. Müller S (2008) Dynamic school network planning in urban areas: a multi-period, cost-minimizing location planning approach with respect to flexible substitution patterns of facilities. Lit, Munster
37. Heckman LB, Taylor HM (1969) School rezoning to achieve racial balance: a linear programming approach. Socio Econ Plan Sci 3(2):127–133
38. Maxfield D (1972) Spatial planning of school districts. Ann Assoc Am Geogr 62(4):582–590
39. Clarke S, Surkis J (1968) An operations research approach to racial desegregation of school systems. Socio Econ Plan Sci 1(3):259–272
40. Church R, Schoepfle OB (1993) The choice alternative to school assignment. Environ Plan B 20(4):447–457
41. Müller S, Haase K, Kless S (2009) A multiperiod school location planning approach with free school choice. Environ Plan A 41(12):2929–2945
42. ESRI (2010) ArcGIS 10. online help. ESRI [cited]. Available from http://help.arcgis.com/en/arcgisdesktop/10.0/help/index.html#//004700000050000000
43. Church RL (2002) Geographical information systems and location science. Comput Oper Res 29(6):541–562
44. Armstrong M, Densham P (2008) Cartographic support for locational problem-solving by groups. Int J Geogr Inf Sci 22(7):721–749
45. Densham PJ (1991) Spatial decision support systems. In: Geographical information systems: principles and applications, vol vol 1, pp 403–412
46. Eldrandaly K (2010) A GEP-based spatial decision support system for multisite land use allocation. Appl Soft Comput 10(3):694–702
47. Ribeiro A, Antunes AP (2002) A GIS-based decision-support tool for public facility planning. Environ Plan B 29(4):553–570
48. Tong D, Murray A, Xiao N (2009) Heuristics in spatial analysis: a genetic algorithm for coverage maximization. Ann Assoc Am Geogr 99(4):698–711

49. Glover F, McMillan C (1986) The general employee scheduling problem. An integration of MS and AI. Comput Oper Res 13(5):563–573
50. Glover F (1989) Tabu search—part I. ORSA J Comput 1(3):190–206
51. Black PE (2013) Greedy algorithm. U.S. National Institute of Standards and Technology (NIST), 2005 [cited 2013]
52. Holland JH (1975) Adaption in natural and artificial systems. University of Michigan Press, Ann Arbor, MI
53. Hosage C, Goodchild M (1986) Discrete space location-allocation solutions from genetic algorithms. Ann Oper Res 6(2):35–46
54. Zhang X, Armstrong MP (2008) Genetic algorithms and the corridor location problem: multiple objectives and alternative solutions. Environ Plan B 35(1):148–168
55. Xiao N, Bennett DA, Armstrong MP (2002) Using evolutionary algorithms to generate alternatives for multiobjective site-search problems. Environ Plan A 34(4):639–656
56. Li X, Liu Z, Zhang X (2009) Applying genetic algorithm and Hilbert curve to capacitated location allocation of facilities. In: Proceedings of the 2009 international conference on artificial intelligence and computational intelligence, pp 378–383

A Space-Time Approach to Reducing Child Pedestrian Exposure to Motor-Vehicle Commuter Traffic

Nikolaos Yiannakoulias and William Bland

Introduction

Pedestrian Injury and Pedestrian Activity

Until the 1990s, pedestrian injury caused by motor-vehicles was one of the leading causes of death in children in North America and Western Europe [1, 2]. Since then the incidence and mortality of child pedestrian injuries has declined in many regions of the world [3–8]. Changes in urban design, traffic engineering, legislation and safety behaviour may have contributed to a noteworthy (though non-linear) decline in traffic-related injury and mortality in recent decades [9]. This is a widespread trend in transportation safety generally, where some of the greatest improvements have benefitted drivers—particularly in the development of motor-vehicle occupant safety technology [10]. Some have argued that the emergence of features of the urban environment—such as single-use zoning, curvilinear street design and hierarchical road plans—may have contributed to a safer pedestrian experience as well [11]. Smaller scale interventions—such as traffic calming infrastructure and intersection controls—also show some promise in reducing the risk of injury and mortality among pedestrians, particularly for children, and have been thought to explain recent declines in pedestrian injury incidence and mortality [12].

An alternative perspective suggests the declines in pedestrian injury risk may simply be a matter of reduced exposure; with less independent pedestrian activity, children are less exposed to the hazards of the transportation environment, and less likely to be harmed in a collision with a motor-vehicle. There is a general consensus

N. Yiannakoulias (✉) • W. Bland
School of Geography and Earth Sciences, McMaster University, 1280 Main Street West, Hamilton, ON, Canada L8S4K1
e-mail: yiannan@mcmaster.ca

J.-C. Thill, S. Dragicevic (eds.), *GeoComputational Analysis and Modeling of Regional Systems*, Advances in Geographic Information Science,
DOI 10.1007/978-3-319-59511-5_18

that children engage in less independent outdoor play than in the past, and are less likely to adopt active transportation options [13, 14]. The consequences of declining physical activity are of considerable interest and importance in the public health community, particularly as obesity prevalence continues to rise among school-aged children [15]. While unravelling the relative contributions of safer environments, motor-vehicle engineering, safety education and reduced exposure is challenging, there is a clear tension between maximizing the safety of children and maximizing their physical activity levels in the current transportation environment.

While this tension has been trending towards lower child physical activity levels, there have been attempts to increase *safe* pedestrian activity within the constraints of existing urban environments—specifically, increasing the levels of activity without an offsetting increase in pedestrian risk. For example, walking school busses—where children, holding hands, walk to school in large, visible groups—have been used in a number of regions to varying levels of success [16, 17]. After successful pilot programs in California and Massachusetts, the Unites States government dedicated federal funding for 'Safe Routes to School' programs emphasizing the importance of active transportation to school. As of 2012, U.S. funding for Safe Routes to School exceeded $1 billion, with similar programs now in Canada, New Zealand and several Western European countries [18]. These programs have signaled an important shift injury prevention strategy—ensuring that children are encouraged to be active, but in a way that does not increase the risk of harm.

Temporal Intervention

Child pedestrian injuries involving collisions with motor-vehicles occur when motor-vehicles and children arrive at the same location in space at the same time. In the language of time geography, these represent bundles of activity, but in a less commonly used sense, since they are unproductive or 'negative bundles', as the outcome of the interaction is undesirable, i.e., a child injury or fatality.

As noted above, small-scale attempts to reduce the negative bundling of child pedestrians and motor-vehicles are thought to have had some success in reducing the risk of child pedestrian injury in recent years. Most of these preventative measures can be classified into one of two types: (1) measures that reduce the spatial convergence of children and motor-vehicles in the transportation system or (2) measures that reduce the speed or design of vehicles to minimize harm when convergence of these agents occurs. Less often considered are measures that reduce the temporal convergence of agents, and in particular, how changes in the timing of trips to school taken by child pedestrians could reduce their exposure to motor-vehicle traffic. Currently, the times children commute to school typically coincide with periods of highest traffic volume, particularly in the morning hours [19–21]. This observation suggests that it may be possible to reduce the risk of pedestrian injuries caused by motor-vehicles by having children travel to school either before or after the peak periods of local traffic intensity. This general strategy would reduce the bundling of travel paths of motor-vehicles and child pedestrians by de-bundling the temporal component, and has the secondary benefit of reducing exposure to motor-vehicle related air pollutants.

Developing such a strategy would have to be based on spatial and temporal information about traffic intensity and child pedestrian routes to school, however some general observations are possible. Assuming a general traffic flow model from suburban areas to downtown areas, suburban motor-vehicle commuters would have to depart from work at earlier times than motor-vehicle commuters living closer to downtown areas in order to arrive on time. It follows that traffic volume would be high in residential suburban areas at earlier times than it would be for downtown areas. Hence, any scheduling of school times that takes area-specific traffic volumes into account should result in spatially patterned optimal school start times—for example, with schools in some areas starting later (after the period of highest traffic volume) and schools in other areas starting earlier (before the period of highest traffic volume). The precise scale and magnitude of these patterns will depend on features of the traffic volume by time.

Earlier work by the authors showed that changing travel times could reduce the negative bundling of pedestrians and motor-vehicles on a real transportation network [22]. However, this study assumed that children take the shortest walking trips to school, something that is often not observed in practice [23]. Given the important role of 'safe' route choices in existing prevention efforts, the purpose of this research is to use simulations of pedestrian activity on a synthesized street network to determine how the *interaction* between route choice and route timing may affect exposure to traffic. Our analysis will help us answer two specific questions: (1) is the timing of a trip less important when the route choice is safe? and (2) do safe routes and safe times combine (in an additive or multiplicative form) to make walking trips safer?

Method

We use the term exposure to describe the negative bundling between motor-vehicles and child pedestrians in space and time. Exposure does not necessarily imply harm, but simply instances in which child pedestrians and motor-vehicles are close enough in spatial and temporal proximity that a child is at risk of harm. The simplest way to reduce (or even entirely eliminate) this exposure is to do away with all child pedestrian travel; however, this is not a preferred option given the physical and psychological benefits of walking. Instead, the modern problem of child pedestrian safety attempts to maximize safety without reducing the frequency of pedestrian trips made overall. With this in mind, our objective is to reduce this risk per child by reducing the number of motor-vehicles children encounter on their walk to school—either at intersections or mid-block. Traffic volume is a predictor of collisions involving pedestrians and motor vehicles at intersections and mid-block locations [24, 25]. We therefore assume that all else being equal, the more motor-vehicles a child is exposed to on a given pedestrian trip, the more likely that the child will be involved in a collision. To conceptualize this, assume that the composite of factors that determine the hazardousness of an instance of exposure at a location on a street network is represented by v. v would include driver distractedness, speed, reaction time, features of the road environment and a variety of other factors. It follows that

$$p_i = \frac{1}{1 + e^{-v_i}}, \tag{1}$$

where p is the probability that a child pedestrian will be struck by a vehicle at this instance of exposure, i, which represents a location on the street network in which a motor-vehicle and child pedestrian bundle in space and time. Furthermore,

$$\varphi = \sum_{i=1}^{n} p_i, \tag{2}$$

where φ the cumulative probability of a child being struck by a car on a pedestrian journey to school with n instances of exposure. Child pedestrian injury strategies involving environmental modification typically target a reduction in the magnitude of some of the factors that comprise v at locations where the probability of collision is considered high. On the other hand, child education strategies target changes in behaviour that would reduce the magnitude of v for many or most instances of exposure on a journey to school.

A space-time intervention augments these other prevention efforts collision by reducing the instances of exposure, n. This can be achieved by de-bundling the temporal schedules of child pedestrians and motor-vehicles; for example, having children walk to school at times when local motor-vehicle traffic activity is lower. In earlier work we proposed that this could be achieved by making small changes to the times that children walk to school [22]. However, it remains unclear how the proposed system would perform in light of safe route to school strategies, which can include changing the routes that children walk to school.

Transportation Model

We address the research questions by analyzing the space-time interactions of child pedestrians and motor-vehicles. The first step is to obtain paths and travel times of motor-vehicle drivers and child pedestrians. At small scales such data can be based on empirical observation—for example, observing sites around schools and taking inventory of the time and location of contact between pedestrians and motor-vehicles. However, at larger scales data on all travel paths is required, and collecting such data would be expensive and time consuming. As an alternative, it is possible to model of the general properties of motor-vehicle and child pedestrian travel, and then use the output of this model to approximate real data.

The general properties of this model are as follows. Our model is comprised of schools, workplaces, a transportation network and households. Households consists of one adult driver and zero or one children. All children are assigned to a school, and all adults are assigned to a workplace. In the model all children are assumed to arrive at school at exactly their school's start time. This approximates the reality that parents usually have a window of only a few minutes during which to leave their children at school; after the bell, the children will be late for school, but if they arrive too early, children may be left unsupervised. Every adult is assumed to work at a workplace outside the home at which they must arrive by a certain work start time. Adult motor-vehicle commuters in households with no children are assumed to drive directly to work, arriving exactly at their work start time. Adults with children may drive directly to work, in which case their children walk to school, or they may drive their children to school before continuing to work. Motor-vehicle and pedestrian trips occur on the transportation network according to some route choice option such that each agent in the system has a time-stamped trip—where the time at each location in the network is known.

Computing Exposure

The transportation model we describe above facilitates a simple method for computing precise exposure to traffic—in space and time—for all children walking to school. Once all pedestrian and all motor-vehicle space-time paths are generated, it is trivial to determine exposure by simply enumerating all instances in which the agents come into contact. Exposure is calculated for intersections and mid-block and summed for each school as well as the system as a whole. Calculating exposure works as follows. First, an instance of exposure at an intersection is defined as an occasion when a motor-vehicle and child pedestrian arrive at an intersection within some interval of time, for example, 30 s. This takes into consideration the role of traffic control measures in extending potential contact between agents. Varying this window size changes the absolute quantity of exposure for a given child, but has no observable effect on the patterns of exposure [22]. An instance of exposure at a midblock location occurs when the two agents cross paths at any location on a road segment. At a micro-scale, children use sidewalks and motor-vehicles use roads such that the paths of these agents do not normally cross. However, given the important role of midblock collisions on child safety [26], we treat these instances of exposure as potentially harmful. We sum up all the instances of exposure for all locations on the network, and also calculate the average time of exposure at each intersection (calculated by summing the times of all exposure for each child at the intersection and dividing it by the instances of exposure). The former is used to set the optimal school schedule and the latter can be used to visualize the space-time pattern of exposure.

Optimization

The final task is to identify a school schedule—comprised of a specific start time for each school—that reduces children's exposure to traffic without also reducing the number of pedestrian trips to school. The problem involves assigning each school, n, one of m possible start times in a way that minimizes total exposure per child pedestrian. For any problem with more than a very small number of schools, the number of possible schedules is m^n, too many to compute exhaustively.

Instead, we reduce the complexity of the problem by treating each school as independent of all other schools. The optimization strategy involves, for each school, assigning each of the m possible times in turn, while assigning all other schools the same default start time (say, 8:30 a.m.). Once this is done for all schools, each school has a start time for which none of the other possible start times result in lower total child pedestrian exposure for that school. The number of schedules under this scheme is nm, few enough that we can compute the total child pedestrian exposure for each school and potential start time. Solutions from the procedure as described are independent of one another, since changing a school's start time should have no effect on the traffic flow at any other school, and therefore, each locally best start time contributes to a globally optimal schedule of school start times. A more complex scenario—where one school's schedule may influence traffic flow near another school, or where school times could affect a parent's decision to drive their child to school or not—could benefit from optimizing the overall schedule rather than at each school individually.

Application

Data Requirements

In order to identify the best school schedule for minimizing exposure in a particular setting, a variety of data are required (Table 1). In previous research we applied the model, exposure calculation and a different optimization procedure to a dataset of schools, street network and journey to work information based on data on the city of Hamilton, Ontario, Canada [22]. While this exercise helped to contextualize the problem, uncertainties associated with some of the data sources left the potential of the findings unclear. Several sources of data required to solve this problem are widely available: digital street network, the location of schools and school catchment areas. Other data are not as commonly available except in geographically aggregated samples: such as the residential locations of child pedestrians and adult motor-vehicle commuters. Yet other data required for the model are generated on assumptions of behaviour—specifically, the modelling of travel paths and the work start times.

Table 1 Data requirements for school schedule optimization model

Model input	Certainty
Locations of schools and school catchment areas	Very high. Based on publicly available data manually geocoded onto a digital transportation network
Digital transportation network	High. Based on digital street data from several sources
Locations of households with adults that drive motor-vehicles to work (including knowing which households have school-aged children)	Moderate. Used census data to estimate these locations as well as the populations of adults and children in households at the census tract level
Locations of work	Moderate. Some data from census on the location of work
Work start time	Low. General estimates are probably reasonable, but would require a survey of working adults to obtain precise information
Paths from households to work	Low. Data from the Census provide start and end points at an aggregate geographic level, but the path actually taken by drivers is unknown. There is evidence that drivers generally prefer to minimize travel time
Paths from households to school	Low. Start and end points are known from municipal data on school locations and census data on child populations, but the path actually taken by child pedestrians is unknown

In order to address the questions specific to this study (is the timing of a trip more or less important when the route choice is safe? and do safe routes and safe times combine to make walking trips safer?) we used a synthesized transportation network as well as synthesized locations of trips. This gives an experimental framework to specifically answer the research questions that would be difficult to address if we were to use real data; specifically, we control for the location of households, location of schools and the structure of the road network by ensuring that these parts of the synthetic data are spatially homogeneous.

Experiment

We generate a simple 'city' with schools, workplaces, households and an hierarchical road network with speeds selected to facilitate efficient traffic flow (Fig. 1). We choose four motor-vehicle road speeds: 40, 50, 60 and 70 Kph. Roads with speeds of 50, 60 and 70 Kph are visible on the map but the 40 Kph are immediately adjacent to schools, and not visible. Motor-vehicle drivers are assumed to take the fastest route to work, reflecting the importance of travel time in route decisions [27]. Pedestrians are assumed to walk along sidewalks adjacent to roads, and therefore are

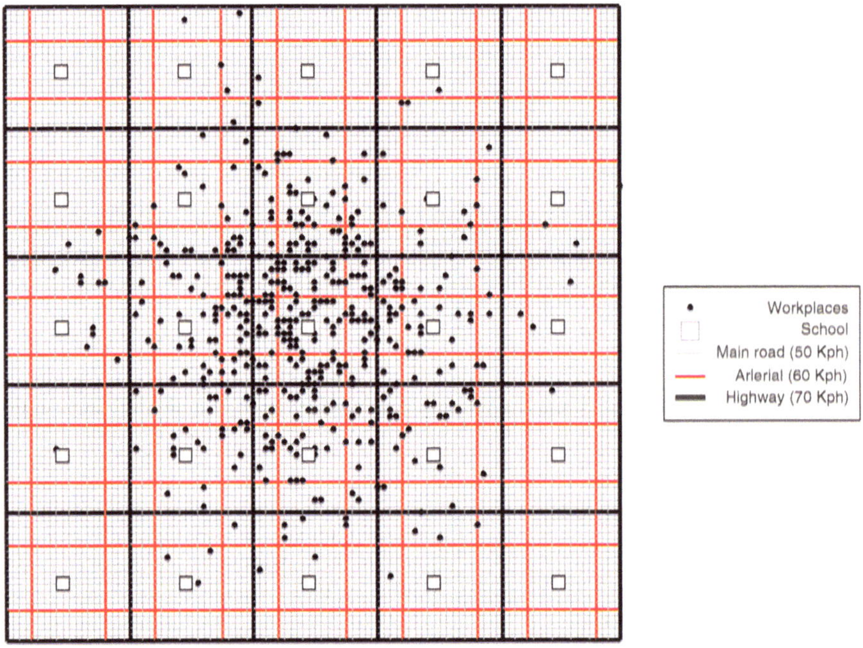

Fig. 1 Infrastructure of the simulated city

assumed to use the same potential paths as motor-vehicles. Child pedestrians are assumed travel at 1.1 m per second based on recommended intersection clearance times for pedestrians less than 65 years of age [28]. Motor-vehicles are assumed to travel at the posted speed limit. In order to observe the effect of route choice on child pedestrian exposure to motor-vehicle traffic under different school scheduling schemes, pedestrian route choices are assumed to be influenced directly by the speed limit of the roadway, where roads with faster speed limits are avoided over roads with slower speed limits. We assign a weight, A, to all roads to determine their attractiveness to a child pedestrian commuter where

$$A = (S/40)^{\beta}, \quad (3)$$

S is the road speed and β is a constant determining the degree to which child pedestrians avoid high speed roadways. The product of A and the length of the road segment determine the travel cost of the road that comprises potential travel paths. The rationale for calculating travel cost as a product of perceived safety and road length is that parents have some underlying awareness that risk is at least partly proportional to exposure—all else being equal, the longer the trip a child takes, the greater the risk of collision. Child pedestrians are assumed to take the path to school with the lowest travel cost, but where cost is a function of distance and some preference for safety. When $\beta = 0$ all road segments are treated equally safe, and children take the shortest path to school. As β increases, higher speed roadways become increasingly unattractive, which may result in longer trips to school to avoid roadways that are perceived as more dangerous. We test out several different values of β in order to observe how safe route choice interacts with the school schedule optimization scheme.

We populate this network with 30,000 households (three at each of the 10,000 intersection locations), 500 workplaces and 25 schools. Households are distributed uniformly across the network. Workplaces are distributed in a cluster around the centre of the network (Fig. 1). Each household is assumed to have a working adult, and 10,000 households (1 at each intersection location) have one child walking to school. The ratio of three drivers to one pedestrian is close to the ratio of drivers to elementary school-aged children in Canada as of 2011. All workers are assumed to attend a workplace, selected at random with replacement, at the fixed time 8:30 a.m. There are 25 schools, each with a population of 400 students. Children go to the school associated with the catchment area in which they reside. The catchment areas boundary are defined by the 70 Kph freeways (Fig. 1). The optimization procedure is limited to finding an optimal schedule where all schools operate between 8:20 and 8:40 a.m.

The model, exposure calculation and optimization routines were programmed in the C++ language.

Results

Figure 2 is a map of average exposure time and frequency by school catchment areas. The shading on the map is used to delineate the times of exposure, and the numeric labels represent the counts of exposure. For this map, all schools are assumed to have the same schedule, and children travel the shortest path to school with no consideration of road speed ($\beta = 0$). There is a clear pattern in time of exposure, where more peripheral areas see exposure earlier in the morning and more central areas see exposure later in the morning. Children in the most central catchment area experience the highest exposure by a considerable margin, but there is little obvious pattern in exposure beyond this.

Figure 3 summarizes the change in total exposure per child for different values of β for both an optimized scheme (found using methods described above) and an homogenous scheme where all schools share the identical start time of 8:30 a.m. Total exposure per child is calculated by dividing the total instances of exposure at midblock locations and at intersections by the number of children walking to school. In spite of the narrow range of times available to select from (8:20 to 8:40 a.m.) the optimized scheme is still superior to the homogenous scheme for all values of β. The optimized scheme and homogeneous schemes are most similar when children have no preference for safe routes ($\beta = 0$). Exposure associated with an optimized school schedule is lowest when $\beta = 0.5$, and increases slightly for larger values of β, though in all cases remains lower than for the homogeneous schedule.

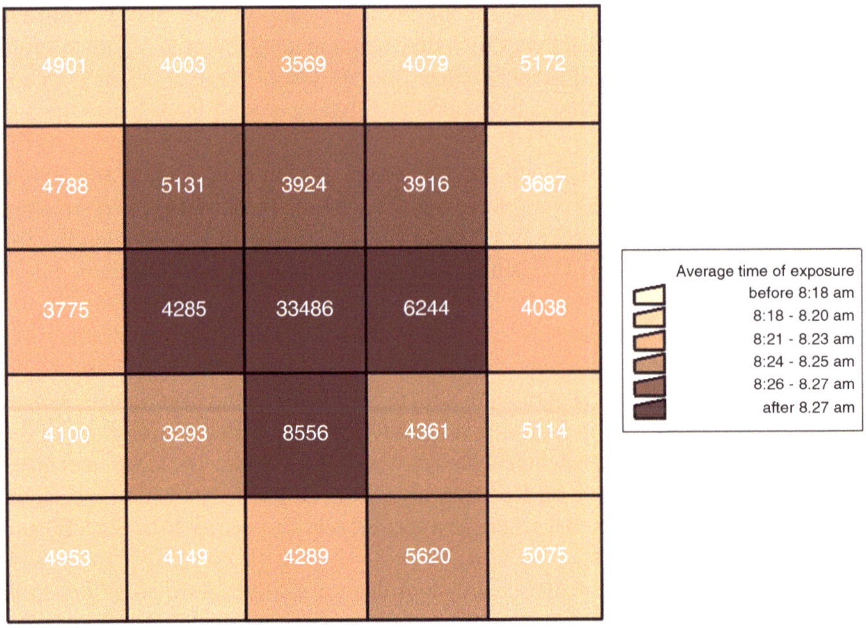

Fig. 2 Map of exposure and average time of exposure for school catchment areas (8:30 a.m. school start times, $\beta = 0$)

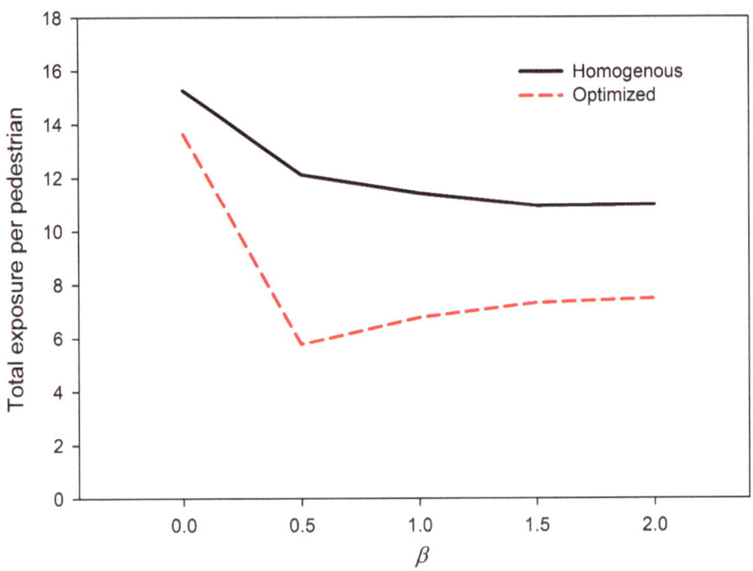

Fig. 3 Total exposure per child pedestrian with changes in the safe route preference, β

Table 2 Exposure and trip time for the optimized scheme as safe route preference (β) increases

	β = 0	β = 0.5	β = 1.0	β = 1.5	β = 2.0
	Midblock				
Mean	5.688	1.647	2.508	3.192	3.743
Standard deviation	3.193	1.960	2.406	3.084	3.735
Maximum	19.038	7.323	9.650	13.943	20.045
Minimum	0.890	0.135	0.335	0.720	1.103
	Intersection				
Mean	7.959	4.136	4.275	4.132	3.739
Standard deviation	12.975	5.602	6.363	5.229	3.152
Maximum	70.003	25.813	34.018	27.895	15.490
Minimum	0.415	0.190	0.403	0.945	1.033
Average pedestrian trip time (min)	15.000	16.439	18.048	19.850	21.865

In Table 2 we aggregate total exposure per child across the 25 schools stratified for midblock and intersection locations, and summarize the mean, standard deviation, minimum and maximum exposure per child for the different values of β. As seen in Fig. 3, the mean exposure per child is highest when β = 0, and lowest when β = 0.5. For intersection locations, as β increases, maximum exposure and variance of exposure per school decline, however for midblock locations all metrics of exposure decline when β = 0.5, but rise as β approaches 2. Interestingly, as β increases, variation in intersection exposure declines, particularly with respect to the maximum value. When β = 2.0, the school with the highest exposure per child is 15.49, less than 25% the school with highest exposure per child when β = 0. As β increases, the average time spent walking to school also increases for child pedestrians, from 15 min when β = 0 to almost 22 min when β = 2.

In Fig. 4A and B we show the spatial pattern of exposure for the homogenous start times in which children travel the shortest route to school (β = 0) and a preference for a safer route (β = 1). Each dark line on the graph defines the boundary of a school catchment area. Dark points on the maps indicate the locations of exposure, and in all cases, represent at least dozens (and in some cases hundreds) of instances of exposure. The spatial patterns do not apparently differ for these two schemes, although as shown in Table 1, there is less exposure when pedestrians prefer safe routes (β = 1) than when they choose the shortest path (β = 0). The pattern of clustering within the catchment areas appears to vary; for the central catchment area, for example, exposure clusters much closer to the school than it does for the peripheral catchment areas.

In Fig. 4C and D, we show the spatial pattern of exposure by time for the optimized schemes in which children travel the shortest route to school (β = 0) and a preference for a safer route (β = 1). Here the spatial patterns differ considerably more across catchment areas. In 4C, the pattern looks somewhat like the patterns for homogeneous time schemes—where exposure clusters around schools—but in some peripheral catchment areas, the pattern is more irregular, with exposure occurring in

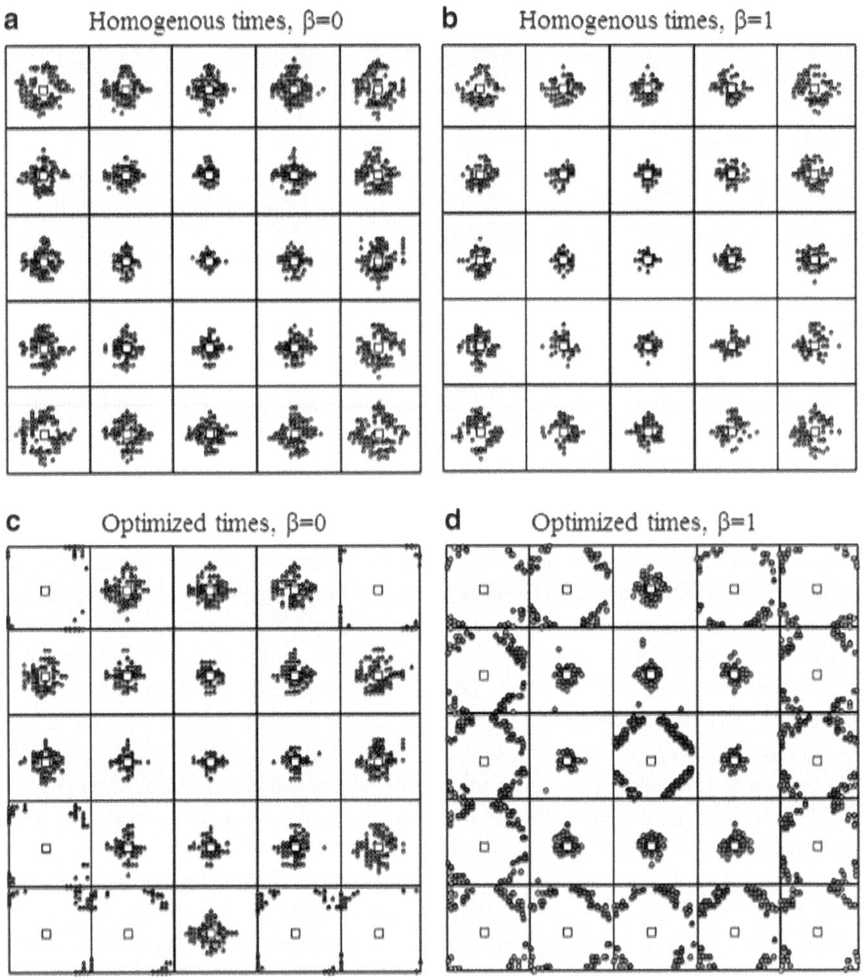

Fig. 4 Locations of exposure for scenarios A–D

more outlining parts of the catchment area. In 4D, there emerges a clearer pattern at two different geographic scales. Within more peripheral catchment areas and the catchment area in the centre of the map, exposure occurs peripherally rather than clustered around schools. Within the eight catchment areas surrounding the centre of the city, exposure is clustered around schools.

Discussion

The well-known association between periods of peak hourly traffic flow and peak hourly child pedestrian injury risk suggests that a judicious school schedule may help reduce exposure to motor-vehicle traffic, and in turn, reduce injury and fatality risk. Since the objective of selecting such school times is to de-bundle the space-time travel paths of pedestrians from the space-time travel paths of motor-vehicles, it seems reasonable to view this strategy as complementary to strategies for finding and/or creating safe walking routes to school. Our results are consistent with the hypothesis that the time of morning a child walks to school and the route they take to school both influence their exposure to motor-vehicle traffic. Our results also suggest that combining safe time scheduling with safe route selection may be the best option for reducing child pedestrian exposure to traffic. Safe route selection can reduce exposure independent of the school schedule, and remains an important part of existing strategies to increase pedestrian safety. Safe time scheduling appears to further reduce exposure to motor-vehicles regardless of whether or not children select safer routes. When combined, these strategies may offer the greatest reduction in exposure to motor-vehicle traffic.

Our analysis is based on synthesized data, so it provides only theoretical support for the idea that changes in school scheduling can lead to measurable changes in exposure. The advantage of this approach is that it allows us to limit the problem to a smaller number of parameters than we would see in the real world. As such, our analysis is specific to the safety and timing of pedestrian trips independent of the many other factors that could influence child safety—such as large and small scale urban design and driver and child behaviour. We now discuss specific observations from our results: (1) the timing of exposure, (2) the safe route paradox, and (3) the emergence of space-time patterns in exposure.

Timing of Exposure

A simple and tempting alternative might be to set all school arrival times to very early in the morning, well before most motor-vehicle commuters are on the road. This would ensure less exposure to traffic, however in high latitude regions, early start times could result in children walking to school before sunrise for certain times of the year. Commuting in poor lighting conditions has been shown to increase risk of collisions in higher latitude regions [29]. This could also expose children to other risks that may concern parents and children (such as fear of strangers) as well as exposure to ‘drowsy drivers’, who are less alert, and more likely to pose a hazard on the road [30]. Uniformly later start times may also seem an attractive alternative, though this could face logistical challenges such as leaving children unsupervised in the home after parents leave for work, or shortening the school day.

A more practical challenge is ensuring that any proposed new schedule is reasonably close to what is already employed in practice. Schools are unlikely to adopt any major change to their existing schedules for various reasons—including maintaining labour agreements with staff and accommodating the habits of local parents. We constrained the feasible school schedules to a 20 min window of time in order to approximate the challenge of identifying practical school start times. In spite of this, we still observed a reduction in exposure when the school schedule is optimized; specifically, we saw a 10–50% reduction in exposure for optimized school time schedules depending on the route choices taken by pedestrians. Whether or not such a reduction would occur in the real world is unclear, but seems likely to be at least partly influenced by the temporal concentration of peak commuter traffic volumes. Specifically, the more temporally compact the peak period of traffic volume, the smaller the change in school time required to separate the two agents in time on the network.

Our results suggest that the timing of peak exposure may not be spatially homogeneous. We synthesized a city with a spatially homogenous distribution of drivers and children and relatively centralized work destinations for motor-vehicle drivers. This resulted in earlier exposure in more outlying regions where drivers needed to embark earlier on their trips to arrive at work on time and later exposure in more central areas where drivers could embark later. Other urban structures would likely result in different patterns; for example, a city with multiple centres of workplace activity could see a more complex pattern of peak exposure times, or even the potential for multi-modal distributions of exposure. This would make the planning of safe travel times a more complex endeavour, but also suggests the importance of local information on exposure for planning school times. It is quite likely that there is no general conceptual model for setting safe school walking times in the real world, but that local information on the flow of traffic and the available walking routes can be vital for planning at the local school catchment level.

A Safe Route Paradox

As noted above, our results suggest that combining safe routes with safe school schedules can reduce child pedestrian exposure to motor-vehicle traffic more than either one of these strategies could do independently. However, our results also highlight a potential challenge in safe route selection; specifically, that the route to school that appears safest at the road level (which determines route choice) may take a child on a longer walking trip, which could actually increase total trip exposure as well as risk of collision. Our experiment used road speed as the determinant of perceived road safety where high speed roads were deemed less safe. As the safe route preference value, β, increases children increase the relative importance of safety over road length in their route choice decisions, and thus total trip time increased. In our experiment we observed that as child pedestrians place increasing importance on the safety of a route to school (as β increases from 0.5 to 2), the total exposure at mid-block locations increased. This change was not observed for intersections, which saw a more systematic decline in exposure with increased safe route selection.

In our experiment, this trade-off emerges because pedestrian exposure is a function of total exposure to traffic for the entire journey to school. So while choosing a relatively safer road decreases the risk of collision on that road compared to another less safe road, choosing this safer road may increase the length of the total journey to school. At some point safer walking trips may become so long as to increase total journey exposure level even if each individual road and intersection may be safe. This results in a curious paradox—where route planning based on avoiding unsafe intersections and roads could result in a less safe journey overall. The point at which this apparent paradox would emerge in the real world is context specific, but the results of our experiment suggest that this problem could be avoided by focussing on a journey-based planning approach rather than education about specific intersections or roads that are deemed unsafe or safe. This would reduce the likelihood of encouraging children to take journeys that are less safe overall in order to avoid specific intersections that may be less safe than other intersections, but part of an overall safer route to school.

Emergence of Space-Time Patterns in Exposure

The emergence of distinct spatial patterns of exposure in Fig. 4A–D illustrates an interesting space-time dynamic in the synthesized environment presented here. Since all motor-vehicle commuters are expected to arrive at work at the same time in our model, commuter traffic emerges as a wave of activity, where the first motor-vehicle drivers to depart their homes are in the peripheral areas, where they have the longest commutes to work. We see this expressed in the timing of exposure—where children are exposed earlier in the morning in peripheral areas, and slightly later in more central areas. As the morning commuting period progresses, centrally located motor-vehicle drivers enter the system up to the point where all arrive at their workplaces on time. Under the homogeneous school scheduling schemes, child pedestrians are behaving in precisely the same manner, but at a smaller geographic scale; the first children to leave home for school are in the periphery of the catchment area, and as the morning progresses, more and more children enter the system until they all arrive at school on time. For the homogeneous school time scheme, exposure is highest clustered around schools, but the level of clustering varies. This indicates that these two waves of commuting activity converge with greatest frequency around school locations, but the convergence is less spatially concentrated for school catchment areas farther from the central region of the city—where workplaces tend not to be located.

Under the optimized school scheduling schemes, the location of exposure is more complex. For the most peripheral school catchment areas, the spatial pattern of exposure emerges in a ring-like pattern near the periphery of the catchment area. For other catchment areas, the spatial pattern of exposure clusters around schools. This is a clear illustration of the impact of time on spatial interaction. For the optimized school scheme, times are selected that minimize the contact between pedestrians and motor-vehicles using information about where motor-vehicles are located at certain times. As such, the optimization routine is attempting to de-synchronize the intersection of these two waves of commuters, and does so in a way that can result in interesting looking spatial patterns.

In a general sense, these observations suggest that safety interventions meant to reduce exposure at specific locations (such as intersection crossing guards and street modifications) should take some consideration of the space-time interaction of motor-vehicles and pedestrians. There is empirical evidence of clustering of child pedestrian injuries near schools [20], and typically, intersection crossing guards, speed controls and other interventions are concentrated at or near these locations. But it could be that for some schools, the highest exposure is elsewhere simply because of the timing of pedestrian and driver commuting waves. Indeed, our model may suggest that in more outlying regions of a city—where drivers have to depart earlier from their homes to get to work on time—exposure may occur farther away from schools than is typically thought. Such spatially diffuse exposure could require an alternative strategy for intervention in some areas. For example, it may recommend placing crossing-guards at more strategic locations away from schools, or moving crossing-guards between locations as the morning commuting period progresses.

Limitations

We use a synthesized environment as a framework for experimentation. While the synthesized city has some general attributes similar to real-world urban environments—a hierarchy of roads and centrally located workplaces—it does not approximate the spatial or other features of any city precisely. As such, it is unclear how meaningful our findings are in any real-world context. We would suggest, however, that the findings are an important exploratory exercise, since any empirical work in this area could be costly, and even risky to the child pedestrian population. This work provides exploratory information based on a simple agent-based model, and as such, resides somewhere between inductive and deductive social science [31]. We hope to use more real world data in real world environments in future models to help test the generalizability and accuracy of our findings.

Our traffic estimation model does not account for congestion. Traffic congestion can affect driver route choice; changes in the routes of motor-vehicle travel due to traffic congestion could affect the estimates of magnitude of exposure, particularly at busy intersections—over-estimating exposure at some intersections. However, it seems unlikely that congestion would have greatly affected the spatial and/or temporal patterns of exposure in a way that would have greatly changed our observations. Another related limitation is that our model does not take into account non-commuter traffic (such as trucks, buses or cars travelling on non work related journeys), or traffic in the city that originates from outside the city. Nevertheless, motor-vehicle commuters represent an important source of risk to pedestrians independent of other traffic, so our findings are unchanged—though the overall impact on the safety of children could be less than what our findings suggest.

Conclusion

Increasing the rates of active transportation to school—either by walking or cycling—may help to reduce rates of child obesity, a public health issue of growing concern, particularly in North America. Pedestrian activity may also be important for child development generally, as it provides children opportunities to build relationships, make decisions and explore their environment, all essential components in their cognitive and emotional development. Parental concerns about the safety of the urban environment may explain observed declines in child pedestrian activity in recent years, and strategies for improving safety need to respond to these somewhat competing realities. Safe route to school are important for ensuring that children have the opportunity for routine pedestrian activity, and based on our findings, trip timing can enhance the effectiveness of safe routes at reducing exposure to motor-vehicle traffic. Our findings also suggest that safe route planning may need to be journey based; rather than identifying the safe walking locations for child pedestrian commuters, emphasis needs to be on identifying trips that minimize exposure to traffic as a whole.

Acknowledgement The research was supported financially by a grant from the Social Sciences and Humanities Research Council of Canada (410-2008-0789).

References

1. Malek M, Guyer B, Lescohier I (1990) The epidemiology and prevention of child pedestrian injury. Accid Anal Prev 22:301–313
2. Rivara FP (1990) Child pedestrian injuries in the United States: current status of the problem, potential interventions and future research needs. Am J Dis Child 114:692–696
3. Cook A, Sheikh A (2003) Trends in serious head injuries among English cyclists and pedestrians. Inj Prev 9:266–267
4. Kypri K, Chalmers D, Langley J, Wright CS (2000) Child injury mortality in New Zealand 1986-1995. J Paediatr Child Health 36:431–439
5. Roberts I (1993) International trends in pedestrian injury. Arch Dis Child 68:190–192
6. Johnston BD, Ebel BE (2013) Child injury control: trends, themes, and controversies. Acad Pediatr 13:499–507
7. Koopmans JM, Friedman L, Kwon S, Sheehan K (2015) Urban crash-related child pedestrian injury incidence and characteristics associated with injury severity. Accid Anal Prev 77: 127–136
8. Kypri K, Chalmers DJ, Langley JD, Wright CS (2001) Child injury morbidity in New Zealand 1987–1996. J Paediatr Child Health 37:227–234

9. van Beeck EF, Borsboom GJJ, Mackenbach JP (2000) Economic development and traffic accident mortality in the industrialized world, 1962-1990. Int J Epidemiol 29:505–509
10. Graham JD (1993) Injuries from traffic crashes: meeting the challenge. Annu Rev Public Health 14:525–543
11. Wang S, Smith PJ (1997) In quest of 'forgiving' environment: residential planning and pedestrian safety in Edmonton, Canada. Planning Perspect 12:225–250
12. Retting RA, Ferguson SA, McCartt AT (2003) A review of evidence-based traffic engineering measures designed to reduce pedestrian-motor-vehicle crashes. Am J Public Health 93: 1456–1463
13. McDonald NC (2007) Active transportation to school: trends among U.S. schoolchildren, 1969–2001. Am J Prev Med 32:509–516
14. van der Ploeg HP, Merom D, Corpuz G, Bauman AE (2008) Trends in Australian children traveling to school 1971-2003: burning petrol or carbohydrates? Prev Med 46:60–62
15. Ogden CL, Carroll MD, Curtin LR, McDowell MA, Tabak CJ, Flegal KM (2006) Prevalence of overweight and obesity in the United States, 1999-2004. JAMA 295:1549–1555
16. Kingham S, Ussher S (2007) An assessment of the benefits of the walking school bus in Christchurch, New Zealand. Transp Res Part A 41:502–510
17. Mendoza JA, Levinger DD, Johnston BD (2009) Pilot evaluation of a walking school bus program in a low-income, urban community. BMC Public Health 9:122
18. Tremblay MS, Gray CE, Akinroye KK, Harrington DM, Katzmarzyk PT, Lambert EV, Liukkonen J, Maddison R, Ocansey RT, Onywera VO, Prista A, REilly JJ, del Pilar Rodriquez Martinez M, Sarmiento Duenas OL, Standage M, Tomkinson G (2014) Physical activity of children: a global matrix of grades comparing 15 countries. J Phys Act Health 11:113–125
19. Calhoun AD, McGwin G, King WD, Rousculp MD (1998) Pediatric pedestrian injuries: a community assessment using a hospital surveillance system. Acad Emerg Med 5:685–690
20. Warsh J, Rothman L, Slater M, Steverango C, Howard A (2009) Are school zones effective? An examination of motor-vehicle versus child pedestrian injury crashes near schools. Inj Prev 15:226–229
21. Yiannakoulias N, Smoyer-Tomic KE, Hodgson MJ, Spady DW, Rowe BH, Voaklander DC (2002) The spatial and temporal dimensions of child pedestrian injury in Edmonton. Can J Public Health 93:447–451
22. Yiannakoulias N, Bland W, Scott DM (2013) Altering school attendance times to prevent child pedestrian injuries. Traffic Inj Prev 14:405–412
23. Buliung RN, Larsen K, Faulker GE, Stone MR (2013) The "path" not taken: exploring structural differences in mapped-versus shortest-network-path school travel routes. Am J Public Health 103:1589–1596
24. Bennet SA, Yiannakoulias N (2015) Motor-vehicle collisions involving child pedestrians at intersection and mid-block locations. Accid Anal Prev 78:94–103
25. Morency P, Gauvin L, Plante C, Fournier M, Morency C (2012) Neighborhood social inequalities in road traffic injuries: the influence of traffic volume and road design. Am J Public Health 102:1112–1119
26. Agran PF, Winn DG, Anerson CL (1994) Differences in child pedestrian injury events by location. Pediatrics 99:284–288
27. Abdel-Aty MA, Kitamura R, Jovanis PP (1995) Exploring route choice behavior using geographic information system-based alternative routes and hypothetical travel time information input. Transport Res Rec 1493:74–80
28. Fitzpatrick K, Brewer MA Turner A (2005) Improving pedestrian safety at unsignalized crossings. Transit Cooperative Research Program D-08, National Cooperative Highway Research Program 3–71
29. Fridstrøm L, Ingebrigtsen S (1991) An aggregate accident model based on pooled, regional time-series data. Accid Anal Prev 23:363–378
30. Landrigan CP (2008) Driving drowsy. J Clin Sleep Med 46:536–537
31. Epstein J (2006) Generative social science: studies in agent-based computational modeling. Princeton University Press, Princeton

Decomposing and Interpreting Spatial Effects in Spatio-Temporal Analysis: Evidences for Spatial Data Pooled Over Time

Jean Dubé and Diégo Legros

Introduction

Spatial dependence and spatial correlation among observations have been suspected for many years [1]. Anselin and Bera [2] define spatial autocorrelation as *the coincidence of value similarity with locational similarity*. As opposed to the unidirectional autocorrelation, such as *comparable sales*, the particularity of spatial autocorrelation lies in its multidirectional effect. Its complexity explains why spatial autocorrelation has received such attention since spatial data is now widely available and used. Spatial autocorrelation among residuals of a statistical model can have various consequences on estimated coefficients and variances, depending on sample size [3–5].

Spatial autocorrelation is also the starting point of the development of spatial econometrics begun at the end of the 70s [6]. The most recent development in the field focuses on spatial panel models and the development of an appropriate estimation procedure [7]. Most of the literature now focuses on how it can be possible to takes into account the spatial characteristics of the data, while using the temporal source of variability as well. Essentially, these approaches apply very well to data representing given geometric delimitation, such as a region, provinces, states or countries. However, there are many databases that do have both dimensions, spatial and temporal, but that are clearly different from the panel case.

J. Dubé (✉)
Université Laval, 2325, rue des Bibliothèques, Pavillon Félix-Antoine-Savard, Québec, QC, Canada G1V 0A6
e-mail: jean.dube@esad.ulaval.ca

D. Legros
Université de Bourgogne, 2, Boulevard Gabriel, Dijon, 24100, France
e-mail: diego.legros@u-bourgone.fr

J.-C. Thill, S. Dragicevic (eds.), *GeoComputational Analysis and Modeling of Regional Systems*, Advances in Geographic Information Science,
DOI 10.1007/978-3-319-59511-5_19

Real estate transactions are different from panel data. In such a case, spatial, or cross-sectional, databases are pooled over time: individual spatial data are usually only observed once[1] (not repeated) over time. This is also the case for business openings (or)/closings, crime location, innovation, and so on. For these cases, the panel procedures cannot be applied. Of course, it is possible to build spatial (pseudo) panel data from such observations, by pooling individual data into a geometric form imposed by the researcher. However, such an approach necessarily implies a spatial aggregation choice which, in turn, can yield three problems: (1) a loss of variability of data, given the fact that the new spatial data express the mean characteristics (a pseudo-panel); (2) the fact that a different spatial aggregation can yield different results (the modifiable areal unit problem—MAUP); and (3) the fact that the inference on individual spatial units is impossible (ecological fallacy).

In short, spatial data collected over time is clearly different from the spatial panel case, but it has received little attention and not much has been done about the modeling strategies for such data. In practice, spatial data pooled over time have been treated has being purely spatial data and the usual spatial econometric methods and models are applied. The introduction of time dummy variables to control for the nominal aspect of the data or for the temporal global trend does not ensure that time dimension is fully adequate, since it only controls for the nominal price evolution. Time dimension can play an even more important role, introducing a spatial autocorrelation pattern respecting temporal directionality: multidirectional spatial effect and unidirectional spatial effect [8].

This chapter endeavors to determine the impact of omitting to decompose the spatial effect (multidirectional and unidirectional) using spatial data pooled over time. This is done by presenting the data generating process (DGP) using a Monte Carlo experiment. The results clearly suggest that neglecting one of the spatial effects generates bias on the spatial (autoregressive) effect, which leads to erroneous interpretation of the marginal effect. This exercise is complemented by an empirical example based on transactions occurring in Paris between 1990 and 2003. The results largely confirm the Monte Carlo results and an out-of-sample prediction shows that the estimation performance of a complete model controlling for spatial multidirectional and unidirectional effect outperforms all the other modeling strategies.

This chapter is divided into five sections. The next section presents the list of the authors that have addressed the question of spatio-temporal modeling using spatial data pooled over time in real estate. Emphasis is placed on the possible ways the spatial autoregressive (SAR) model can be extended to account for a complete decomposition of the spatial effect considering the temporal dimension. The third section presents a brief discussion on the estimation method in the spatio-temporal framework. The fourth section presents a Monte Carlo framework used to evaluate the impact of neglecting the decomposition of the spatial effect using spatial data pooled

[1]Or very few times. It is common, when one point is repeated twice, to assume that this recurrence is strictly related to hazard.

over time with particular emphasis placed on the interpretation of the marginal effect in such context. The fifth section presents an empirical example based on apartment transactions in Paris between 1990 and 2003 and compares the out-of-sample prediction power of the different specifications before turning to the conclusion.

Spatial and Spatio-Temporal Modeling in Real Estate Literature

Hedonic pricing models (HPM) have been used extensively since the formal work of Rosen [9] and the formalization of the hedonic theory. Many empirical applications are based on HPM. In statistical terms, the HPM usually expresses the (log) price of a complex good *i*, such as a real estate good, sold in time period *t*, noted $\mathbf{y}_{it}$, as a function of all its characteristics, intrinsic and extrinsic, stocked in a vector $\mathbf{Z}_{it}$ (Eq. 1). Given the fact that transactions occurred in time, it is usual practice to control for the nominal evolution of real estate price by including a set of dummy variables, in a vector noted $\mathbf{D}_{it}$, indicating the time period in which the house was sold.

$$\mathbf{y}_{it} = \alpha\boldsymbol{\iota} + \mathbf{D}_{it}\boldsymbol{\delta} + \mathbf{Z}_{it}\boldsymbol{\beta} + \boldsymbol{\varepsilon}_{it} \tag{1}$$

Transactions databases consist of a set of individual cross-sectional layers (*i* is different for all observations), while the spatial layers of data are pooled over time (see Fig. 1). Thus, both subscripts are necessary, but the interpretation is different from the panel data case because the subscript *i* is never (or rarely) repeated. Recurrences of house sales are usually treated as random events. In such a case, the total number of observations is noted $N_T = \sum_t N_t$ where N_t is the total number of observations in one time period *t*: for $t = 1, 2, \ldots, T$.

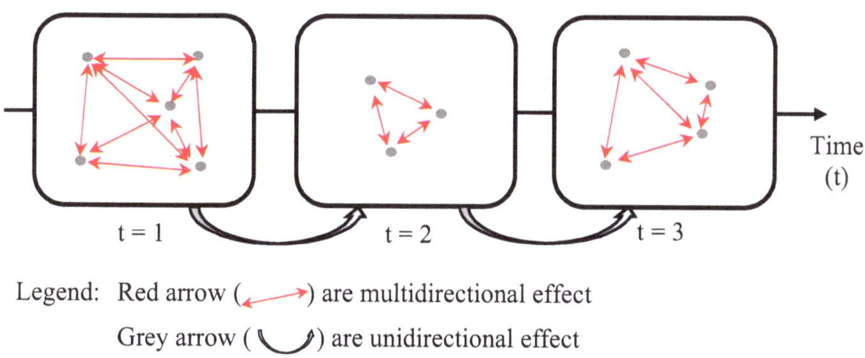

Fig. 1 Distinction between multidirectional and unidirectional spatial effect. *Red arrow* are multidirectional effect. *Grey arrow* are unidirectional effect

In consequence, the vector $\mathbf{y}_{it}$ is of dimension $(N_T \times 1)$, the vector $\boldsymbol{\iota}$ is of dimension $(N_T \times 1)$, the matrix $\mathbf{D}_{it}$ is of dimension $(N_T \times (T-1))$, where T is the total number of time periods, and the matrix $\mathbf{Z}_{it}$ is of dimension $(N_T \times K)$, where K is the total number of independent variables. The vectors of parameters $\boldsymbol{\delta}$ and $\boldsymbol{\beta}$ are, respectively, of dimension $((T-1) \times 1)$ and $(K \times 1)$ and the parameter α is a scalar. The vector of parameter $\boldsymbol{\delta}$ allows control for the nominal aspect of the price (and recompose a general price index), while the vector of parameter $\boldsymbol{\beta}$ expresses the (mean) implicit price of the individual amenities of the house. Finally, the vector of perturbations $\boldsymbol{\varepsilon}_{it}$ is of dimension $(N_T \times 1)$, assumed of homogenous variance and not spatially correlated. However, these last assumptions are rarely satisfied in practice.

Since 1990, it is widely recognized that the error terms of the hedonic pricing model of real estate studies are spatially correlated [10–12]. Various techniques have been developed to deal with spatial autocorrelation among residuals, such as geo-statistical techniques [13–15], coefficient expansion method [16, 17], local regression techniques [18–21] and spatial econometric models [22–24]. If there is still debate about how spatial dimension should be taken into account (geostatistical models or spatial econometric models), there is no doubt that spatial dimension is important in the data generating process of real estate data.

Spatial Econometrics and Hedonic Pricing Models

Spatial econometrics directly addresses the problem of spatial autocorrelation among residuals of ordinary least squares (OLS) models by proposing an autoregressive specification to be incorporated in the HPM. There is still much debate on which specification should be used. Some argue that both processes are not necessarily related to any theory [25], while others argue that spatial econometrics is clearly an appropriate way to deal with issues [26]. Without entering this debate, it must be noted that spatial econometrics are largely related to the way the spatial weights matrix, $\mathbf{W}$, is constructed.

One popular model in spatial econometrics is the spatial autoregressive (SAR) model (Eq. 2). In such a case, it is assumed that the price of a house is related to and explained by the other sale prices occurring in the vicinity ($\mathbf{W}\mathbf{y}_{it}$). This may be largely related to what real estate professionals refer to as being the "*comparable sales*" approach. This approach is also popular among regional scientists since it captures the (spatial) spillover effect. Thus, price is not only determined by the individual characteristics of the house, but also by a spatial effect related to environmental amenities, market conditions, or any other phenomenon that is internalized through the sale price of other houses.

$$\mathbf{y}_{it} = \rho\mathbf{W}\mathbf{y}_{it} + \alpha\boldsymbol{\iota} + \mathbf{D}_{it}\delta + \mathbf{Z}_{it}\beta + \boldsymbol{\varepsilon}_{it} \tag{2}$$

Notwithstanding, the spatial autoregressive (SAR) specification[2] is based on the definition of the general elements, w_{ij}, of the spatial weights matrix **W** of dimension ($N_T \times N_T$). Usually, the general elements are uniquely based on spatial distances. The particularity with such specification is the fact that a single parameter β_k can no longer determine the marginal effect [4], except if the effect comes from a strictly pecuniary effect [27].

Spatio-Temporal Hedonic Pricing Models

The spatial specification of the HPM neglects the fact that data are collected over time and that the spatial multidirectional effect is subject to temporal constraints [28]. There has been some work trying to incorporate both dimensions, spatial and temporal, in hedonic pricing models through the construction of a spatio-temporal weights matrix (Table 1). In all cases, the construction of the weights matrix is usually based on *a priori* temporal chronological ordering, ensuring that the first line of the database corresponds to the oldest transactions, while the last line of the database corresponds to the latest transactions. With such a specification, it is possible to decompose the weights matrix in its triangular parts [28, 43]. This has been the main idea underlying the seminal work of Pace et al. [30, 31], which has influenced the way many empirical analyses have been conducted [33, 34, 36, 40].

Table 1 List of spatio-temporal modeling strategies in HPM

Authors	Years	City	Sample size (N_T)
Can and Megbolugbe [29]	1990	Miami	944
Pace et al. [30]	1966–1991	Fairfax County	70,822
Pace et al. [31]	1984–1992	Baton Rouge	5243
Gelfand et al. [32]	1985–1995	Baton Rouge	1327
Tu et al. [33]	1992–2001	Singapore	2950
Sun et al. [34]	1990–1999	Singapore	54,282
Smith and Wu [35]	2004–2005	Philadelphia	400
Nappi-Choulet and Maury [36]	1991–2005	Paris	2587
Huang et al. [37]	2002–2004	Calgary	5000
Nappi-Choulet and Maury [38]	1991–2005	Paris	220,418
Thanos et al. [39]	1995–2001	Athena	1613
Dubé and Legros [28]	1993–1998	Lucas County	25,237
Liu [40]	1997–2007	Dutch Randstad	437,734
Dubé and Legros [41, 42]	1990–2001	Paris	10,000

[2]It is also the case for most of the geostatistical and local models.

The spatio-temporal model developed by Pace et al. [30], and the extensions proposed, have the advantage of controlling for the spatial and the temporal dimensions using spatial (**S**) and temporal (**T**) weights matrices taking both a lower triangular specification. The spatio-temporal dimension is isolated using a cross-product of the two matrices (**ST** and **TS**), which act as a filter to control for the complexity of the interaction effects [31]. However, the interpretation of the cross-product of the spatial and temporal weights matrices is not trivial and is only an indirect measure of the spatio-temporal effect [35], which makes it hard to isolate the unidirectional and multidirectional spatial effects[3] (Fig. 1).

For these reasons, other spatio-temporal models have been based on the extension of the SAR model (Eq. 2), taking into account the temporal constraints related to spatial relations. One of the first attempts in such a context was proposed by Can and Megbolugbe [29], who developed a weights matrix based on the transactions occurring 6 month before. The weights matrix is lower triangular and allows to capture the effect of the "*comparable sales*", since it only accounts for the effect of past transactions on actual sales. The relations among observations are no longer multidirectional, but subject to temporal constraints: the past observations can influence the actual realizations, but the inverse is not possible (unidirectional spatial effect). In this case, the spatio-temporal model is simply an adaptation of the SAR model where the weights matrix expresses the spatial links from previous observations to actual observations and where ψ capture the (dynamic) effect spatially located (Eq. 3). In real estate applications, such dynamic effect can be seen as a "*comparable sales*" effect on price determination process.

$$\mathbf{y}_{it} = \psi \underline{\mathbf{W}\mathbf{y}}_{it} + \alpha \boldsymbol{\iota} + \mathbf{D}_{it}\delta + \mathbf{Z}_{it}\beta + \boldsymbol{\varepsilon}_{it} \tag{3}$$

A similar approach is used by Gelfand et al. [32] for transactions occurring between 1985 and 1995 in Baton Rouge, by Smith and Wu [35] for transactions occurring between January 2004 and September 2005 in Philadelphia, and by Thanos et al. [39][4] for transactions occurring between January 1995 and December 2001 in Athena. This idea was also reintroduced by Des Rosiers et al. [44], but in a different way: accounting for the mean sale price in a predefined neighborhood in the previous quarter.

Smith and Wu [35] and Huang et al. [37] have extended this concept and have proposed a general framework to develop a unique spatio-temporal weights matrix, **W**, simultaneously integrating spatial and temporal distances and constraints using the Hadamard product ($\square$) between a (full) spatial weights matrix, **S**, and a temporal weights matrix, **T**. The term-by-term operation ($\mathbf{W} = \mathbf{S} \square \mathbf{T}$) indicates that a general element of the matrix is defined by $w_{ij} = s_{ij} \times t_{ij}$.

[3]This can partly explain why Nappi-Choulet and Maury [38] have proposed a specification using only the spatial, **S**, and temporal, **T**, weights matrices.

[4]The authors explicitly consider the temporal distances by weighting the spatial elements by the inverse of time elapsed between sales.

The temporal elements, t_{ij}, can be seen as a filtering process, allowing the isolation of different spatial effects related to different temporal specifications [8]. Numerous scenarios are possible: defining the t_{ij} elements as taking a value of one if observation *i* is collected before observation *j* and zero otherwise gives a spatio-temporal weights matrix where only the lower triangular part have non-zero elements, and thus measure the unidirectional spatial effect ψ (Eq. 3); defining the t_{ij} elements as taking a value of one if observation *i* is collected during the same time period as observation *j* and zero otherwise gives a block diagonal (spatial) weights matrix allowing to measure multidirectional spatial effect, ρ (Eq. 4) [42].

$$\mathbf{y}_{it} = \rho \underline{\mathbf{S}}\mathbf{y}_{it} + \alpha \boldsymbol{\iota} + \mathbf{D}_{it}\delta + \mathbf{Z}_{it}\beta + \boldsymbol{\varepsilon}_{it} \tag{4}$$

Of course, the two spatial effects, multidirectional and unidirectional, can be isolated by introducing the lower triangular weights matrix, $\underline{\mathbf{W}}$, and the block diagonal weights matrix, $\underline{\mathbf{S}}$, in the same equation (Eq. 5). Such functional form avoids the pitfall of multiple weights matrices for the dependent variable [45] since the term $\underline{\mathbf{W}}\mathbf{y}_{it}$ expresses the mean sale price for houses sold previously within a predefined spatial zone of influence. This new variable is thus completely exogenous from the point of view of the transaction occurring in the actual time period.

$$\mathbf{y}_{it} = \rho \underline{\mathbf{S}}\mathbf{y}_{it} + \psi \underline{\mathbf{W}}\mathbf{y}_{it} + \alpha \boldsymbol{\iota} + \mathbf{D}_{it}\delta + \mathbf{Z}_{it}\beta + \boldsymbol{\varepsilon}_{it} \tag{5}$$

The advantage of the latter specification is that the spatial effect can be expressed as the sum of two distinct components.[5] The spatial multidirectional effect is similar to what is usually captured through a SAR model estimated on cross-sectional data, while the spatial unidirectional effect is similar to the dynamic effect in time series analyses, except that it is spatially localized. This approach shows promising avenues by correctly isolating the spatial effect in the actual time period, as well as the dynamic spatial effect measured through observations collected in the previous time period. These effects have been shown to be significant for transactions occurring in Lucas County (Ohio) between 1993 and 1998 [8] and in Paris between 1990 and 2001 [41].

To summarize, spatio-temporal models for spatial data pooled over time can be based on a simple extension of the spatial econometric specification, by accounting for the temporal dimension in the construction of the weights matrix. The representation of the DGP for spatial data pooled over time underlines the pertinence, as mentioned by the literature, of thinking about how the database structure should be analyzed before doing any mechanical construction of the weights matrix. According to some authors, the weights matrix should be based on

[5]The construction of a spatio-temporal weights matrix can also be done with the construction of a temporal weights matrix. However, the decomposition presented here can simplify the exercise by introducing constraints on the spatial weights matrix through a block diagonal decomposition ([42], Chap. 5).

a priori theoretically-defendable knowledge [46, 47]. This idea is also supported by the work of Pinkse and Slade [48] who suggest that approaches should be developed from an empirical perspective, and by McMillen [49] who stresses that what really matters when working with spatial data is the relative position of the observations. To some extent, these restrictions to modeling strategies were previously highlighted by what Legendre [50] calls the *new paradigm* of spatial autocorrelation. In short, the DGP for spatial data pooled over time is different from the spatial case, where all relations are multidirectional.

Estimation Methods

In spatial econometrics, a general step before estimating the models is to row-standardize the final form of the weights matrix (Eqs. 2–5). The row-standardization procedure is a common practice that ensures the comparability of the usual statistic tests as well as the autoregressive estimated coefficients [42]. There is also a computational advantage of row-normalizing the weights matrix (see [4, 51]).

For functional forms based solely on the unidirectional spatial effect (Eq. 3) or introducing multidirectional and unidirectional spatial effects (Eq. 5), an adjustment is necessary. Since the weights matrix is based on a lower (block diagonal) triangular specification, the first N_1 elements are set to 0. A simple way to avoid the issue related to false 0 values is to drop the first N_1 observations from the database and not use them in the estimation process. Consequently, the model is not estimated using the whole sample size, N_T, but instead is estimated using N_{T-1} observations, where $N_{T-1} = N_T - N_1$. In the end, the final sample size is reduced by the total number of observations in the first time period, N_1 (see [52]).

The same procedure is used when dealing with dynamic spatial panel data. Since it is impossible to have information on the initial time period ($t = 0$), this transformation is necessary to avoid false zero values in the time lag variable for $t = 1$ and introduce potential bias on the ψ parameter. Thus, the vectors of variables are now of dimension ($N_{T-1} \times 1$) and the weights matrices are of dimension ($N_{T-1} \times N_{T-1}$).

After reducing the total sample size, models using only the lower triangular specification (Eq. 3) can be estimated by OLS (or generalized least squares—GLS) method,[6] while the other specifications (Eqs. 2, 4, and 5) can be estimated using the maximum likelihood (ML) method[7] [54, 55], two step estimation process [56], method of moments [57], or the new HAC method [58–60].

---

[6]The model uses a variable based on realizations recorded one period before, $\underline{\mathbf{Wy}}$, and thus assumed exogeneity in the actual time period (see [31]).

[7]See LeSage [53] for a complete presentation of such methods using MatLab software.

A Monte Carlo Experiment

Given the complete decomposition of the spatial effect from spatial data pooled over time, a Monte Carlo experiment is conducted to see what happens to the estimated parameters when the modeling strategies are omitted to account for one or the other spatial effects. To do this, it is assumed that the DGP is constructed using the full decomposition of the spatial effects (multidirectional and unidirectional—Eq. 6) while the different specifications proposed in the literature (Eqs. 2–5) are estimated using the adequate specifications of the weights matrices.

$$\mathbf{y} = (\mathbf{I}-\rho\underline{\mathbf{S}}-\psi\underline{\mathbf{W}})^{-1} \quad [\alpha\iota + \mathbf{z}\beta + \boldsymbol{\varepsilon}] \tag{6}$$

Where $\mathbf{y}$ is the vector of the dependent variable to be constructed of dimension $(N_T \times 1)$, $\mathbf{z}$ is a vector of the independent variable of dimension $(N_T \times 1)$ and can be seen as resulting from a principal component analysis (PCA) summarizing all the pertinent information in a unique variable, and $\boldsymbol{\varepsilon}$ is a vector of an independent and identically distributed error term of dimension $(N_T \times 1)$. $\mathbf{I}$ is the identity matrix, and $\underline{\mathbf{S}}$ and $\underline{\mathbf{W}}$ are row-standardized weights matrices, all of dimension $(N_T \times N_T)$. The $\underline{\mathbf{S}}$ matrix is a spatial weights matrix accounting for the spatial multidirectional relations occurring in the same time period (block diagonal), and the $\underline{\mathbf{W}}$ is a spatio-temporal weights matrix accounting for the spatial unidirectional (lower triangular) relations occurring from the observations collected in the previous time period.

The ρ coefficient represents the usual spatial spillover (multidirectional) effect, the ψ coefficient measures the *comparable sales* (unidirectional) spatial effect, the α coefficient represents the constant term, and the β coefficient measures the effect of the independent variable on the dependent variable. For simplicity's sake, all coefficients are scalars in this framework.

The Monte Carlo Set up

To conduct a Monte Carlo experiment,[8] we must first fix the variables and the parameters related to the DGP (Table 2). The first step is to build the individual spatial units and declare where, in time, these observations are collected. The individual spatial units are drawn from a square spatial grid of dimensions 10×10. The spatial units can be seen as kilometers for instance and or obtained through a uniform law (0, 10). The temporal dimension is set in a similar way. The t variable is expressed in a continuous way, varying from 0 to 10, and is simulated using a uniform law (0, 10). These three variables are fundamental to build the spatio-temporal weights matrices, $\underline{\mathbf{S}}$ and $\underline{\mathbf{W}}$. The weights matrix controlling for observations collected in the same

[8]See Adkins and Gade [61] for an interesting discussion about how to conduct Monte Carlo experiments and Dubé and Legros [62] for an application.

Table 2 Set up of the Monte Carlo experiment

Variables	Distribution
$\mathbf{z}$	N(0.9)
ε	N(0.1)
X	U(0.10)
Y	U(0.10)
t	U(0.10)
Parameters	Values
α	0.5
β	1
ρ	0.2; 0.4; 0.6
ψ	0.2; 0.4; 0.6
N_T	2500
# of draws	1000
Weights matrices	Cut-off criteria
Inverse distance	$d_c \leq \mu_{d(i)}$
Negative exponential	$d_c \leq \mu_{d(i)}$
Nearest neighbors	25

time period ($\underline{\mathbf{S}}$) controls for observations occurring in $\Delta t \leq |0.25|$ surroundings temporal units, while the weights matrix controlling for previous observations ($\underline{\mathbf{W}}$) considers all observations collected before this temporal window ($\Delta t > 0.25$) and the temporal weight gives more weight to temporally close observations. Moreover, the spatial weights are built using three specifications: (1) an inverse distance matrix with cut-off distance criteria; (2) an exponential negative distance matrix with cut-off distance criteria; and (3) a 25-nearest neighbors matrix.

Since the DGP depends on the value of the independent variable, $\mathbf{z}$, and the error term, $\boldsymbol{\varepsilon}$, these two variables are generated using a normal distribution (Table 2). The simulations are based on 1000 repetitions of 2500 observations ($N_T = 2500$). Since the objective of the paper is to explore the effect of neglecting the temporal dimension of the DGP on the autoregressive parameter, the parameter β is set to 1, while the parameter α is set to 0.5. Only the autoregressive parameters, ρ and ψ, can vary in the simulations. The values of the parameters ρ and ψ are set to 0.2; 0.4; and 0.6. Thus, using the values of the parameters and the value of the independent variable as well as the value of the error term, it is possible to recompose the value of the dependent variable.

In the end, the values of the dependent variables expressed in Eq. (7) are recomposed and all the models proposed in the literature (Eqs. 2–5) are estimated using the different specification of the weights matrices. The resulting parameters are stored and the distribution is compiled and compared to the true values postulated by the experiment.

Monte Carlo Results

Two statistics are used to evaluate the impact of using one or the other autoregressive specification on the estimated coefficients: (1) the bias (Eq. 7); and (2) the mean square error (Eq. 8). Here, θ indicates the true value of the parameter (fixed in the Monte Carlo set up), while b represents the estimated values of the parameter. These statistics are calculated for each coefficient and each specification.

$$\text{Bias} = \text{E}(\text{b}) - \theta \tag{7}$$

$$\text{MSE} = \text{E}(\text{b} - \theta)^2 \tag{8}$$

The main concern here lies with regards to the possible bias on the β coefficient, as well as the different autoregressive parameters, ρ and ψ. For the β parameter, the biases are small for all values of the ρ and ψ parameters, except for the specification using the nearest neighbor's weights matrix (Table 3). The MSE are very low for all specifications and slightly higher for specifications based on the nearest neighbor's weights matrix (Table 4). Thus, independent of the specification of the weights matrix, the bias appears to be negligible for the β parameters. This is good news since the main interest in econometric specifications lies on the β parameters.

Concerning the autoregressive parameters (ρ and ψ), the bias and the MSE are somewhat more pronounced, given the form of the weights matrix used. This is particularly true for the model using the strictly spatial weights matrix (first three columns at the bottom of the tables). In such a case, bias is positive and increases with the value of ψ, as do the MSE. This results can be explained by the fact that in such a case, the two spatial effects are amalgamated in only one single statistic capturing the multidirectional and the unidirectional spatial effect. Moreover, this specification of the weights matrix can introduce some over-connectivity problems, introducing a bias in the estimated autoregressive coefficient [63].

For the two other specifications using the multidirectional spatial effect (second three columns at the bottom of the tables) or the unidirectional spatial effect (last three columns at the bottom of the tables), the bias is less pronounced. However, when the omitted spatial effect (ρ or ψ) is high, the bias increases. Once again, the omitted spatial effects are then internalized, at least partly, through the other parameter. This problem is more pronounced using the nearest neighbors to build the spatial relations. In this case, the bias is noted, except when ρ and ψ are low ($\rho = \psi = 0.2$).

Why do these results matter? Because omitting to decompose the spatial effect into multidirectional and unidirectional effects leads to potential bias and erroneous interpretation of the source of spatial variability of the phenomenon under study. For the multidirectional spatial effect, the marginal effect is usually decomposed into two components: the direct and the indirect effect. Both effects compose the total marginal effect and are expressed, for a row-stochastic matrix, by $(1-\rho)^{-1}\beta_k$ ([4], p. 38), where β_k is the coefficient associated with the kth independent variable.

Table 3 Bias on the estimated coefficients of α and β

	w_{ij}	SAR specification with **S**			STAR specification with <u>**S**</u>			STAR specification with <u>**W**</u>		
		$\psi = 0.2$	$\psi = 0.4$	$\psi = 0.6$	$\psi = 0.2$	$\psi = 0.4$	$\psi = 0.6$	$\psi = 0.2$	$\psi = 0.4$	$\psi = 0.6$
		for β			for β			for β		
$\rho = 0.2$	$(1/d_{ij})$	0.0001	0.0004	0.0000	0.0002	0.0005	0.0005	0.0002	0.0006	0.0008
	$\exp(-d_{ij})$	0.0001	0.0004	0.0000	0.0003	0.0005	0.0005	0.0002	0.0007	0.0010
	$\min(d_k)$*	0.0015	0.0018	−0.0039	0.0049	0.0085	0.0073	0.0041	0.0099	0.0171
$\rho = 0.4$	$(1/d_{ij})$	−0.0001	0.0001	0.0005	0.0003	0.0007	0.0022	0.0000	0.0005	0.0018
	$\exp(-d_{ij})$	−0.0001	0.0002	0.0008	0.0004	0.0009	0.0038	0.0000	0.0006	0.0027
	$\min(d_k)$*	−0.0010	0.0055	0.0105	0.0106	0.0226	0.0384	0.0049	0.0155	0.0336
$\rho = 0.6$	$(1/d_{ij})$	−0.0006	−0.0005	0.0022	0.0005	0.0012	0.0033	−0.0004	0.0001	0.0028
	$\exp(-d_{ij})$	−0.0008	−0.0006	0.0055	0.0007	0.0020	0.0129	−0.0008	−0.0003	−0.0021
	$\min(d_k)$*	−0.0123	−0.0016	0.0189	0.0171	0.0420	−0.0116	−0.0061	0.0028	0.0125
		for ρ			for ρ			for ψ		
$\rho = 0.2$	$(1/d_{ij})$	0.1058	0.2586	0.4266	−0.0068	0.0091	0.0191	0.0091	0.0122	0.0064
	$\exp(-d_{ij})$	0.1606	0.3469	0.5348	0.0018	0.0329	0.0855	0.0163	0.0243	0.0278
	$\min(d_k)$*	0.1616	0.3368	0.5134	0.0145	0.0725	0.1829	0.0279	0.0434	0.0677
$\rho = 0.4$	$(1/d_{ij})$	0.0900	0.2494	0.4139	−0.0122	0.0082	−0.0759	0.0420	0.0415	−0.0700
	$\exp(-d_{ij})$	0.1881	0.3527	0.5098	0.0053	0.0549	−0.0262	0.0738	0.0942	−0.0210
	$\min(d_k)$*	0.1630	0.3279	0.4900	0.0236	0.1159	0.2607	0.1022	0.1509	0.2138
$\rho = 0.6$	$(1/d_{ij})$	0.1056	0.2566	0.3692	0.0023	0.0343	0.0770	0.1548	0.2059	0.1395
	$\exp(-d_{ij})$	0.2148	0.3365	0.3998	0.0303	0.0778	0.3904	0.2464	0.3137	0.5210
	$\min(d_k)$*	0.1616	0.3060	0.4000	0.0530	0.2001	0.4000	0.2627	0.3578	0.4804

*k = 25

Table 4 Mean square error (MSE) on the estimated coefficients of α and β

	w_{ij}	SAR specification with **S**			STAR specification with **S**			STAR specification with **W**		
		$\psi = 0.2$	$\psi = 0.4$	$\psi = 0.6$	$\psi = 0.2$	$\psi = 0.4$	$\psi = 0.6$	$\psi = 0.2$	$\psi = 0.4$	$\psi = 0.6$
		for β			for β			for β		
$\rho = 0.2$	$(1/d_{ij})$	0.0000	0.0000	0.0000	0.0000	0.0000	0.0001	0.0000	0.0000	0.0000
	$\exp(-d_{ij})$	0.0000	0.0000	0.0001	0.0000	0.0000	0.0001	0.0000	0.0000	0.0000
	$\min(d_k)$*	0.0001	0.0001	0.0001	0.0001	0.0001	0.0001	0.0001	0.0001	0.0003
$\rho = 0.4$	$(1/d_{ij})$	0.0001	0.0001	0.0001	0.0000	0.0001	0.0001	0.0000	0.0001	0.0001
	$\exp(-d_{ij})$	0.0001	0.0001	0.0001	0.0001	0.0001	0.0001	0.0001	0.0001	0.0001
	$\min(d_k)$*	0.0001	0.0001	0.0003	0.0002	0.0006	0.0017	0.0001	0.0003	0.0012
$\rho = 0.6$	$(1/d_{ij})$	0.0001	0.0001	0.0001	0.0000	0.0000	0.0001	0.0000	0.0001	0.0001
	$\exp(-d_{ij})$	0.0001	0.0001	0.0002	0.0000	0.0001	0.0004	0.0001	0.0001	0.0001
	$\min(d_k)$*	0.0002	0.0001	0.0087	0.0003	0.0019	0.0177	0.0001	0.0001	0.0099
		for ρ			for ρ			for ψ		
$\rho = 0.2$	$(1/d_{ij})$	0.0214	0.0776	0.1946	0.0026	0.0033	0.0053	0.0030	0.0027	0.0022
	$\exp(-d_{ij})$	0.0383	0.1296	0.2939	0.0023	0.0049	0.0158	0.0027	0.0029	0.0028
	$\min(d_k)$*	0.0266	0.1138	0.2639	0.0004	0.0057	0.0344	0.0010	0.0021	0.0048
$\rho = 0.4$	$(1/d_{ij})$	0.0202	0.0738	0.1821	0.0026	0.0034	0.0063	0.0058	0.0065	0.0055
	$\exp(-d_{ij})$	0.0467	0.1317	0.2639	0.0024	0.0072	0.0034	0.0101	0.0148	0.0028
	$\min(d_k)$*	0.0271	0.1078	0.2402	0.0008	0.0139	0.0694	0.0109	0.0233	0.0465
$\rho = 0.6$	$(1/d_{ij})$	0.0248	0.0751	0.1392	0.0021	0.0023	0.0161	0.0361	0.0629	0.0405
	$\exp(-d_{ij})$	0.0551	0.1162	0.1599	0.0031	0.0085	0.1543	0.0765	0.1177	0.2760
	$\min(d_k)$*	0.0265	0.0938	0.1600	0.0030	0.0406	0.1600	0.0701	0.1293	0.2309

*k = 25

In consequence, a bias in the autoregressive coefficient is of prime importance if the goal is to have a full consideration of the total marginal effect, including spatial spillover (feedback loops). Thus, an overestimation of the ρ coefficient leads to overestimation of the total marginal effect.

For the unidirectional effect, there is a problem related to the calculation of the marginal effect on the short- and the long-term. In such a case, the marginal effect of a one-unit change in x_k leads to a change of β_k in time *t*. The same one-unit change in x_k in time period *t*, leads to a change of $\psi\beta_k$ in time $t + 1$. Thus, bias in the coefficient ψ can lead to an erroneous conclusion about the effect on the next time period. Moreover, the marginal effect on the long-term is equal to $\beta_k/(1-\psi)$, or, equivalently, to $(1-\psi)^{-1}\beta_k$. Once again, an overestimation of the ψ parameter leads to an overestimation of the short- and long-term marginal effect.

In all cases, the omission of the decomposition of the spatial effects into its multidirectional and unidirectional components can lead to the erroneous interpretation of the marginal effect on the spatial equilibrium, as well as on the temporal equilibrium. Allowing only for a block-diagonal or a lower triangular specification of the weights matrix does not solve all the problems: the omission of the unidirectional spatial effect introduces bias on the estimated autoregressive coefficients when the omitted spatial effect is high. The consideration of the two effects (multidirectional and unidirectional) is thus a necessity for spatial data pooled over time to ensure a correct interpretation: a direct, indirect and total marginal effect is usually captured through the spatial multidirectional effect (based on the estimation of the ρ—[4]), while a dynamic effect (short- and long-run) is captured through the unidirectional effect (based on the estimation of the ψ coefficient—[64]).

An Empirical Application

To see the impact of the different specifications on the results obtained through an empirical application, an HPM model is estimated based on transactions of apartments in Paris between 1990 and 2003. The transactions come from the *Base d'Informations Economiques Notariales* (BIEN), compiled by French notaries. The database has recently been used by Dubé and Legros [65] for a similar exercise.[9] The full database consists of 294,768 observations and contains the exact address of the property sold in Paris. The database also contains information about the characteristics of the dwelling: type of dwelling, date of sale, living area (in m^2), date of construction, number of rooms, mean area/room, number of bathrooms, number of garages or parking spaces, and for apartments, floor level and presence of an elevator, and number of service rooms.

[9]However, the transactions only include those occurring between 1990 and 2001.

To compare the impact of the choice of a given spatio-temporal specifications, we have built two sub-samples: (1) one that contains 15,000 observations and serves to estimate the coefficients; and (2) another one that contains 7500 observations and serves for an out-of-sample prediction exercise. Six models are estimated: (1) the usual OLS specification (Eq. 1); (2) the SAR specification based only on spatial relations with no regard to temporal constraints (Eq. 2); (3) the SAR model accounting only for transactions occurring in the same month (Eq. 4); (4) the STAR model, controlling for comparable sales approach using transactions occurring one quarter before (Eq. 3); and (5) the complete STAR model that distinguishes between multidirectional spatial effect and unidirectional spatial effect (Eq. 5).

The spatial weights matrix is built using the negative exponential transformation, whilethe spatial relations are limited using cut-off criteria based on the mean distance for each of the observations. All the past transactions are accounted for in the lower triangular specification ($\underline{\mathbf{W}}$), but a higher weight is given to transactions occurring in a close temporal window. This is achieved using an inverse temporal distance transformation, where the time distance is calculated using the number of months that have passed between two transactions (see [66]).

The comparison shows some major divergence in the autoregressive coefficients (Table 5). For the SAR specification using the full spatial weights matrix (Eq. 2), a high coefficient is obtained, even after introducing some distance cut-off criteria. Thus, the empirical results support the fact that the lack of constraints on the individual weights can potentially lead to a bias in the estimated autoregressive coefficient. As compared to the autoregressive coefficient using the block diagonal specification ($\underline{\mathbf{S}}$–Eq. 3), the amplitude is highly reduced (0.2950 vs. 0.9760). The difference is even larger when the unidirectional and the multidirectional spatial effects are accounted for (0.0639).

A quick comparison with the specification using only the unidirectional effect (Eq. 3) proposes that the main part of the spatial effect can be attributable to this "*comparable sales*" effect since the value of the coefficient is 0.8140. A comparison with the full specification supports this assumption since the coefficient associated with the unidirectional spatial effect is high (0.7667) and highly significant.

It is assumed that a technological change, such as the development of a new mass transit system, impacts the sale price of houses located within a distance of 500 m of the station. This effect is estimated to be 5% ($\beta_k = 0.05$). The estimation results obtained in Table 5 suggest that the total marginal effect[10] using the SAR model will result in a change of more than 208% in housing prices for those located within 500 m. In comparison, the total marginal effect using the specification in Eq. (4) would be 7 and 5.3% for the specification using the complete decomposition of the spatial effect (Eq. 5). Of course, the specification using the unidirectional effect gives no spatial spillover effect for the same time period since spatial effect is assumed to come from the previous time period.

[10]All the effects are obtained using the formula identified in section "A Monte Carlo Experiment": $(1-\rho)^{-1} \times \beta$.

Table 5 Estimation results, transactions occurring in Paris between 1990 and 2003

	Equation (1)		Equation (2)		Equation (4)		Equation (3)		Equation (5)	
	OLS		SAR with **S**		SAR with <u>**S**</u>		STAR with <u>**W**</u>		STAR with <u>**S**</u> and <u>**W**</u>	
	Coefficient	*t-stat*	Coefficient	*t-stat*	Coefficient	*t-stat*	Coefficient	*t-stat*	Coefficient	*t-stat*
Constant	6.8514	*92.09*	−3.8718	*−52.70*	3.5592	*32.31*	−2.2510	*−16.24*	−2.4356	*−45.99*
Log surface	1.0918	*143.03*	1.0294	*159.18*	1.0756	*148.51*	1.0301	*156.51*	1.0301	*157.86*
Lift	0.1001	*14.75*	0.0499	*8.65*	0.0868	*13.49*	0.0471	*8.05*	0.0473	*8.19*
No. of bathroom (log)	0.1635	*13.91*	0.1401	*14.06*	0.1544	*13.86*	0.1431	*14.23*	0.1423	*14.23*
Terrace	0.1100	*7.78*	0.1090	*9.10*	0.1126	*8.40*	0.1067	*8.83*	0.1075	*8.93*
Garage	0.0196	*2.94*	0.0326	*5.76*	0.0228	*3.61*	0.0364	*6.36*	0.0361	*6.34*
Communal heating	−0.0107	*−0.93*	0.0045	*0.47*	−0.0040	*−0.37*	0.0073	*0.75*	0.0077	*0.79*
Built before 1850	*Reference*		*Reference*		*Reference*		*Reference*		*Reference*	
Built between '50-'13	−0.1786	*−11.24*	−0.1516	*−11.22*	−0.1706	*−11.27*	−0.1416	*−10.41*	−0.1420	*−10.48*
Built between '14-'47	−0.1821	*−11.03*	−0.1674	*−11.94*	−0.1796	*−11.43*	−0.1483	*−10.51*	−0.1497	*−10.65*
Built between '48-'69	−0.2090	*−12.51*	−0.1811	*−12.75*	−0.2042	*−12.84*	−0.1465	*−10.23*	−0.1491	*−10.46*
Built between '70-'80	−0.2167	*−12.54*	−0.1476	*−10.04*	−0.1945	*−11.81*	−0.1177	*−7.93*	−0.1186	*−8.04*
Built between '81-'91	−0.0992	*−4.99*	−0.0365	*−2.16*	−0.0883	*−4.66*	−0.0130	*−0.76*	−0.0156	*−0.92*
Built between '92-'00	0.1273	*6.63*	0.1973	*12.08*	0.1411	*7.73*	0.2222	*13.49*	0.2197	*13.41*
Built after 2001	0.0850	*3.37*	0.1507	*7.03*	0.1029	*4.30*	0.1941	*8.98*	0.1916	*8.91*
Basement	*Reference*		*Reference*		*Reference*		*Reference*		*Reference*	
Floor 1	0.0424	*3.98*	0.0513	*5.68*	0.0413	*4.09*	0.0483	*5.30*	0.0477	*5.26*
Floor 2	0.0678	*6.36*	0.0809	*8.94*	0.0665	*6.58*	0.0791	*8.67*	0.0781	*8.60*
Floor 3	0.0803	*7.43*	0.0937	*10.22*	0.0804	*7.84*	0.0887	*9.59*	0.0882	*9.58*
Floor 4	0.0727	*6.43*	0.0888	*9.27*	0.0737	*6.88*	0.0796	*8.24*	0.0794	*8.25*
Floor 5 and more	0.0587	*5.69*	0.0784	*8.96*	0.0627	*6.41*	0.0691	*7.83*	0.0694	*7.89*
Apartment of type 1	−0.0176	*−0.28*	−0.0476	*−0.87*	−0.0020	*−0.03*	−0.0535	*−1.01*	−0.0480	*−0.92*

Apartment of type 2	−0.0811	*−0.56*	−0.0834	*−0.68*	−0.0947	*−0.68*	−0.0740	*−0.60*	−0.0774	*−0.63*
Apartment of type 3	0.0798	*1.27*	−0.0025	*−0.04*	0.0813	*1.24*	−0.0078	*−0.14*	−0.0024	*−0.05*
Apartment of type 4	0.0628	*0.98*	0.0376	*0.67*	0.0840	*1.26*	0.0332	*0.61*	0.0395	*0.74*
Sold in 1990	*Reference*		*Reference*		*Reference*		*Reference*		*Reference*	
Sold in 1991	0.0719	*2.97*	0.0578	*2.81*	0.0773	*3.35*	0.0732	*3.53*	0.0743	*3.60*
Sold in 1992	−0.0399	*−1.77*	−0.0548	*−2.86*	−0.0024	*−0.11*	−0.0382	*−1.98*	−0.0302	*−1.57*
Sold in 1993	−0.1137	*−5.16*	−0.1223	*−6.53*	−0.0827	*−3.93*	−0.1012	*−5.37*	−0.0952	*−5.07*
Sold in 1994	−0.1220	*−5.69*	−0.1330	*−7.30*	−0.0909	*−4.44*	−0.0949	*−5.17*	−0.0897	*−4.91*
Sold in 1995	−0.1917	*−8.57*	−0.1982	*−10.43*	−0.1381	*−6.46*	−0.1479	*−7.73*	−0.1388	*−7.29*
Sold in 1996	−0.2685	*−12.65*	−0.2733	*−15.15*	−0.2175	*−10.72*	−0.2180	*−12.00*	−0.2098	*−11.60*
Sold in 1997	−0.2704	*−12.75*	−0.2832	*−15.71*	−0.2085	*−10.27*	−0.2125	*−11.71*	−0.2024	*−11.20*
Sold in 1998	−0.2501	*−11.81*	−0.2599	*−14.43*	−0.2034	*−10.05*	−0.1867	*−10.29*	−0.1803	*−9.98*
Sold in 1999	−0.2008	*−9.77*	−0.2132	*−12.20*	−0.1663	*−8.47*	−0.1321	*−7.50*	−0.1286	*−7.33*
Sold in 2000	−0.0381	*−1.94*	−0.0490	*−2.93*	−0.0449	*−2.39*	0.0301	*1.78*	0.0246	*1.47*
Sold in 2001	0.0790	*3.78*	0.0426	*2.40*	0.0911	*4.57*	0.1194	*6.68*	0.1197	*6.73*
Sold in 2002	0.1655	*8.22*	0.1311	*7.66*	0.1480	*7.69*	0.2050	*11.91*	0.1989	*11.60*
Sold in 2003	0.2552	*12.57*	0.2276	*13.19*	0.2086	*10.74*	0.2935	*16.89*	0.2812	*16.25*
Department 75	0.7369	*74.81*	0.3345	*41.09*	0.6174	*63.42*	0.3832	*39.54*	0.3778	*47.57*
Department 92	0.4805	*47.50*	0.0010	*0.12*	0.3379	*33.35*	0.0740	*7.22*	0.0667	*8.26*
Department 93	*Reference*		*Reference*		*Reference*		*Reference*		*Reference*	
Department 94	0.2423	*22.09*	0.0454	*4.91*	0.1897	*18.12*	0.1300	*13.67*	0.1251	*13.47*
Dynamic effect							0.8140	*73.91*	0.7667	*144.96*
ρ			0.9760	*295.54*	0.2950	*43.33*			0.0639	*182.86*
N_{T-1}	14,916		14,916		14,916		14,916		14,916	
R^2	0.7844		0.1809		0.7819		0.8423		0.8414	
Log-Likelihood			3449,4123		1712,9397				3402,9686	

The same changes lead to price movement in the next time periods if the spatial effect is unidirectional (and dynamic).[11] For the specification using only the unidirectional effect (Eq. 3), the increase of the house price in the next time period is 4.1%, while it is equal to 3.8% when using the specification in Eq. (5). Similarly, these two models also suggest an important appreciation of prices in the long-run, equivalent to 26.8% with the specification using only the multidirectional effect (Eq. 3) and equivalent to 21.4% with the specification using the full decomposition of the spatial effect (Eq. 5). The difference does not appear so large however, applied to the total housing stock, the difference in total added value, in dollars, can be quite considerable.

In the end, only the specification using the multidirectional and the unidirectional spatial effect (Eq. 5) is able to decompose the spatial effect of the price determination process over time. The consideration of both spatial effects correctly addressing the question of the spatial (spillover and dynamic) effects suggests that the total price appreciation is of 5.3% in the initial time period, of 4.1% in the next time period, and returns a total impact of 22.9% in the long-run.[12]

The advantage of the full STAR specification including multidirectional and unidirectional spatial effects is also revealed through the out-of-sample performance of the different models (Table 6). The correlation between the true values and the predicted values is over 0.9 for the specification using only the lower triangular part (0.9158) and the lower triangular part and the block diagonal part, while accounting for distinct effects on both matrices (0.9150). Thus, the full spatio-temporal specification not only helps in decomposing the spatial effect, but clearly helps improving the predictions of the models, reinforcing the necessity to account for both effects.

Thus, there is a need, in a spatio-temporal context, to decompose the spatial effect to account for the temporal reality of cross-section data pooled over time.

Table 6 Out-of-sample performance of the different specifications, Paris 1990–2003

	Out-of-sample performance index	
	$\rho_{y,\hat{y}}$	$(\hat{y}\text{-}y) < \sigma_y$
y_{OLS}	0.8863	96.42%
$y_{SAR(\mathbf{S})}$	0.8294	86.63%
$y_{SAR(\underline{\mathbf{S}})}$	0.8762	86.54%
$y_{STAR(\underline{\mathbf{W}})}$	0.9158	99.81%
$y_{STAR(\underline{\mathbf{S}} + \underline{\mathbf{W}})}$	0.8502	98.96%
$y_{STAR(\underline{\mathbf{S}}\ \&\ \underline{\mathbf{W}})}$	0.9150	99.22%

[11] All the effects are obtained using the formulas identified in section "A Monte Carlo Experiment": for the short-run ($\psi \times \beta$); and the long-run $[(1-\psi)^{-1} \times \beta]$.

[12] These effect include the spatial spillover effect. The short run effect is obtained from the formula $\psi \times [(1-\rho)^{-1} \times \beta]$, while the long-run effect is obtained from the formula $[(1-\psi)^{-1} \times \{(1-\rho)^{-1} \times \beta\}]$.

Conclusion

In this chapter, we have proposed a method/framework to decompose the spatial effect into two different components for spatial data pooled over time by extending the spatial autoregressive (SAR) model specification: (1) a spatial multidirectional effect; and (2) a spatial unidirectional effect. Both spatial effects can be isolate by building appropriate and distinct weights matrices that correctly expresses the different spatial relations. Monte Carlo results clearly show that isolating both spatial effects through appropriate weights matrix has a major impact on the calculation of the marginal effects.

The results clearly conclude that using a spatial specification can lead to a potential problem, while only controlling for spatial multidirectional effects or unidirectional spatial effects does not help in solving the problem. Since spatial econometrics modeling is gaining in popularity and because many software programs now offer packages to perform tests and estimations, these conclusions are fundamental for empirical analysis.

Even if there is no evidence of a large bias on the β parameters according to the choice of the functional form with or without taking into account the spatial unidirectional and multidirectional effects, the interpretation of the spatial effect is different regarding the form of the weights matrix used. Assuming that the spatial effect is multidirectional (unidirectional), using a block diagonal (lower triangular) weights matrix leads to a completely different interpretation. In such a case, the spatial effect is assumed to be simultaneous (dynamic), with the actual (past) observations depicting influence on the actual observations. An empirical analysis on real estate transactions in Paris between 1990 and 2003 is in line with the conclusions drawn from the Monte Carlo experiment. Using an out-of-sample prediction, the results show that the decomposition of the spatial effect, through multidirectional and unidirectional effects, provides better performance.

Given the fact that the goal of the econometric modeling remains the same—identifying the marginal mean effect of a given variable on a particular outcome—the conclusions should be of primal importance for those who work with spatial data pooled over time, such as in real estate analysis, crime detection, business start-ups and closings and so on. However, what is less clear at this stage is how the temporal dimension should be treated. As is the case with spatial data (the Modifiable Areal Unit Problem—MAUP), the effect of the temporal aggregation, in time period equivalent to a month or a quarter, on the results needs to be explored in detail since this can also have a potential effect on the estimated coefficients. It is possible to treat temporal dimension in a different way, based on time windows, considering a given number of days before and after a particular transaction as a better tool to isolate spatial multidirectional effect.

Acknowledgement This research was funded by the Fonds de recherche québécois sur la société et la culture (FRQSC).

References

1. Student (1914) The elimination of spurious correlation due to position in time or space. Biometrika 5:351–360
2. Anselin L, Bera AK (1998) Spatial dependence in linear regression models with an introduction to spatial econometrics. In: Ullah A, Giles D (eds) Handbook of applied economic statistics. Marcel Dekker, New York, pp 237–289
3. Griffith DA (2005) Effective geographic sample size in the presence of spatial autocorrelation. Ann Assoc Am Geogr 95:740–760
4. LeSage J, Pace RK (2009) Introduction to spatial econometrics. Taylor & Francis Group, Boca Raton
5. LeSage J, Pace RK (2004) Models for spatially dependent missing data. J Real Estate Financ Econ 29(2):233–254
6. Anselin L (2010) Thirty years of spatial econometrics. Pap Reg Sci 89:3–25
7. Arbia G (2011) A lustrum of SEA: recent research trends following the creation of the spatial econometrics association (2007–2011). Spat Econ Anal 6(4):376–395
8. Dubé J, Legros D (2013) Dealing with spatial data pooled over time in statistical models. Lett Spat Resour Sci 6:1–18
9. Rosen S (1974) Hedonic prices and implicit markets: product differentiation in pure competition. J Polit Econ 82:34–55
10. Can A (1992) Specification and estimation of hedonic housing price models. Reg Sci Urban Econ 22:453–474
11. Dubin RA (1998) Spatial autocorrelation: a primer. J Hous Econ 7:304–327
12. Dubin RA, Sung C-H (1987) Spatial variation in the price of housing: rent gradients in non-monocentric cities. Urban Stud 24:193–204
13. Krige DG (1966) Two-dimensional weighted moving average trend surfaces for ore valuation. J South Afr Inst Min Metall 67:13–38
14. Trigg DW, Leach AG (1967) Exponential smoothing with an adaptive response rate. J Oper Res Soc 18:53–59
15. Widrow B, Hoff ME (1960) Adaptive switching circuits. In: IRE WESCON Convention Record, Part 4. IRE, New York, pp 96–104
16. Casetti E (1972) Generating models by the expansion method: applications to geographical research. Geogr Anal 4:81–91
17. Casetti E (1997) The expansion method, mathematical modeling, and spatial econometrics. Int Reg Sci Rev 20:9–33
18. Cleveland WS, Devlin SJ (1988) Locally weighted regression: an approach to regression analysis by local fitting. J Am Stat Assoc 83:596–610
19. Fotheringham AS, Brunsdon C, Charlton M (2002) Geographically weighted regression: the analysis of spatially varying relationships. Wiley, London
20. Fotheringham AS, Charlton ME, Brunsdon C (1998) Geographically weighted regression: a natural evolution of the expansion method for spatial data analysis. Environ Plann A 30: 1905–1927
21. McMillen DP (1996) One hundred fifty years of land values in Chicago: a nonparametric approach. J Urban Econ 40:100–124
22. Anselin L (1988) Spatial econometrics: methods and models. Springer, Boston
23. Dubin R, Pace RK, Thibodeau TG (1999) Spatial autoregression techniques for real estate data. J Real Estate Lit 7:79–96
24. Ord K (1975) Estimation methods for models of spatial interaction. J Am Stat Assoc 70: 120–126
25. Gibbons S, Overman HG (2012) Mostly pointless spatial econometrics? J Reg Sci 52:172–191
26. Corrado L, Fingleton B (2012) Where is the economics in spatial econometrics? J Reg Sci 52:210–239

27. Small KA, Steimetz SSC (2012) Spatial hedonics and the willingness to pay for residential amenities. J Reg Sci 52:635–647
28. Dubé J, Legros D (2013) A spatio-temporal measure of spatial dependence: an example using real estate data. Paper Reg Sci 92:19–30
29. Can A, Megbolugbe I (1997) Spatial dependence and house price index construction. J Real Estate Financ Econ 14:203–222
30. Pace RK, Barry R, Clapp JM, Rodriquez M (1998) Spatiotemporal autoregressive models of neighborhood effects. J Real Estate Financ Econ 17:15–33
31. Pace RK, Barry R, Gilley OW, Sirmans CF (2000) A method for spatial-temporal forecasting with an application to real estate prices. Int J Forecast 16:229–246
32. Gelfand AE, Ecker MD, Knight JR, Sirmans CF (2004) The dynamics of location in home price. J Real Estate Financ Econ 29:149–166
33. Tu Y, Yu SM, Sun H (2004) Transaction-based office price indexes: a spatiotemporal modeling approach. Real Estate Econ 32:297–328
34. Sun H, Tu Y, Yu SM (2005) A Spatio-temporal autoregressive model for multi-unit residential market analysis. J Real Estate Financ Econ 31:155–187
35. Smith TE, Wu P (2009) A spatio-temporal model of housing prices based on individual sales transactions over time. J Geogr Syst 11(4):333–355
36. Nappi-Choulet I, Maury T-P (2009) A spatiotemporal autoregressive price index for the paris office property market. Real Estate Econ 37(2):305–340
37. Huang B, Wu B, Barry M (2010) Geographically and temporally weighted regression for modeling spatiotemporal variation in house prices. Int J Geogr Inf Sci 24(3):383–401
38. Nappi-Choulet I, Maury T-P (2011) A spatial and temporal autoregressive local estimation for the paris housing market. J Reg Sci 51(4):732–750
39. Thanos S, Bristow AL, Wardman MR (2012) Theoretically consistent temporal ordering specification in spatial hedonic pricing models applied to the valuation of aircraft noise. J Environ Econ Policy 1(2):103–126
40. Liu X (2013) Spatial and temporal dependence in house price prediction. J Real Estate Financ Econ 47:341–369
41. Dubé J, Legros D (2014) Spatial econometrics and spatial data pooled over time: towards an adapted modelling approach. J Real Estate Lit 22(1):101–125
42. Dubé J, Legros D (2014) Spatial econometrics using microdata. Wiley, London
43. Dubé J, Baumont C, Legros D (2013) Matrices de pondérations et contexte spatio-temporel en économétrie spatiale. Revue Canadienne de science régionale 36(1/3):57–75
44. Des Rosiers F, Dubé J, Thériault M (2011) Do peer effects shape property values? J Property Invest Financ 29(4/5):510–528
45. LeSage J (2014) What regional scientists need to know about spatial econometrics. Rev Reg Stud 44:13–32
46. Getis A, Aldstadt J (2004) Constructing the spatial weights matrix using a local statistic. Geogr Anal 36:90–104
47. Haining RP (2009) Spatial autocorrelation and the quantitative revolution. Geogr Anal 41: 364–374
48. Pinkse J, Slade ME (2010) The future of spatial econometrics. J Reg Sci 50:103–117
49. McMillen DP (2010) Issues in spatial data analysis. J Reg Sci 50:119–141
50. Legendre P (1993) Spatial autocorrelation: trouble or new paradigm? Ecology 74(6): 1659–1673
51. Pace RK, Barry R (1997) Quick computation of spatial autoregressive estimators. Geogr Anal 29:232–246
52. Dubé J, Legros D, Thériault M, Des Rosiers F (2014) A spatial difference-in-differences estimator to evaluate the effect of change in public mass transit systems on house prices. Transp Res B 64:24–40
53. LeSage J (1999) Applied econometrics using Matlab. www.spatial-econometrics.com
54. Anselin L, Florax R (1995) New directions in spatial econometrics. Springer, New York

55. Lee L-F (2004) Asymptotic distributions of quasi-maximum likelihood estimators for spatial autoregressive models. Econometrica 72(6):1899–1925
56. Kelejian HH, Prucha IR (1998) A generalized spatial two-stage least squares procedure for estimating a spatial autoregressive model with autoregressive disturbance. J Real Estate Financ Econ 17(1):99–121
57. Kelejian HH, Prucha IR (1999) A generalized moments estimator for the autoregressive parameter in a spatial model. Int Econ Rev 40(2):509–533
58. Kelejian HH, Prucha IR (2007) The relative efficiencies of various predictors in spatial econometric models containing spatial lags. Reg Sci Urban Econ 37(3):363–374
59. Kelejian HH, Prucha IR (2007) HAC estimation in a spatial framework. J Econ 140(2007): 131–154
60. Kelejian HH, Prucha IR (2010) Specification and estimation of spatial autoregressive models with autoregressive and heteroskedastic disturbances. J Econ 140(1):53–130
61. Adkins LC, Gade MN (2012) Monte Carlo experiments using stata: a primer with examples. Working paper. http://learneconometrics.com
62. Dubé J, Legros D (2015) Modeling spatial data pooled over time: schematic representation and Monte Carlo evidences. Theor Econ Lett 5:132–154
63. Smith TE (2009) Estimation bias in spatial models with strongly connected weight matrices. Geogr Anal 41:307–332
64. Hamilton JD (1994) Time series analysis. Princeton University Press, Princeton
65. Dubé J, Legros D (2014) Spatial econometrics and the hedonic pricing model: what about the temporal dimension? J Prop Res 31(4):333–359
66. Dubé J, Baumont C, Legros D (2011) Utilisation des matrices de pondérations en économétrie spatiale: Proposition dans un contexte spatio-temporel, Documents de travail du Laboratoire d'Économie et de Gestion (LEG), Université de Bourgogne, e2011–01, 33p

An Open Source Spatiotemporal Model for Simulating Obesity Prevalence

Jay Lee and Xinyue Ye

Introduction

Obesity is an exceedingly complex public health problem with hypothesized causes at multiple interacting levels that are embedded in the very structure of society [1, 2]. This complexity appears to be the reason that most one-dimensional preventive or therapeutic interventions have not been very successful. For example, the Foresight causal map prepared by UK Government Office illustrates the inherent complexity of obesity as a public health problem [3]. The Foresight map was built around energy balance and mammalian physiology, but the model rapidly expanded to include individual and collective physical activity, the built environment, individual and collective psychology, industrial food production, and population food consumption. Even with the expanded list of variables, obesogenic policy determinants of the relevant environments were excluded which seems to limit the validity of that approach. Obesity, per se, is only a small part of a larger public health problem that includes obesogenic policy, environments, and population characteristics. These population characteristics include unhealthy dietary habits and sedentary behavior, a high prevalence of obesity, high obesity-related morbidity and mortality, and high rates of diabetes or cardiovascular diseases among historicallydisadvantaged

J. Lee
Department of Geography, Kent State University, Kent, OH, 44242, USA

College of Environment and Planning, Henan University, Kaifeng Shi, Henan Sheng, China
e-mail: jlee@kent.edu

X. Ye (✉)
Department of Geography, Kent State University, Kent, OH, 44242, USA
e-mail: xinyue.ye@gmail.com

J.-C. Thill, S. Dragicevic (eds.), *GeoComputational Analysis and Modeling of Regional Systems*, Advances in Geographic Information Science,
DOI 10.1007/978-3-319-59511-5_20

groups. Thus the obesity problem includes long-standing area disparities in health. Addressing these disparities, their spatio-temporal components, and their determinants requires new approaches.

Obesity prevalence has been predicted by using statistical models and simple dynamic models. However, they predicted only the size of the obese population as a whole without further distinguishing the population to various levels of obesity [4]. Such models over-generalized the movements of subpopulations between different levels of obesity. In addition, the simple models from current literature (e.g., [5, 6]) are often too simplified in the following ways: modeling future trends of obese population at a geographic scale that is often too coarse to be useful in revealing area disparities. Finally, most models, in order to accommodate the statistical and simple dynamic modeling structure, often miss important factors, such as death rates, birth rates of the population, and more importantly; lumping all levels of normal weight/overweight/obese/extremely obese subpopulations into one.

As such, the results of statistical analysis and predictions have limited practical use in assisting policy-making process by public health districts when designing and implementing more geographically- and temporally-focused intervention programs. Auchincloss and Roux [7] pointed out the weaknesses of traditional epidemiologic approaches when dealing with complex multilevel data with spatio-temporal components. They noted that traditional regression-based approaches to analyzing multi-level exposures and health disparities are limited by a variety of assumptions. These assumptions include the requirements that realizations of each independent variable do not influence one another, and that there are no feedback loops to address the interactions among variables. These requirements do not fit well with the complex realities of obesogenic policy, environments, and population characteristics where dependencies and feedback loops are common.

Obesity may be the single most challenging example for a condition with causes and consequences at multiple levels and with multiple feedback loops among the causes. New approaches are obviously needed. The principal research question of our work is: can we develop a prototype for a comprehensive simulation mechanism for estimating obesity prevalence and obesity-related disease or disparities that (1) addresses obesogenic policy, environments, and population characteristics; and (2) is calibrated against obesity-related morbidity and mortality?

Obesity studies have been, and continue to be challenged by dealing with temporal trend of geographic patterns and spatial dynamics of health development. There is an imperative need for effective and efficient methods to represent and examine the coupled space-time attributes of obesity phenomena in the comparative context. As a multi-dimensional and multi-scale phenomenon, obesity studies witness the role of geography and the awakening emphasis on space among public health practitioners. As discussed above, it is clear that a space-time perspective has become increasingly relevant to our understanding of public health dynamics. To this end, we argue that an open source solution is needed to systematically integrate space and time so to share and promote any advances in this direction. Though rich conceptual frameworks have highlighted the complexity of obesity dynamics, the gap has been widening between empirical studies and theories. Hence, the most crucial step is to systematically understand obesity dynamics data from the theoretical and policy context. Thus, the availability of codes and tools to support space-time data analysis are vital in the adoption of such a perspective in obesity studies.

An Open Source Approach to Obesity Simulations

The prevalence of obesity among adults and children in the United States has increased dramatically in recent decades [8]. This is a public health issue as obesity causes many other chronic health conditions, such as, hypertension, cardiovascular disease, type II diabetes, among others. Increasing obesity prevalence in a region affects the life expectancy and quality of its residents. It also increases social costs in many ways.

The basic cause of obesity is the imbalance between the amount of energy taken in through eating and drinking and the amount of energy expended through metabolism and physical activity [9]. To offset excessive energy intake, increased physical activity is encouraged as a way to keep energy in balance. However, energy imbalances appear to be encouraged by features of the physical, social, and economic environments. Lee et al. [10] found that the density of fitness centers and non-fresh food outlets are related to the prevalence of obesity, and that an analysis of smaller geographic units provides more details regarding area disparities in health than analyses carried out with larger geographic units.

Most of the obesity studies that have looked at the food environment have concentrated on the hypothesized effect of non-fresh food (fast-food, packaged food, pre-processed food, etc.) consumption on people's diet and public health. With today's fast-paced life styles and intensive marketing of various types, non-fresh food outlets have become an important part in people's daily diet because of convenience, price, distance and other cultural factors [11]. The literature in this area suggests a positive correlation between regularly consuming non-fresh food and the prevalence of obesity unless daily physical activities are performed on a regular basis [12]. Positive correlation means, the more frequently one eats from non-fresh food outlets over time, the higher are the chances of being obese [13].

A study on non-fresh food consumption and obesity among Michigan adults suggested that regular fast food consumption was higher among younger adults and men [8]. In that study, the prevalence of obesity increased consistently with frequenting non-fresh food outlets, from 24% of those going less than once a week to 33% of those going three or more times per week. The predominate reason for choosing fast food was convenience. Another study found that youths 11–18 years old ate at non-fresh food outlets an average of twice per week [14], which also points to the alarming possibility of increasing obesity rates among young people.

Non-fresh food consumption has been found to be highly correlated with the prevalence of obesity. Reasons that may affect the consumption of non-fresh food are the price of the food, the walking or driving distance, and various cultural, behavioral, or environmental factors [8, 15]. In addition, marketing campaigns of non-fresh food outlets could play a significant role in the consumption of unhealthy food [9]. If marketed well, non-fresh food outlets can attract a significant number of customers, which can later lead to increases of overweight and obese people. Most often non-fresh food outlets are unhealthy because of the way foods are cooked and the high calories per "serving". The increased supply of non-fresh food outlets has a significant impact on obesity. Frequently eating at non-fresh food outlets is becoming an important issue in the public health literature because of the apparent health effects.

Physical activity and the distribution of fitness centers can have a significant impact on the prevalence of obesity if exercise is taken regularly [16]. Over the last few years, there have been studies focused on the relationship between the built environment and physical activity [16]. However, there were no other studies besides Lee et al. [10] that examine the relationship between distances from fitness centers and obesity rates by using small geographical units such as tracts or block group. The proximity of fitness centers could change the prevalence of overweight and obesity in some neighborhoods. A relevant study in New Zealand neighborhoods found evidence of a relationship between beach access and body mass index (BMI) and physical activities [17]. Several other studies reported a positive association between the recreational environment and physical activity for both adults and children [18, 19]. Going to recreational centers regularly increased physical activity; therefore, lower rates of obesity and overweight can be expected in neighborhoods with sufficient access to fitness centers. Mobley et al. [20] found there is a lower average BMI in areas with more fitness centers. In addition, Boehmer et al. [21] reported that having fewer fitness centers within close proximity was associated with higher likelihood of obesity among women but not men.

Furthermore, being obese was found to be significantly associated with perceived absence of sidewalks, unpleasant communities, lack of interesting sites, and presence of garbage [21]. Several studies show that people tend to increase their frequencies of visiting fitness centers when the distance between home and facilities decreases [22]. For long-term health benefits, people should focus on improving fitness by increasing physical activity rather than relying only on diet for weight control [23]. It should be noted, however, that going to fitness centers maybe a critical behavior, but there are multiple factors that may discourage or encourage this key behavior (such as the price of membership, geographical (distance), time required for finding a parking space, etc.)

Our review of the literature in obesity suggests that a comprehensive computation model of obesity-related disparities with extensive calibration is possible. Some basic components of the model have been developed, but key components of a comprehensive model have been omitted from prior work. Calibration is also insufficient. As far as we know, no one has developed a comprehensive model of obesity and related area disparities with extensive calibration against obesity-related morbidity and mortality. Our innovative project has scientific merit because of the breadth of the proposed model and the possible calibration of the simulation against hard outcomes including obesity-related morbidity and mortality. A strength of our approach is that it may be possible to use a multi-year sample of geocoded individual inpatient discharge data from all hospitals in a representative urban-suburban county (such as Summit County, Ohio) where the simulation will be anchored as well as a corresponding sample of geocoded death certificates, US Census data, and geocoded environmental data from Summit County Public Health, the Ohio Department of Health, and other sources. Use of real world geocoded individual health outcome data in this research project will provide more robust tests of a given modeling strategy in nearly all circumstances.

In terms of obesity simulations, there have been various attempts discussed in obesity literature. In their review of obesity simulations, Levy et al. [24] list two agent-based models (ABM) and seven Markov models. Burke and Heiland's ABM [25] looks at the obesity epidemic in terms of food prices and social norms, while the Hammond and Epstein [26] ABM looks at obesity in terms of the physiology of dieting and socially influenced weight changes. More recently, Auchincloss et al. [27] models residential segregation, income disparities, and diet quality; while Yang et al. [28] models disparities and walking behaviors in an urban setting. While these obesity simulations achieved the objectives of estimating obesity prevalence in some ways, they all fell short of allowing more detailed classification of population (e.g., grouping populations into normal/overweight/obese/extremely obese) and allowing movements between subpopulations. Furthermore, the geographic units of these simulations are mostly too big to have practical uses in assisting policy-making processes for intervention programs.

Overall, from many of the analyses we reviewed, they showed that obesity ratios are indeed affected by educational attainment, income level, and unemployment level (see reviews in [10]). In addition, obesity ratios also show the expected relationships with densities of fitness centers and non-fresh food outlets. While such relationships are all statistically significant, it is important for us to explore in more detail where inside the county we can expect such relationships to be stronger or weaker. This is so that, when making policies on how to promote health and allocating funding to different areas in the county. For example, area disparities in health can be incorporated for more effective outcomes at neighborhood level.

In terms of implementing a software tool for simulating obesity prevalence, we argue that both space and time are critical components in such simulations. Spatial turn in many socioeconomic theories has been noted in many disciplines, encompassing both social and physical phenomena [29–31]. This intellectual and technological change has yielded important insights on physical sciences, social sciences and the humanities, with an explosion of interest across disciplines [32]. During the past several decades, a number of efforts have been witnessed on the development and implementation of spatial statistical analysis packages, which continues to be an active area of research [33]. Meanwhile, spatial public health analysis is increasingly being supported by the emergence of advanced analytical methods in space-time data analysis and data visualization. The interactive spatial data analysis has motivated, if not directly provoked, new queries on spatial public health theories. Therefore, the current research implements the new methodological advances in an open source environment for exploring data that has both temporal and spatial dimensions, which lend support to the notion that space and time cannot be meaningfully separated.

The fast growth of spatial public health analysis is increasingly seen as attributable to the availability of spatio-temporal datasets. By contrast, most public health geographers have been slow to adopt and implement new spatially explicit methods of data analysis due to the lack of extensible software packages, which becomes a major impediment to promoting spatial thinking in public health studies.

ABM is not new to public health inequality studies, whereas an open source solution would give better support for the scientific investigation and management of data sets, including its description, representation, analysis, visualization, and simulation. Additionally, comparative space-time analysis enables access to a much wider thinking that addresses the role of space at different stages and thus identifies the research gaps and opportunities for more in-depth study.

Obesity Prevalence Simulator: A Case Study of Summit County, Ohio

Timely and rigorous analysis of obesity will open up a rich empirical context for the social sciences and policy interventions. The Obesity Prevalence Simulator (ObPSim) was developed in Python programming language with funding provided by the Summit County Public Health District of Summit County, Ohio. Python is a versatile language that is free to acquire, install, and use. Python is also a cross-platform programming language, which means a python script can be used by computers with one platform of operating system and be usable in other operating system platforms. In addition, many libraries that process GIS and other forms of data have been developed and are freely available in public domain. This allows further improvements and updates for existing codes to be carried out easily. The open source environment offers a straightforward way of benefiting wider community.

While Lee et al. [10] used Summit County, Ohio as a case study because of the availability of key data and the project's funding, their findings may be applicable to many other geographic locations since demographic and socio-economic profiles in this area are very close to the national average in the US.

The objective of the study reported here is to model known multiple parameters associated with changes in body mass index (BMI) classes and to establish conditions under which obesity prevalence will plateau. Following Thomas et al. [4], a differential equation system is adopted that predicts population-wide obesity prevalence trends. The equation system is complex but very logical and practical. Interested readers can find the equation set in Thomas et al. [4].

The model considers both social and non-social influences on weight gain, incorporates other known parameters affecting obesity trends, and allows for country specific population growth. With 2011 data from American Community Survey (Census Bureau, 2011) and the 2008–2013 BMI data from the Bureau of

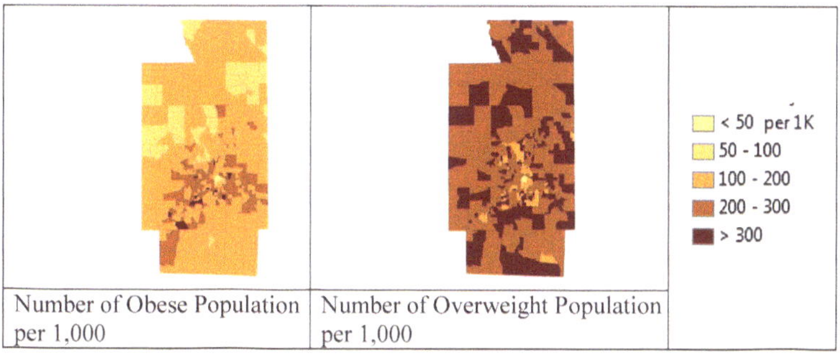

Fig. 1 Obesity and overweight ratios in Summit County, Ohio based on driver licenses. *Data sources: BMI data from Ohio Bureau of Motor Vehicles, 2008–2013; population data from American Community Survey of the US Census Bureau, 2011*

Motor Vehicles, Summit County has 452 census block groups with a wide spectrum of obesity ratios (ranging from 16 per 1000 population to 549 per 1000 population) and overweight ratios (ranging from 32 per 1000 population to 541 per 1000 population).

As can be seen in the two maps in Fig. 1, (1) obese population, though still are in lower ratios than those of overweight population, does seem to have a geographic clustering patterns in the county, (2) overweight population prevails in most of the county with exceptions of only a few census block groups, and (3) the use of census block groups as the unit for geographic analysis indeed reveals more detail of how obesity prevails in the county than using the entire county as an analytic unit.

We adopted the concept of the susceptible, infected, and recovered (SIR) framework to divide a population into subpopulations categorized as normal weight, overweight, obese, and extremely obese by BMI data. To estimate the population moving between these categories, we use a simulation approach that allow analysts to specify the ratios that subpopulations change in between categories. The relationships and potential movements between subpopulations are shown in the diagram in Fig. 2 below:

In each neighborhood (i.e., census block group in this project), population is categorized into six (6) subpopulations:

- Normal weight (*S_T*),
- Overweight (*1_T*),
- Obese (*2_T*),
- Extremely Obese (*3_T*),
- Exposed (*E_T*, or *S_T* → *1_T*), and
- Recovered (R_T, or *1_T* → *S_T*).

The ratios that define how subpopulations move in between categories are

- α_1 *(1_T → 2_T)*,
- α_2 *(2_T → 3_T)*,

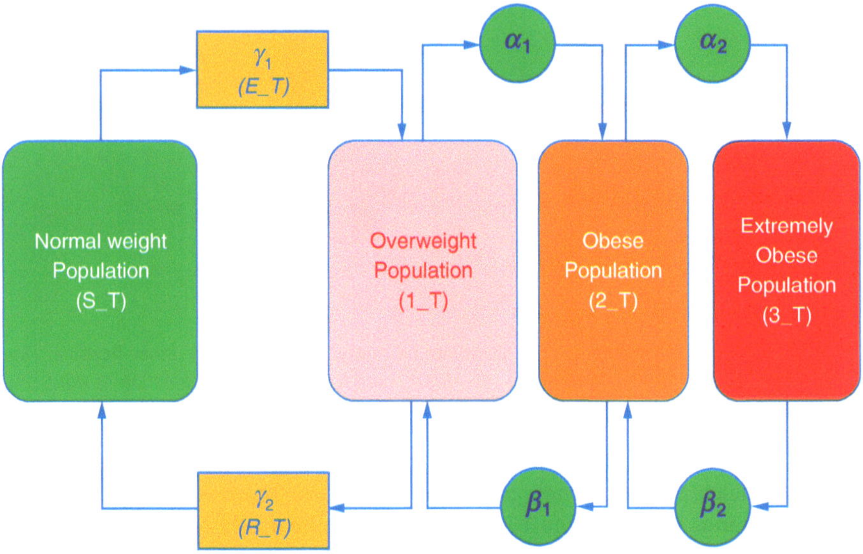

Fig. 2 The susceptible, infected, and recovered (SIR) framework for obesity prevalence simulation

- β_1 *(3_T ➔ 2_T),*
- β_2 *(2_T ➔ 1_T),*
- Υ_1 *(S_T ➔ 1_T),* and
- Υ_2 *(1_T ➔ S_T).*

Following Thomas et al. [4]:

- Total population at $time_0$ (*TotalPopulation*) = *S_T* + *1_T* + *2_T* + *3_T* + *E_T* + *R_T*
- The exposed subpopulation (*E_T*) are individuals who are exposed to either social or non-social influences that lead to weight gain and these individuals will eventually become overweight.
- The subpopulation (*R_T*) are individuals who have weight loss under social or non-social influences.
- Social interactions between compartments are governed by the law of mass action and modeled by multiplying the population numbers in each class.
- Estimated subpopulations at $time_1$ can be derived as solutions for α_1, α_2, β_1, β_2, Υ_1, and Υ_2 from a set of differential equations as proved in Thomas et al. [4].
- For the purpose of modeling and simulations, initial values for model parameters are estimated from publications in the obesity literature:
- The probability of being born in obesogenic environment is set to be ***0.55*** of females of reproductive age who are overweight or obese, based on Balcan et al. [34].
- Birth rate is set to be ***0.0144***, based on Jacobson et al. (2007).

- Baseline prevalence rates are set to be ***0.32*** for overweight, ***0.22*** for obese, ***0.03*** for strictly obese, based on Flegal et al. [35].
- Social influence by overweight and obese are set to be ***0.4*** for overweight subpopulation and ***0.2*** for obese subpopulations, both are based on fitting to initial trends as discussed in Flegal et al. [35].
- Spontaneous rate of weight gain to each class are set to be: exposed (***0.05***), overweight (***0.14***), obese (***0.08***), and extremely obese (***0.014***), also based on Flegal et al. [35].
- Rate of weight loss to each class are set to be: extremely obese to obese (***0.05***), obese to overweight (***0.03***), and overweight to normal weight (***0.033***), also based on Flegal et al. [35].
- Rate of weight regainers transitioning from normal weight to overweight is set to be ***0.04***, also beased on Flegal et al. [35].
- Death rate of obese and extremely obese populations is set to vary between ***16.5*** to ***22*** per 1000 population as suggested by Oizumi [36].

ObPSim comes with a sample data file in shapefile format (ESRI, Inc., Redlands, California). Users of the ObPSim can use it to work with any customized shapefile data. The only requirement for the shapefiles is to have the following columns in the attribute table:

- ***S_T***: the number of people in each neighborhood who are in normal weight range (BMI < = 25)
- ***1_T***: the number of people in each neighborhood who are considered overweight (20 < BMI < = 30)
- ***2_T***: the number of people in each neighborhood who are considered obese (30 < BMI < = 40)
- ***3_T***: the number of people in each neighborhood who are considered extremely obese (BMI > 40)
- ***E_T***: the number of people in each neighborhood who are exposed to possibility of changing from normal weight to overweight
- ***R_T***: the number of people in each neighborhood who may have weight loss so to return from overweight to normal weight.

In the obesity prevalence folder of the sample data set, a shapefile subfolder holds a set of shapefiles, entitled SummitBG. This can be used to test run the Obesity Prevalence Simulator. Please note that the boundary data for block group polygons were downloaded from http://www.esri.com. Data for the *S_T*, *1_T*, *2_T*, and *3_T* subpopulations were calculated using height/weight data derived from drivers' license data from the Ohio Bureau of Motor Vehicles. *E_T* and *R_T* data were derived from geographically weighted regression of the following relationships:

$$ET = \mathit{function}\ \ (ST, \text{density non-fresh food outlets})$$

$$RT = \mathit{function}\ \ (1T, \text{Distance to nearest fitness centers})$$

It should be noted that estimations for *E_T* and *R_T* with the above regression are provided here purely for the purpose of demonstrating the usage of ObPSim. Additional studies and analysis may be needed in order to derive better or more precise estimates.

The estimates for *E_T* and *R_T* should be done so each neighborhood has its own estimates. The examples included in the sample shapefile were derived using the relationships

- between *S_T* and the density of non-fresh food outlets in each neighborhood for estimating *E_T* and
- between *I_T* and the distance to the nearest fitness centers from the neighborhood center for estimating *R_T*.

A simulation control panel, entitled Simulation, shows the various simulated year, parameters, and the Update button as below:

Please note that the parameters in Fig. 3 are set to their initial values (default values), which can be changed in simulation runs. Please note that parameters such as birth rates and death rates are assumed to be the same across the entire county. This is because a county is a small geographic area and there wasn't any such data available for any geographical units inside a county. Other parameters may be formulated such that local conditions (i.e., unique parametric values for census blockgroups) can be reflected by the different values describing each neighborhood's unique characteristics.

Needless to say, any of the parameter values in this model can be changed to reflect the conditions of the simulated area. Essentially, we implemented the model described by Thomas et al. [4] for each neighborhood (census block groups) in Summit County. We developed ObPSim by using years as the temporal unit of analysis. The modeling process as described in Thomas et al. [4] was repeated for each neighborhood. With this approach, ObPSim allows users to

- Observe the spatial distribution of obesity prevalence at any given year.
- Observe the changes in each neighborhood's obesity prevalence over time.
- Observe the spatio-temporal patterns by neighborhoods by changing one or more parameter values.
- Each round of simulation will generate an output file.

For example, Fig. 4 below shows the simulated obesity prevalence by neighborhoods from 2013 to 2019. As shown in this table, obesity prevalence does seem to plateau into future years. As can be seen in this series of maps, Summit County was simulated to evolve from having many neighborhoods (census blockgroups) seeing fast growth of obesity ratios (shown in bluish colors) in 2013–2015 to having much slowed growth of obesity ratios (shown in reddish colors 0 in 2016–2019. When obesity ratios are increasing (or growing) fast, the obesity prevalence is high. On the other hand, when obesity ratios are already high and only change little, the obesity prevalence is plateaued.

The advantage of using ObPSim to estimate obesity prevalence is the ability to change values of model parameters by holding all others constant while varying

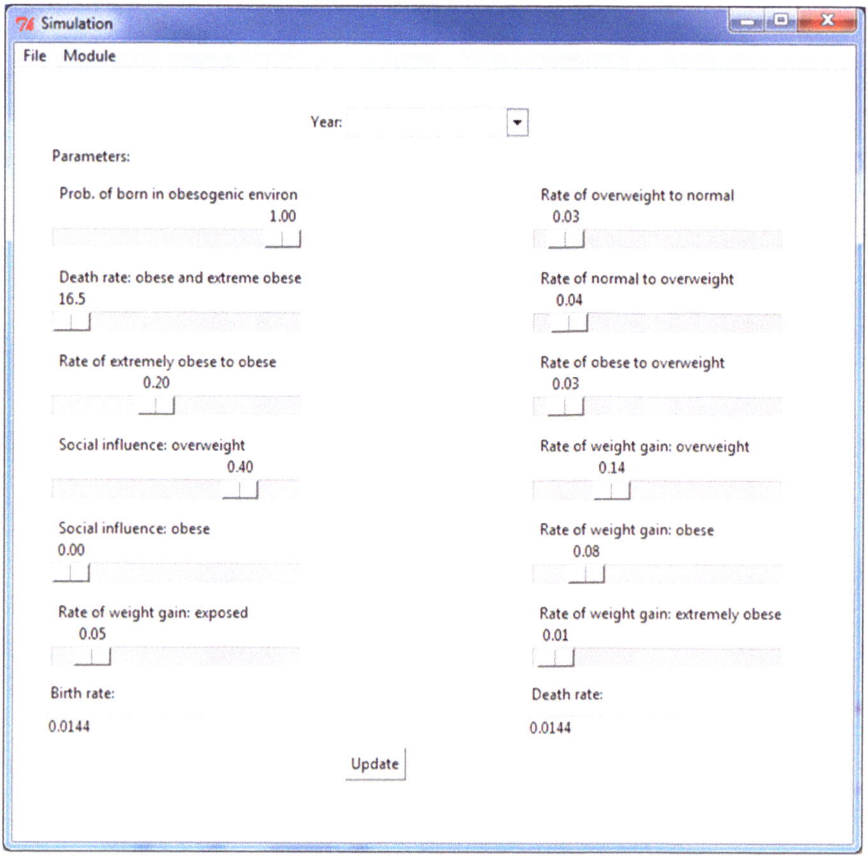

Fig. 3 Control panel for simulation parameters in Obesity Prevalence Simulator

only one or only a few parameter values in simulation runs. In Fig. 5 below, obesity prevalence is simulated for year **2018**, by setting social influence value to be **0.2**, **0.3**, **0.4**, and **0.5**.

As can be seen in the progressive changes of obesity prevalence by increasing social influences on overweight and holding that influence on obese constant, above figure shows that higher levels of social influence seem to be important in shaping simulated obesity prevalence. As a comparison, Fig. 6 below shows the insensitivity of social influence on obese subpopulation while that influence on overweight is held constant at **0.20**. Figures 5 and 6 are listed here to demonstrate the influence of model parameters in the simulated pace of obesity prevalence.

The concept of exploratory space-time data analysis is strongly associated with visualization because graphical presentation enables the analyst to open-mindedly explore the structure of the data set and gain some new insights. Shneiderman [37] argues that exploratory data analysis can be generalized as a three-step process: "overview first, zoom and filter and then details-on-demand". More importantly, it

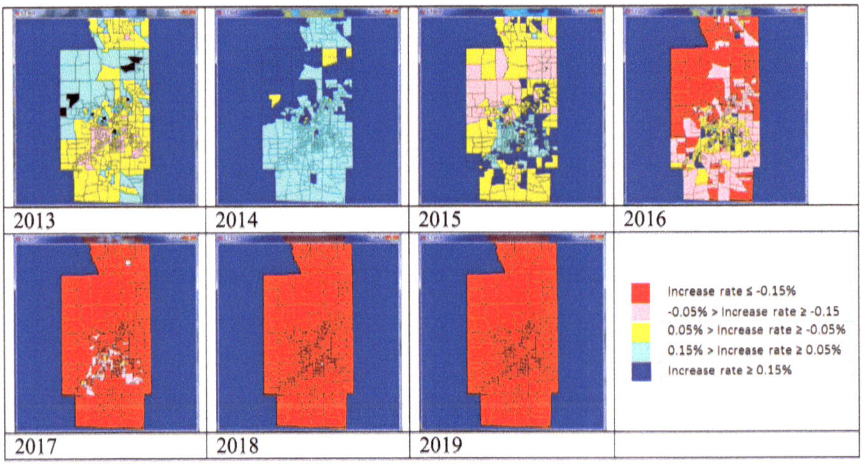

Fig. 4 Example runs of obesity prevalence simulations. *Note: Increase rates are calculated with reference to baseline figures in 2013*

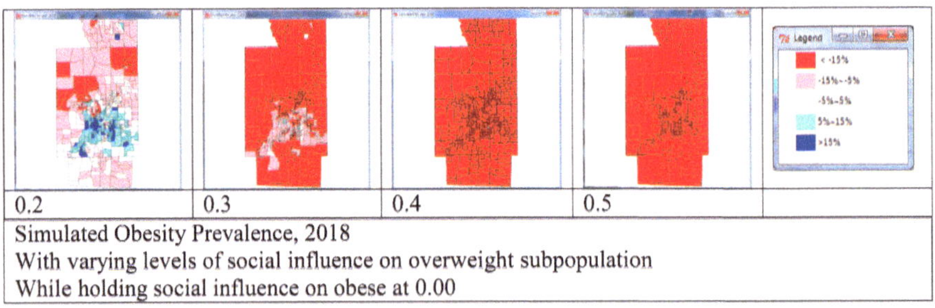

Fig. 5 Effects of social influence changes in obesity prevalence

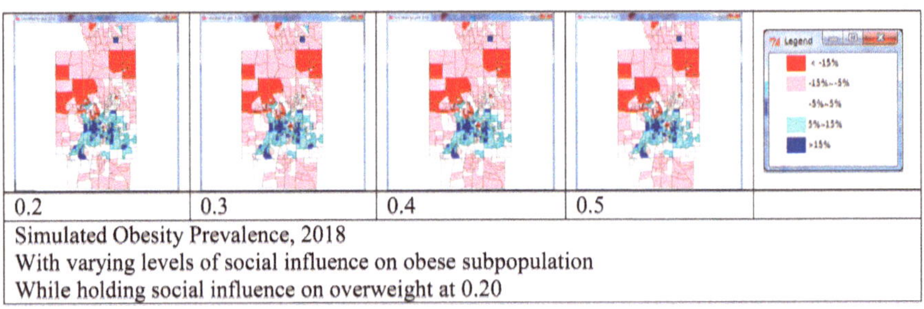

Fig. 6 Insensitivity of social influence on obese subpopulation while that influence on overweight is held constant at **0.20**

is worth noticing that this process should be iterative, and the methods implemented in the current research addressed the challenge. To explain the observed patterns and trends, a follow-up research is needed on collecting determinants of economic growth.

As the last, but the most important step in an analysis such as using ObPSim to investigate spatio-temporal changes in obesity prevalence is the calibration of the model. If (and when) actual data are available for simulated years, it is possible to run the simulations retroactively for a target year and then calibrate the model parameters by incorporating actual data. For example, one can first simulate obesity prevalence in 2012 by using 2000 data and then calibrate the model with actual 2012 data. Such calibration would help to derive a set of parametric values that best approximates simulated results to actual trends in 2012. Understandably, the calibration processes can be tedious and repetitive, they are, however, necessary steps in ensuring simulations are meaningful and applicable.

Concluding Remarks

This paper explores the potential for the new open source tool to function in obesity studies. In other words, the current work is mainly from an exploratory perspective, which can motivate scholars to design a series of analysis questions and formulate new hypotheses from theoretical and policy perspectives. This space-time work provides an important contribution to the current literature, which lacks in comparative space-time studies. Although this comparative study stems from the analysis of obesity dynamics, it broadly aims to analyze the role of geography and location in public health phenomena. In addition, the methods are built in open source environments and thus easily extensible and customizable.

Obesity is an exceedingly complex public health problem with hypothesized causes at multiple interacting levels that are embedded in the very structure of society. This complexity appears to be the reason that one-dimensional preventive or therapeutic interventions are not very successful. The traditional epidemiologic approaches fail to address complex and multilevel data with spatial components. These simplifications do not fit well with the complex realities of obesogenic policy, environments, and population characteristics where dependencies and feedback loops are common. Hence, the reported research extends traditional regression-based approaches to multi-level exposures through a set of differential equation system. This project also integrates the following elements: spatial components, the influence among realizations of each independent variable, as well as feedback loops between outcomes and independent variables.

Given this, new approaches are needed to fully understand the complexities associated with obesity. ObPSim developed in this project is a new, more comprehensive, decision support tool for policy makers. The implementation of policies that effectively combat obesity would improve the health and well-being of a high percentage of the population, including both adults and children, as well as greatly reducing associated economic costs to society such as obesity-related health care expenses and loss of productivity. Based on the susceptible, infected, and recovered (SIR) framework, ObPSim is featured by categorizing the population into subpopulations of normal weight, overweight, obese, and extremely obese.

Furthermore, ObPSim allows population to be moved between subpopulations. Such movements can be defined by any reasoning from the various physical environments, food environment, built environment, and socio-economic environments of the neighborhoods.

Beyond the features of categorizing a population to subpopulations and allowing people to move between subpopulations, ObPSim also allows users to set a suite of model parameters in estimating future obesity prevalence. These parameters do affect how estimations are calculated. However, the parameters as defined by the local conditions allow the simulations to be executed with spatial variations and with localized conditions. Finally, ObPSim provides a means of studying obesity prevalence at a very fine geographic scale. By using census block groups as neighborhoods, ObPSim goes beyond the conventional approaches of studying obesity prevalence at the scale of census tracts. The additional details reveal by using smaller geographic units certainly allow us to better understand spatial patterns and processes of obesity prevalence.

Beyond the scope of this project, studies that compare how simulated obesity prevalence levels react to different values of the model's parameters would be valuable to engage. By fixing all but one parameter to vary in simulations, estimated obesity prevalence patterns can be used to related to how that particular parameter changes. If desired, multiple parameters can be allowed to change simultaneously so observations can be made to see how they affect obesity prevalence as a whole. This paper thus demonstrates an example to interface public health analysis with the open source revolution, which is among the burgeoning efforts seeking the cross-fertilization between the two fast-growing communities.

The ObPSim package is entirely open source, which can promote collaboration among researchers who want to improve current functions or add extensions to address specific research questions. Based on the strength of scientific visualization techniques, this paper stresses the need to study the space-time dimension underlying obesity data sets. Finally, a new interactive tool is suggested and demonstrated as providing an explanatory framework for space-time data. On this basis, the sincere hope here is that this dialogue between public health scholars and geographers will embrace the real world challenges of inequality issues.

Acknowledgement This work is partially supported by the National Science Foundation under Grant No. 1416509, project titled "Spatiotemporal Modeling of Human Dynamics Across Social Media and Social Networks". Any opinions, findings, and conclusions or recommendations expressed in this material are those of the author and do not necessarily reflect the views of the National Science Foundation.

References

1. Feng J, Glass TA, Curriero FC, Stewart WF, Schwartz BS (2010) The built environment and obesity: a systematic review of the epidemiologic evidence. Health Place 16(2):175–190
2. Morland KB, Evenson KR (2009) Obesity prevalence and the local food environment. Health Place 15(2):491–495

3. Butland B, Jebb S, Kopelman P, McPherson K, Thomas S, Mardell J, Parry V (2012) Tackling obesities: future choices: project report, 2nd edn. Foresight, United Kingdom Government Office for Science, London
4. Thomas DM, Weedermann M, Fuemmeler BF, Martin CK, Dhurandhar NV, Bredlau C, Heymsfield SB, Ravussin E, Bouchard C (2013) Dynamic model predicting overweight, obesity, and extreme obesity prevalence trends. Obesity. doi:10.1002/oby.20520
5. Finkelstein EA, Khavjou OA, Thompson H, Trogdon JG, Pan L, Sherry B et al (2012) Obesity and severe obesity forecasts through 2030. Am J Prev Med 42(6):563–570
6. Wang Y, Beydoun MA, Liang L, Caballero B, Kumanyika SK (2008) Will all Americans become overweight or obese? Estimating the progression and cost of the US obesity epidemic. Obesity 16(10):2323–2330
7. Auchincloss AH, Roux AVD (2008) A new tool for epidemiology: the usefulness of dynamic-agent models in understanding place effects on health. Am J Epidemiol 168(1):1–8
8. Anderson B, Rafferty AP, Lyon-Callo S, Fussman C, Imes G (2011) Fast-food CONSUMPTION and obesity among Michigan adults. Prev Chronic Dis 8(4):A71
9. Sonya AG, Mensinger G, Huang SH, Kumanyika SK, Stettler N (2007) Fast-food marketing and children's fast-food consumption: exploring parents' influences in an ethnically diverse sample. J Publ Policy Mark 26(2):221–235
10. Lee RE, Mama SK, Medina AV, Ho A, Adamus HJ (2012) Neighborhood factors influence physical activity among African American and Hispanic or Latina women. Health Place 18(1):63–70
11. Nielsen SJ, Siega-Riz AM, Popkin BM (2002) Trends in food locations and sources among adolescents and young adults. Prev Med 35:107–113
12. Cummins S, Macintyre S (2005) Food environments and obesity-neighborhood or nation? Int J Epidemiol 35:100–104
13. Prentice AM, Jebb SA (2003) Fast foods, energy density and obesity: a possible mechanistic link. Obes Rev 4(4):187–194
14. Paeratakul S, Ferdinand DP, Champagne CM, Ryan DH, Bray GA (2003) Fast-food consumption among U.S. adults and children: dietary and nutrient intake profile. J Am Diet Assoc 103(10):1332–1388
15. McEntee J, Aygeman J (2009) Towards the development of a GIS method for identifying rural food deserts: geographic access in Vermont, USA. Appl Geogr 30:165–176
16. Gebel K, Bauman AE, Petticrew M (2007) The physical environment and physical activity: a critical appraisal of review articles. Am J Prev Med 32(5):361–369
17. Witten K, Hiscock R, Pearce J, Blakely T (2008) Neighbourhood access to open spaces and the physical activity of residents: a national study. Prev Med 47:299–303
18. Davison KK, Lawson CT (2006) Do attributes in the physical environment influence children's physical activity? A review of the literature. Int J Behav Nutr Phys Act 3:19
19. Owen N, Humpel N, Leslie E, Bauman A, Sallis J (2004) Understanding environmental influences on walking; review and research agenda. Am J Prev Med 27(1):67–76
20. Mobley LR, Root ED, Finkelstein EA, Khavjou O, Farris RP, Will JC (2006) Environment, Obesity, and cardiovascular disease in low-income women. Am J Prev Med 30(4):327–332
21. Boehmer TK, Hoehner CM, Despande AD, Brennan Ramirez LK, Brownson RC (2007) Perceived and observed neighborhood indicators of obesity among urban adults. Int J Obes (Lond) 97(3):486–492
22. Giles-Corti B, Timperio A, Bull F, Pikora T (2005) Understanding physical activity environmental correlates: increased specificity for ecological models. Exerc Sport Sci Rev 33(4): 175–181
23. Lee CD, Blair SN, Jackson AS (1999) Cardiorespiratory fitness, body composition, and all-cause and cardiovascular disease mortality in men. Am J Clin Nutr 69:373–380
24. Levy D, Mabry P, Wang Y, Gortmaker S, Huang TK, Marsh T, Moodie M, Swinburn B (2011) Simulation models of obesity: a review of the literature and implications for research and policy. Obes Rev 12(5):378–394
25. Burke MA, Heiland F (2007) Social dynamics of obesity. Econ Inq 45(3):571–591

26. Hammond R, Epstein J (2007) Exploring price-independent mechanisms in the obesity epidemic. Center on Social and Economic Dynamics Working Paper
27. Auchincloss AH, Riolo RL, Brown DG, Cook J, Diez Roux AV (2011) An agent-based model of income inequalities in diet in the context of residential segregation. Am J Prev Med 40(3):303–311
28. Yang Y, Diez Roux AV, Auchincloss AH, Rodriguez DA, Brown DG (2011) A spatial agent-based model for the simulation of adults' daily walking within a city. Am J Prev Med 40(3):353–361
29. Goodchild MF, Glennon A (2008) Representation and computation of geographic dynamics. In: Hornsby KS, Yuan M (eds) Understanding dynamics of geographic domains. CRC, Boca Raton, FL, pp 13–30
30. Krugman P (1999) The role of geography in development. Int Reg Sci Rev 22(2):142–161
31. Ye X, Wu L (2011) Analyzing the dynamics of homicide patterns in Chicago: ESDA and spatial panel approaches. Appl Geogr 31(2):800–807
32. Ye X, Rey S (2013) A framework for exploratory space-time analysis of economic data. Ann Reg Sci 50(1):315–339
33. Rey S, Ye X (2010) Comparative spatial dynamics of regional systems. In: Páez A et al (eds) Progress in spatial analysis. Springer, Berlin, pp 441–463
34. Balcan D, Goncalves B, Hu H, Ramasco JJ, Colizza V, Vespignani A (2010) Modeling the spatial spread of infectious diseases: the GLobal Epidemic and Mobility computational model. J Comput Sci 1(3):132–145. doi:10.1016/j.jocs.2010.07.002. Epub 2011/03/19
35. Flegal KM, Carroll MD, Ogden CL, Curtin LR (2010) Prevalence and trends in obesity among US adults, 1999–2008. JAMA 303(3):235–241. doi:10.1001/jama.2009.2014. Epub 2010/01/15 2009.2014
36. Oizumi R, Takada T (2013) Optimal life schedule with stochastic growth in age-size structured models: Theory and an application. J Theor Biol 323:76–89. doi:10.1016/j.jtbi.2013.01.020. Epub 2013/02/09
37. Shneiderman B (1996). The eyes have it: a task by data type taxonomy for information visualizations. In: Visual languages, 1996. Proceedings., IEEE Symposium on. IEEE, pp 336–343

MIX
Papier aus verantwortungsvollen Quellen
Paper from responsible sources
FSC® C105338

If you have any concerns about our products,
you can contact us on
ProductSafety@springernature.com

In case Publisher is established outside the EU,
the EU authorized representative is:
Springer Nature Customer Service Center GmbH
Europaplatz 3, 69115 Heidelberg, Germany

Printed by Libri Plureos GmbH
in Hamburg, Germany